INSIDE Pro/ENGINEER®

The Professional User's Guide to Designing with Pro/ENGINEER®

Third Edition

James Utz and W. Robert Cox, P.E.

Revised by Dennis Steffen

ONWORD® PRESS

INSIDE Pro/ENGINEER ®

Third Edition

By James Utz and W. Robert Cox, P.E.

Third Edition revised by Dennis Steffen

Published by:

OnWord Press

2530 Camino Entrada

Santa Fe, NM 87505-4835 USA

SAN 694-0269
Third Edition, 1997

10 9 8 7 6 5 4 3 2 1

Printed in the United States of America

Cataloging-in-Publication Data

Utz, James; Cox, W. Robert

INSIDE Pro/ENGINEER

Includes index.

1. Pro/ENGINEER (computer software) 2. Mechanical CAD

I. Title

95-069618

ISBN 1-56690-114-6

Trademarks

Warning and Disclaimer

About the Authors

James Utz is a senior mechanical engineer for Digital Equipment Corporation in the Portable Computers Engineering Department of the PC Business Unit. He earned a B.S. degree in Mechanical Engineering from the University of Texas at Austin in 1989. James previously worked for COMPAQ Computer Corporation as a Mechanical Engineer in the Portables Division. He currently uses Pro/ENGINEER to design notebook computers and related options.

W. Robert Cox, P.E. is an application engineer for Information Decisions, Inc., a value-added reseller of workstations and CAE software, including Pro/ENGINEER. He has a Master's of Engineering degree from Texas A&M University. He has been the technical operations manager for a valve manufacturing company, and the mechanical CAE productivity tools manager for COMPAQ Computer Corporation.

Dennis Steffen is a senior mechanical design engineer for Digital Lightwave, Inc., of Clearwater, FL. He has 20 years of design experience—nearly half of that in the Pro/Engineer environment. He also serves on the editorial advisory board of *Pro/E: The Magazine* (a publication of ConnectPress, a High Mountain Press company).

Thanks for the Help

Thanks to my wife, Sue Ann, for understanding and believing in me. Thanks to Blaze, my son, for smiling at me and making every problem go away for an instant. I would also like to thank all my family and friends, especially Jeff Keaton and T. Gordon Wu, for making this project a reality.

James Utz

I would like to thank my wife, Sharon, and my kids, Nicole, Ashley, Randy, and Brooke, for their patience while I worked on this edition of *INSIDE Pro/ENGINEER*. I would also like to thank Gerald Goodwin for his help during this time. Finally, I wish to acknowledge my special remembrance of Robynn.

W. Robert Cox

OnWord Press Credits

OnWord Press is dedicated to the fine art of professional documentation. In addition to the authors who developed the material for this book, other members of the OnWord Press team contributed their skills to make the book a reality. Thanks to the following people and other members of the OnWord Press team who contributed to the production and distribution of this book.

Dan Raker, President
Gary Lange, Vice President
Janet Leigh Dick, Associate Publisher; director, Channel Marketing
Daniel Clavio, Director, Market Development and Strategic Relations
David Talbott, Director of Acquisitions
Dave Surette, Senior Manager, Administration and Finance
Rena Rully, Senior Manager, Production and Editorial
Barbara Kohl, Senior Editor
Daril Bentley, Senior Editor
Carol Leyba, Senior Production Manager
Cynthia Welch, Production Editor
Lauri Hogan, Marketing Services Manager
Kristie Reilly, Assistant Editor
Lynne Egensteiner, Cover designer, Illustrator

Table of Contents

Part 2
Working with Parts and Assemblies 153

Chapter 6
Designing a Larger Part 155

Chapter 7
Designing a Part with a Complex Shape 219

Introduction

A Brief Overview and Explanation of Conventions

Pro/ENGINEER's method of design is quite different from traditional, solid modeling Computer Aided Design (CAD) systems. The data structure and intuitive design methodology work together to provide a state-of-the-art design environment that is capable of not only capturing your design, but also your design intent.

Understanding the Advantages of Linked Data

The parts you create in Pro/ENGINEER are the core database for all other Pro/ENGINEER applications. The other application modules such as Pro/DETAIL (for making drawings) and Pro/ASSEMBLY (for assembling two or more parts) *reference* the parts they use. This means that the applications do not copy the part(s) into their local database; they simply create a link to the original data. One obvious advantage is that concurrent engineering is supported and easily maintained by Pro/ENGINEER's data structure.

The term *bi-directional* means that design changes can be initiated from any application, and the results will be shown in all other applications. For example, you can change the length of the part and see the drawing update, or you can change the length dimension in the drawing and see the part, as well as any assemblies that *reference* the part, automatically update. It is your choice. This bi-directional trait saves you time by automatically updating the related data files for the other Pro/ENGINEER applications.

Understanding Parts and Features

The commands available in Pro/ENGINEER allow you to design complex parts using a simple, step-by-step process. The way you build parts in Pro/ENGINEER is by creating features. Features can be additions of material, subtractions of material, rounds, chamfers, and even datum planes or points. New features can be created from conceptual sketches, or copied, mirrored, and patterned from existing features. This method is called feature-based modeling and allows you to design in a fast, easy, and intuitive manner. Almost everything in your part's design will be created as a feature.

Many of the features in Pro/ENGINEER begin with a sketched section. A section is a group of two-dimensional (2D) elements (lines, arcs, circles) that control the size and shape of your three-dimensional (3D) features. When you need to create a section, Pro/ENGINEER opens a Sketcher window, and you simply draw the geometry using a free-hand technique. Pro/ENGINEER then behaves like an electronic sketch pad, with the mouse as your pencil. You simply sketch your shape without worrying about the accuracy or size of the lines.

Who Should Read This Book

Parametric Technology Corporation's Pro/ENGINEER is the premier electronic tool for professional designers of mechanical parts and assemblies. Simply stated, Pro/ENGINEER is powerful. With just a little input from you, Pro/ENGINEER can automate design tasks that would take hours, days, or even weeks to perform manually. Pro/ENGINEER is more than a solid modeling CAD system. It is a family of applications that revolve around a single, 3D database that you design. Pro/ENGINEER includes applications that help you create, analyze, document, and produce your designs.

INSIDE Pro/ENGINEER is a friendly and easy-to-follow guide for both experienced and beginning mechanical engineers to learn and use Pro/ENGINEER. Even experienced users of Pro/ENGINEER can find valuable hints within these pages. *INSIDE Pro/ENGINEER* saves you time and effort and teaches you how to use Pro/ENGINEER in a language you can understand.

What This Book Is About

INSIDE Pro/ENGINEER begins by teaching you about the Pro/ENGINEER design environment. You will learn how a typical design session progresses and see how Pro/ENGINEER works. You will learn everything about controlling the environment. You will also learn about the Pro/ENGINEER Sketcher utility and how to use it to input your designs. *INSIDE Pro/ENGINEER* also guides you through Part mode, Assembly mode, and Drawing mode while you design, build, and detail your own toy sports car. You will then be shown how to customize your environment and take full advantage of Pro/ENGINEER's parametric capabilities.

How This Book Is Structured

INSIDE Pro/ENGINEER is divided into four sections. Section 1 introduces Pro/ENGINEER as a design tool and shows you how to design a part. Section 2 shows you how to design more complex parts and how to create an assembly. Section 3 shows you how to create finished drawings of your design. Section 4 tells you how to import and export designs and how to customize your workstation.

In addition to the four sections of the book, there are two appendices. Appendix A provides information on how to use the Pro/

TABLE editor. Appendix B describes all Pro/ENGINEER modules. There is also a complete index at the end of the book.

Typographical Conventions Used in This Book

Consistent typographical standards are used throughout *INSIDE Pro/ENGINEER.*

Menus and Commands

Menu titles such as MAIN and PART are shown in all capital letters. Menu commands such as **Regenerate** and **Done** are shown in bold characters. Capitalization matches the commands as shown on the menus.

Command strings, which are series of menu selections, are shown in the same fashion, but each item is separated by arrow (➡) and pipe (|) characters. Arrows indicate that you move from one menu to another, whereas the pipes separate choices you make within one menu. The following line shows a typical command string:

Create ➡ Solid ➡ Cut ➡ Extrude | Solid | Done ➡ Single | Done ➡ One Side ➡ Thru All | Done

Computer Prompts and Key-ins

When you need to answer a computer prompt, you will see the prompt in normal type and your response in boldface, as shown in the following example:

`Enter Part name [PRT0001]:` **`wheel`** ↵

Notes, Tips, and Other Conventions

There are several other conventions you will see used throughout this book.

➾ **NOTE:** *Notes are used to show important ideas and explain Pro/ENGINEER's behavior.*

➾ **DESIGN NOTE:** *Design Notes are used to point out design decisions (design intent). They are also used to explain the results of your design decisions.*

✓ **TIP:** *Tips show shortcuts and hints that help you be more productive with Pro/ENGINEER.*

✗ **WARNING:** *Warnings are used to make sure you understand the consequences of a particular action. These will prevent you from losing work or making major mistakes.*

All file names, such as *config.pro,* are written in lower-case italics.

To indicate that it is a good time to save your work, a special icon is placed within the text. The save icon appears below.

How To Use This Book

Depending on your familiarity with Pro/ENGINEER, there are several ways to read this book.

New Pro/ENGINEER Users

This book is written for you. If you start at the beginning and work your way through the first two sections, you will learn the most important lessons about Pro/ENGINEER. After becoming familiar with parametric design, you can move on to sections three and four to learn about creating and plotting drawings. The last chapter focuses on optimizing your Pro/ENGINEER configuration.

Experienced Users

Pay special attention to the design process in the sample design sessions of sections one and two. This book will help you avoid some of the nasty parent/child relationships you may already have experienced. Section three contains helpful information about making finished production drawings.

System Administrators

You should pay special attention to section four because most of your time is spent handling configuration problems and answering questions about plotting. Sections one, two, and three will help you to instruct new users about basic design and drawing practices.

Companion Disk Installation

The diskette inside the back cover of this book contains assorted Pro/ENGINEER files created by the examples in this book. If you have trouble with any of the described features or procedures, you might want to take a look at these example files.

The diskette is a high-density, 1.44-Mb DOS-format diskette. There are two directories on the diskette. The first directory, `UNIX`, holds a compressed tar file that contains sample files for users who run Pro/ENGINEER in a UNIX environment. The second directory, `WINNT35`, contains an install utility and a self-extracting file of samples for users who run Pro/ENGINEER in a Windows NT environment. All Pro/ENGINEER data files are from Version 14 of Pro/ENGINEER.

Installation on a UNIX Workstation That Can Read a DOS Diskette

Some UNIX workstations have a floppy drive that can read DOS diskettes. To install on such systems, follow these steps:

1. Insert the diskette into your floppy drive.

2. Mount the floppy drive if it is not already mounted.
3. Create a directory called `ipe` in your home directory. Change your current working directory to the new directory.
4. Copy the file called `IPEFILES.Z` from the `\UNIX` directory on the diskette to the new directory.
5. Check the file name of the copied file. If the extension is lower case `(.z)`, rename the file to change the extension to upper case `(.Z)`.
6. Use the **Uncompress** command at the UNIX prompt to uncompress the file.
7. Use the **Tar** command at the UNIX prompt to extract the data files from the tar archive. The data files are now ready to use.

Consult your manual or system administrator on the specific commands for these operations if you are uncertain about anything.

Installation on a UNIX Workstation That Cannot Read a DOS Diskette

If you have a workstation that cannot read DOS diskettes but is connected to a network with a PC, you can transfer the files from the PC to the workstation.

1. Insert the diskette into your floppy drive on the PC.

Copy the file `IPEFILES.Z` from the diskette to a temporary directory on the PC.

2. Create a directory called `ipe` in your home directory on the workstation. Change your current working directory to the new directory.
3. Transfer the file to the `ipe` directory on your workstation via TCP/IP connection, network mail, or any other available method.
4. Check the file name extension of the copied file. If the extension is lower case `(.z)`, rename the file to change the extension to upper case `(.Z)`.
5. Use the **Uncompress** command at the UNIX prompt to uncompress the file.
6. Use the **Tar** command at the UNIX prompt to extract the data files from the tar archive. The data files are now ready to use.

Consult your manual or system administrator on the specific commands for these operations if you are uncertain about anything.

Installation on a Windows NT Workstation

Follow these steps to install the files:

1. Insert the diskette into your floppy drive.
2. Open a Command Prompt window from Program Manager.
3. Change your current working directory to the `A:\WINNT35` directory.

4. Type `INSTALL drive:\path`, where `drive` is the drive letter of the destination drive and `path` is the desired directory path on the destination drive. The install utility will automatically create an `ipe` directory at the location you specify. The data files will be copied, and will be ready to use.

Consult your manual or system administrator on the specific commands for these operations if you are unsure about anything.

File Descriptions

Parts

wheel.prt The wheel part example.

chassis.prt The chassis part example.

body.prt The car body part example.

master.prt The part that contains the default datums and the view names.

Assemblies

car.asm The assembly of the wheels, chassis, and car body.

Drawings

wheel.drw The drawing of the wheel part.

chassis.drw The drawing of the chassis part.

car.drw The drawing of the car assembly.

Drawing Formats

btitle.frm The drawing format that contains a custom title block.

bassy.frm The drawing format for the assembly drawing.

btab.frm The drawing format for the tabulated wheel drawing.

Assorted Text Files

ipe_conf.pro An example *config.pro* configuration file.

ipe_menu.pro An example *menu_def.pro* file that adds a custom item to the menu.

ipe_draw.dtl An example drawing setup file.

master.txt A trail file that brings up the master part and renames it.

Part 1

Understanding Pro/ENGINEER

Chapter 1
A Pro/ENGINEER Design Session

A Quick View of Pro/ENGINEER as an Integrated Design Tool

Introduction

Say you work as a designer or engineer for a company that makes coffee mugs, and you are charged with the task of designing a new mug. As part of the job, you must design the mug, create a custom coaster, and provide a detailed drawing.

How would you perform this task using Pro/ENGINEER? That is what will be illustrated in this chapter, an introduction to some of the basic principles of working in Pro/ENGINEER through a simplified design session. In this example, you will see how to design a part, fit (or assemble) the part to another part, and make a drawing. You will also see how to use the program's capability to alter the specifications with a fraction of the effort you otherwise would use.

Note that this is not a step-by-step exercise. You are not shown all of the commands you would use to create the mug. The purpose of this chapter is to acquaint you with how Pro/ENGINEER "thinks" and "behaves" so that when you begin to design with Pro/ENGINEER in later chapters, the environment and terminology are familiar to you. Just sit back for a few minutes and ignore your workstation.

When this design session overview is finished, you should understand the following concepts:

- How feature-based modeling helps the design process
- How parametric features behave
- How a single part, database, drawing, and assembly are linked
- How Pro/ENGINEER captures the design intent, not just the design

Beginning Your Design

Pro/ENGINEER uses feature-based modeling to create a part. This means you design a part by creating all of the features that make up the part. Typically you begin with a base feature, or solid object, roughly the shape of the part you want to create. Then you add features (cuts, slots, holes, chamfers, and others) that shape the base object into the part you want. One of the most powerful aspects of Pro/ENGINEER is its ability to create extremely complicated geometry as a combination of simple steps.

Creating the Base Feature

In that most coffee mugs look like a cylinder, it makes sense to start with a cylinder as the base feature. When creating cylindrical features it is best to create a default datum plane system first. You will find these datums useful later on for sketching-plane setup, assembly references, and drawing-view orientation. How do you make a cylinder? The process for making most features in Pro/ENGINEER includes the following steps:

1. Select menu options that describe the general type of feature you want to create.
2. Establish a sketching plane by selecting opposing datum planes from the default datum plane system.
3. Sketch the basic shape of the feature.
4. Add dimensions to the sketched shape.

Once you have created the feature, you typically modify and refine it. There are many ways to make a cylinder, but the most intuitive is to extrude a circle. The first step is to issue a command to Pro/ENGINEER indicating that you want to make a solid feature that is an extrusion.

NOTE: *Menu selections that perform the most commonly used operations in Pro/ENGINEER appear as default selections (red background) in the menus.*

The previously mentioned command brings up Pro/ENGINEER's Sketcher. Here you sketch a circle (the shape of a cylinder in the end view) and add dimensions. Adding dimensions is a very important step. Pro/ENGINEER will not generate the feature until it has dimensions that are geometrically complete. To create a cylinder by extruding a circle, Pro/ENGINEER needs two dimensions: the diameter of the circle and the length of the extrusion.

NOTE: *Adding dimensions to a sketch is called "constraining the geometry." This is a key step and deserves careful consideration, as these dimensions will be used later to describe the part in the drawing.*

The following illustrations show a circle with a diameter dimension and a cylinder with dimensions. The cylinder is the base feature for the mug.

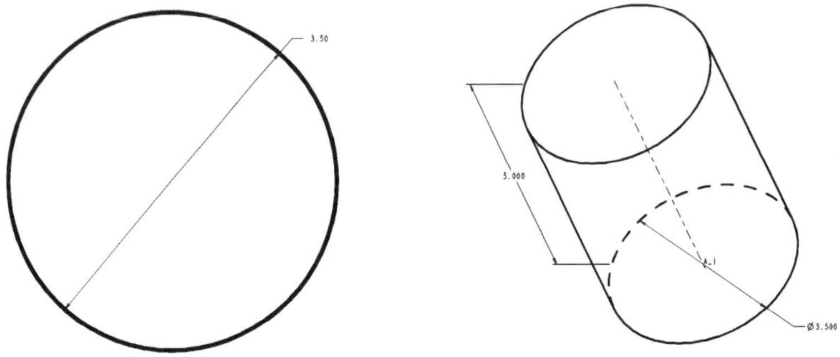

Figure 1-1
A sketched circle with its diameter dimension (left) and the cylinder with dimensions.

Turning the Base Feature into a Mug

Now you have a solid cylinder. To turn the cylinder into a mug, you need to hollow out the cylinder and add a handle.

Hollowing Out the Cylinder

The next step in creating the mug is to remove the inside material of the cylinder so that it can hold liquid. In Pro/ENGINEER, the best way to make a part with a constant wall thickness is to use the Shell feature. Alternatively, a blind slot or blind cut may be used. For the shell feature, you need to tell Pro/ENGINEER which sides to "hollow out," and to supply the program with a desired wall thickness. The following illustration shows the result of hollowing out the cylinder.

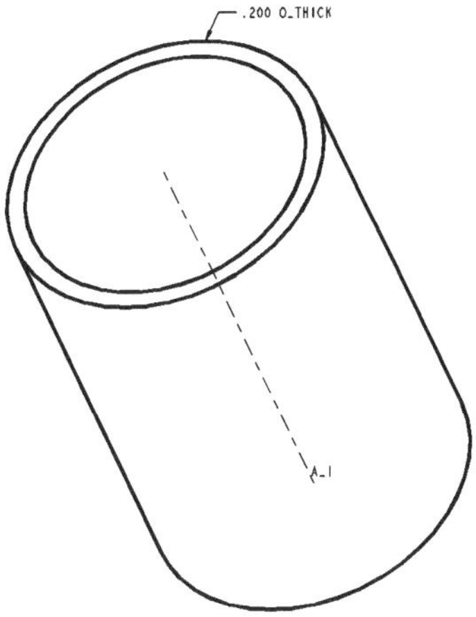

Figure 1-2
Shelled cylinder with thickness dimension.

Adding a Handle

Next, you need to add a handle so that you can pick up the mug. A swept protrusion is best for this type of feature. Sweeps require a sketched section for a trajectory and a sketched section for the profile. The next illustration shows the handle created this way.

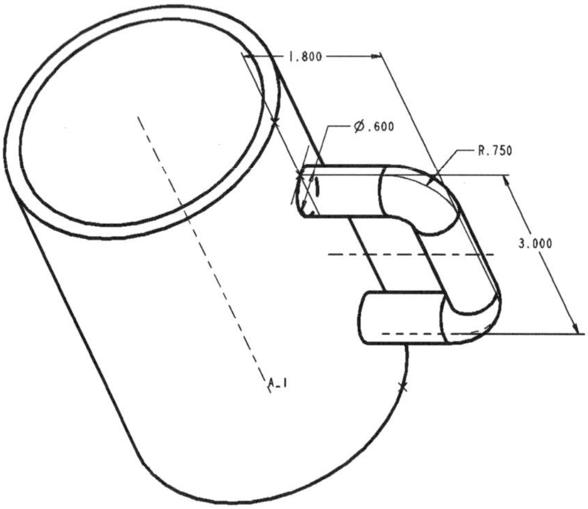

Figure 1-3
Handle with constraining dimensions.

Refining the Mug Design

Now that you have the mug roughed in, you need to add a few finishing details. To make the mug handle more robust, you should fillet the areas where the handle meets the mug. The next illustration shows the handle fillets.

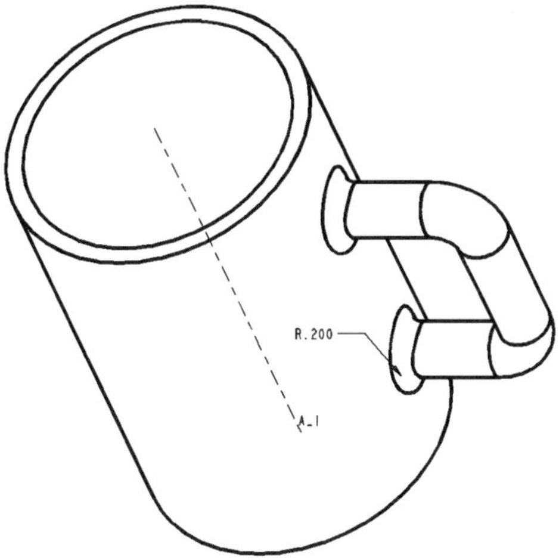

Figure 1-4
Handle fillets with dimension.

Finally, you should round the top lip of the cup so that you may safely drink from it.

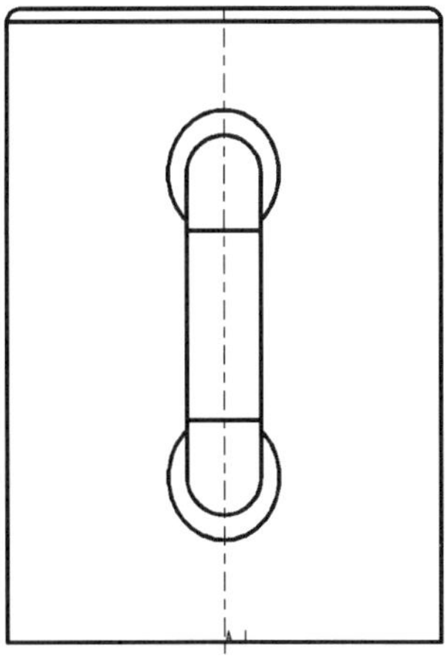

Figure 1-5
Mug with rounded lip.

Summary of the Part Creation Process

The mug example shows a good approach to tackling designs in Pro/ENGINEER. A summary of the methodology follows:

1. Create the basic shape of the part.
2. Add the necessary features.
3. Add the final details.

✓ **TIP:** *We started with the major shape and ended with the minor details. Your designs should be performed in the same manner. Simply break your designs down into their most basic features. Start by modeling the basic shapes and finish with the details.*

Making an Assembly

Now that you have finished the initial design of your coffee mug, you can add a coaster. Pro/ENGINEER's Assembly mode lets you bring multiple parts together with logical constraints so that you can see how your parts fit. For purposes of this section, we assume that the coaster has already been created.

Begin assembly mode by determining which component should be the first component of the assembly. Components can be either a part or another assembly. In most cases, the first component of the assembly should be the foundation of your design. Once you determine which component of your assembly should be first, you can then assemble the other components to it. In this example, there are only two parts. Assume the coaster is an existing part already purchased by the company, and therefore that it is more important that the mug fit the coaster than vice versa.

So, to begin this assembly, we will bring in the coaster as the first component and then assemble the coffee mug to it. A small window containing the new component will appear each time you choose to add a component to the assembly.

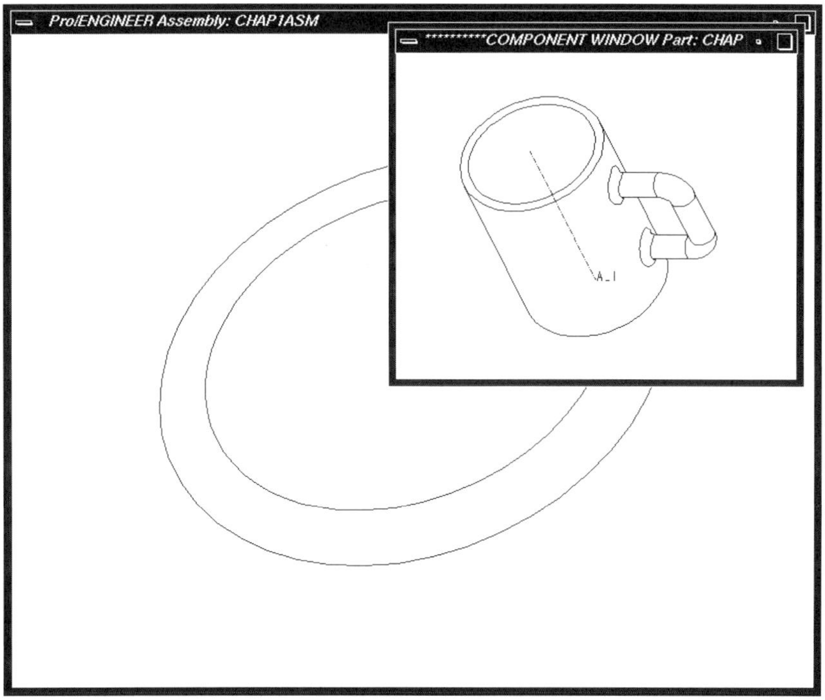

Figure 1-6
Creating an assembly.

Specifying Assembly Constraints

At this point you begin specifying assembly constraints. These are the rules that Pro/ENGINEER will follow when building the assembly. The constraints relate the geometry of one component to the geometry of another component. To assemble the coffee mug, you will use two simple constraints. One constraint tells Pro/ENGINEER which surfaces of the components to mate (the surfaces that will "face" each other). In the case of the mug, you are placing the bottom surface of the coffee mug on the upper surface of the coaster.

The next constraint tells how to position the mug on the coaster. In this case you want to center the coffee mug on the coaster coaxially (by aligning the axes of the parts). These constraints are shown in the following illustration. These same results can also be achieved by aligning datums, if they were used in the construction of the parts, or by simply aligning the diameters of each part. Experiment with various assembly methods to find the ones best suited to your situation and the project at hand.

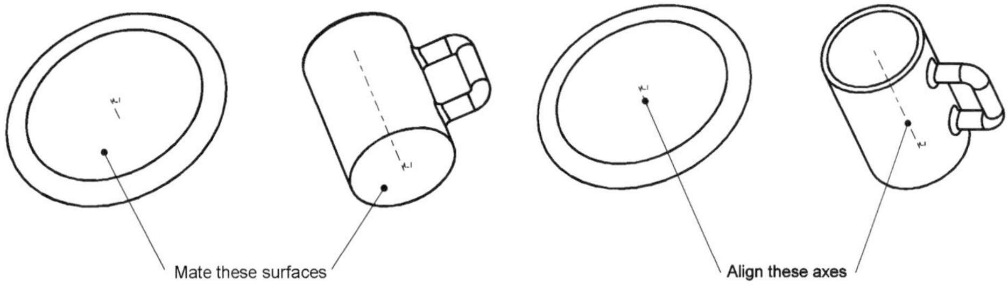

Figure 1-7
The mate (left) and align (right) assembly constraints.

Summary of the Assembly Process

A methodology for building assemblies follows:

1. Select the base component of the assembly.
2. Assemble the other components using assembly constraints.

The constraints you use to assemble these parts will govern the assembly as the parts change. If the coaster changes thickness, the coffee mug will move accordingly, and vice versa. If the coffee mug changes diameter, it will remain centered on the coaster.

Pro/ENGINEER's constraints let you put intelligence into your assemblies so that you do not have to recreate your design every time you change a component. This dynamic relationship among components is the essence of parametric design. In addition, Pro/ENGINEER's modes are linked so that an alteration made in one mode to any component will automatically appear whenever you work in other modes.

Creating a Drawing

Inevitably, somebody you work with will need a drawing of your design. Much of the world still relies heavily on paper, and Pro/ENGINEER's Drawing mode provides tools that make 2D drawing creation of parts or assemblies quite painless. When you begin a new drawing, Pro/ENGINEER asks you for the name of the object you intend to detail. In this example, you will make a drawing of the coffee mug.

Selecting Views for Your Drawing

The new drawing sheet will be blank. Start by placing the first view.

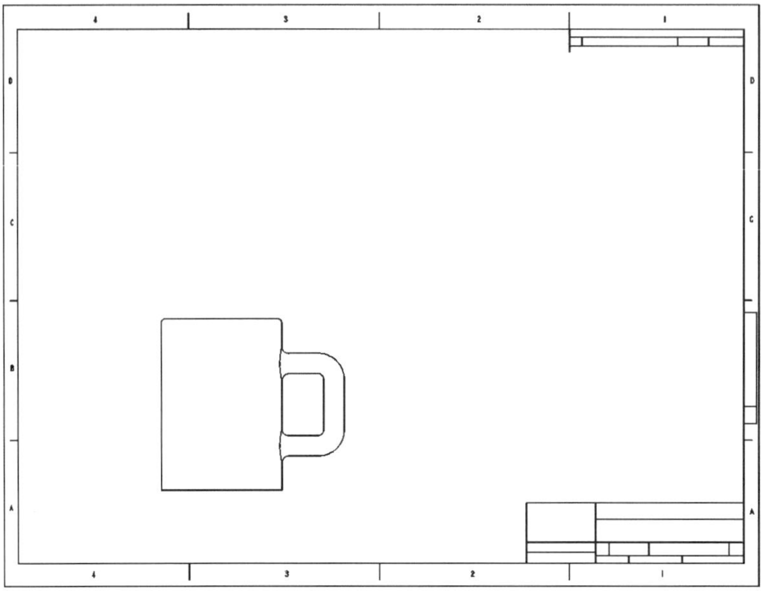

Figure 1-8
Starting a drawing.

When you have the view placed in the desired orientation, you can easily create more projected and section views by picking the location of the new views.

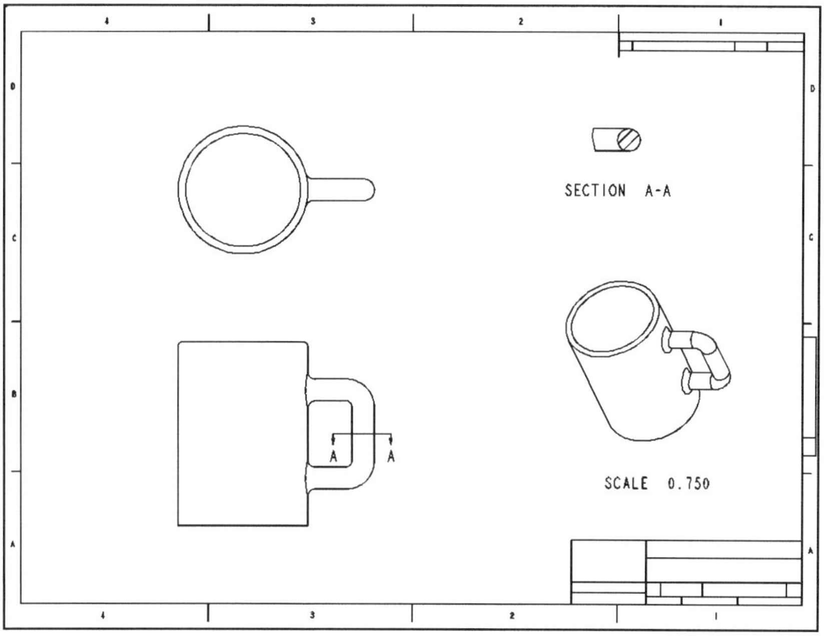

Figure 1-9
Adding other views to the drawing.

Specifying Dimensions

As you created the part, you provided the dimensions needed to fully constrain each feature. The same dimensions can be used in the drawing, and all you have to do is display them.

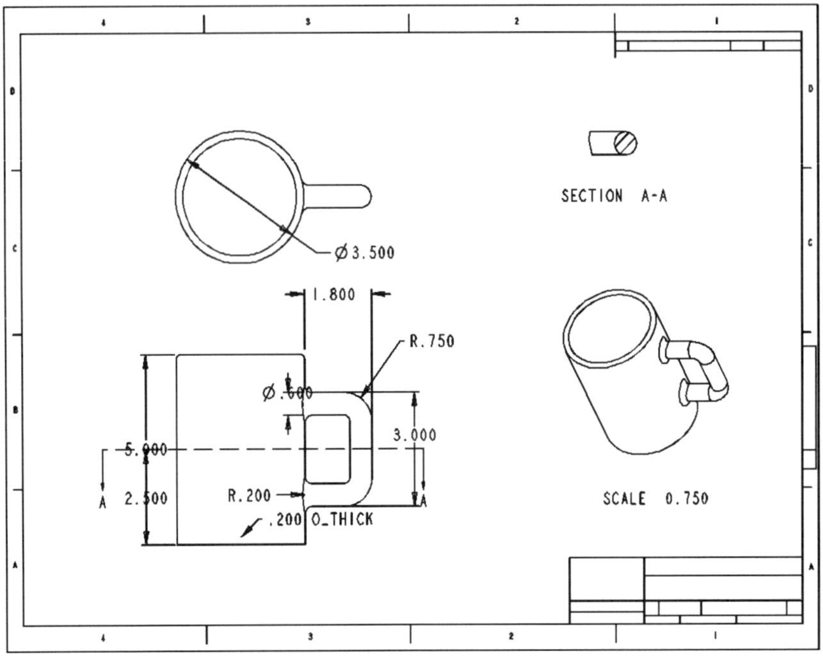

Figure 1-10
Dimensioned drawing sheet prior to cleanup.

Now you can rearrange the dimensions and add notes.

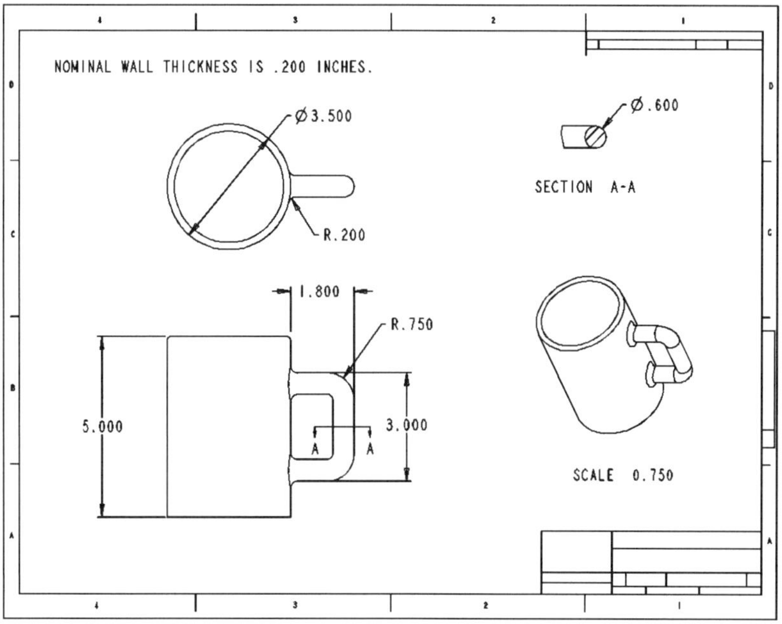

Figure 1-11
The finished drawing.

Summary of Drawing Methodology

To create a drawing from a part or assembly, you typically follow these general steps:

1. Place your first view.
2. Add other views.
3. Show dimensions.
4. Rearrange the dimensions and add notes.

Of course, this drawing is linked to the part. Any changes you make to the part will automatically show in the drawing. You can also modify dimensions in the drawing and the part will change.

Changing the Coffee Mug

A design session would not be typical without a few changes. Unfortunately, in this design session, you must complete your design before the end of the day, and no one else at your company has approved it yet. You quickly arrange a meeting among the management, marketing, and engineering departments. To summarize the meeting, they think your design stinks. Here is what they said:

Management: "Well, I like my mugs to hold about 20 ounces of coffee. I get tired of going back and forth to the coffee pot all day long. I also need about 1/2 inch of extra height so I don't spill any coffee when I carry it."

Marketing: "We wanted a handle rounder than the one you designed. The handle you designed looks like it belongs on a briefcase, not a coffee mug. Oh, by the way, we think the coffee mug is too heavy. See what you can do to make it a little lighter than other manufacturers' coffee mugs."

Believe it or not, these changes are not difficult. The following material takes you through each change so that you can see how Pro/ENGINEER helps you.

Making the Coffee Mug Lighter

This change is the easiest of the three. If you make the mug thinner, it will reduce the weight. All you need to do is modify the wall thickness dimension.

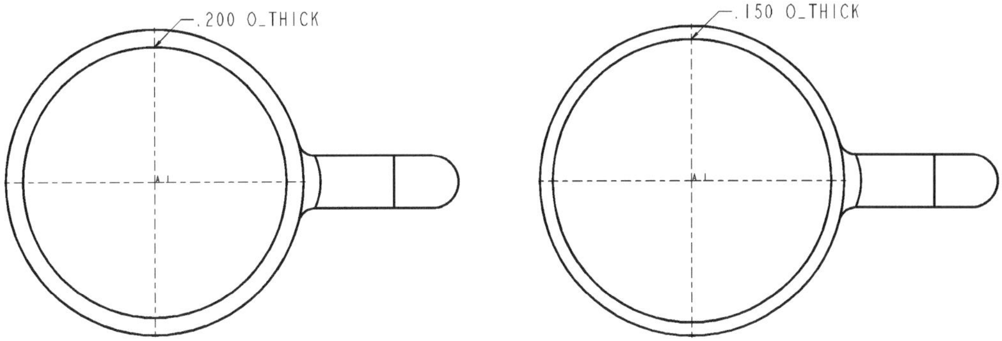

Figure 1-12
Wall thickness before and after modification.

Notice how the round on the upper edge of the coffee mug changed. This is because of the parametric relationship between the round and the wall thickness. The radius of the round is determined by the thickness of the wall.

Changing the Handle Shape

To change the shape of the handle, you need to redefine the trajectory section of the handle feature.

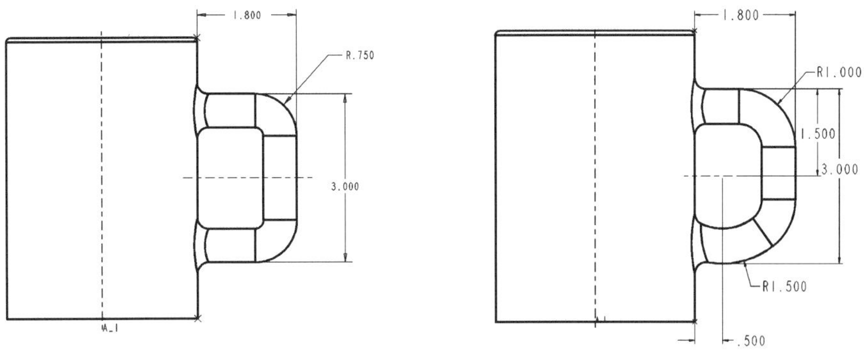

Figure 1-13
Handle shape before and after modification.

As seen here, you can actually replace entire sections without losing your work. Note that the fillets at the intersection of the handle and mug remain.

Modifying Mug Capacity

This change requires some thought. Your boss wants to control the capacity of the mug. Although you do not have a capacity dimension or variable to modify, you *can* make one. In fact, you can govern the shape of the mug with this capacity variable. In that the diameter of the mug is controlled by the diameter of the coaster (recall that it already exists), you can change the height to vary the capacity. A relation can be written that will change the height to meet the capacity requirements.

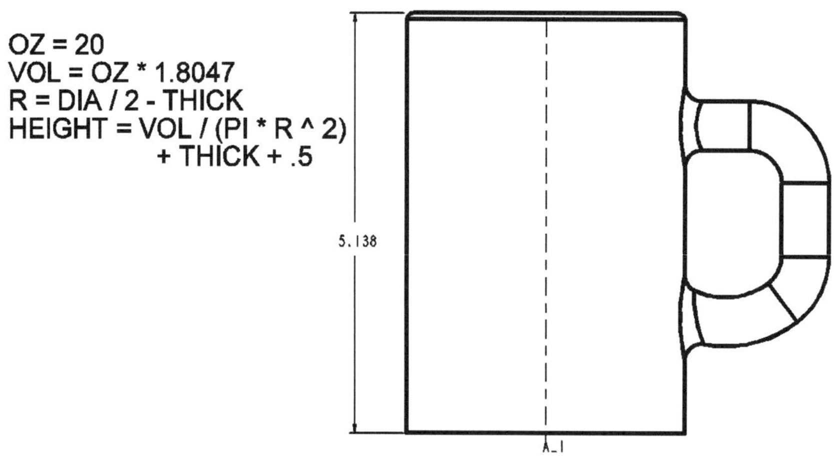

Figure 1-14
Mug after incorporating the equation for its height.

Now if you vary the wall thickness, the diameter, or the desired capacity, the height will adjust accordingly. This is *real* design intent. Even if your boss changes her mind and wants a 14-ounce mug, all you need to do is edit the capacity variable and the mug will update.

The End of the Day

Your coffee mug design is finished. You have a detailed drawing of it (because the drawing and design modes are associated) and you have shaped it to a custom coaster. Congratulations!

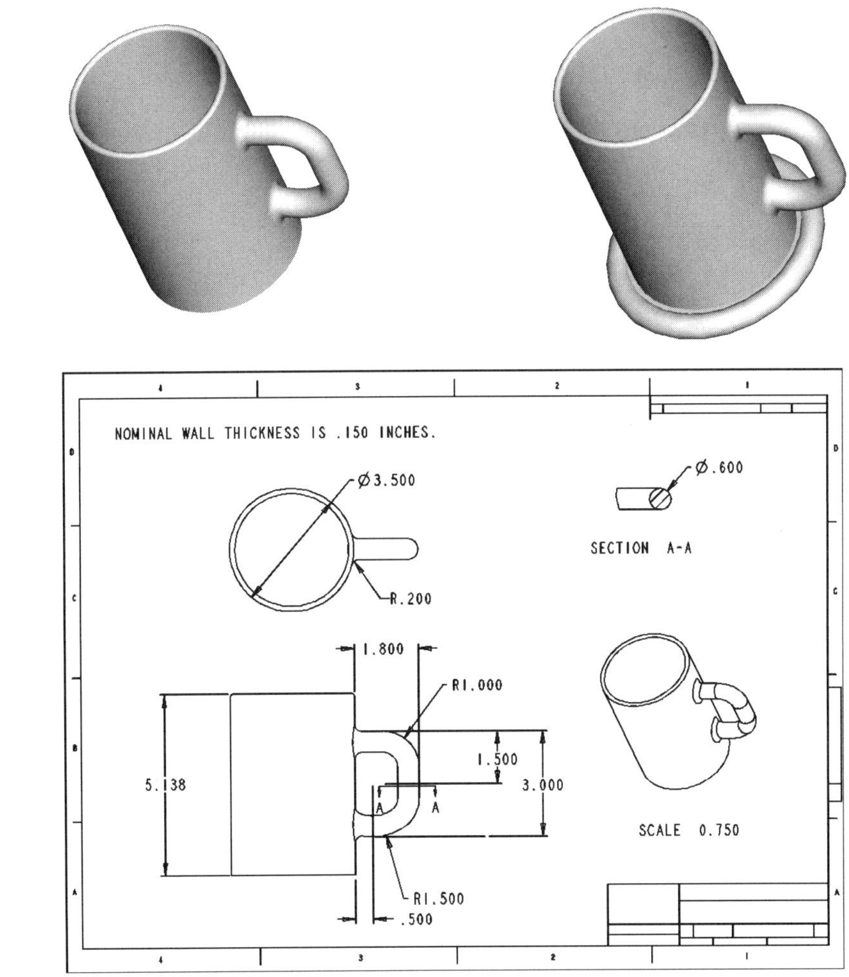

Figure 1-15

The result of this desisgn: a part, an assembly, and a drawing.

Summary

This simple design session has introduced you to some of Pro/ ENGINEER's most important features. By now you should have a basic understanding of the following concepts:

1. Parametric feature behavior
2. Dimensional constraints and relations
3. Assembly techniques
4. Bi-directional associativity between Pro/ENGINEER modules
5. How Pro/ENGINEER captures your design intent

Now that you have a basic overview of how Pro parts and assemblies are developed and interact, in Chapter 2 you will explore the facets of part creation in the Pro/ENGINEER environment.

Chapter 2
Working in the Pro/ENGINEER Environment

Mouse Control, Windows, Menus, Views, and Files

Introduction

In the previous chapter you were introduced to Pro/ENGINEER and supplied with illustrations of some basic features of the Pro/ENGINEER environment. In this chapter you will take a closer look at the environment. You will find descriptions of the mouse and mouse buttons, various windows that make up the Pro/ENGINEER interface, the menu structure, view commands, and file management.

Starting Pro/ENGINEER

The procedure for starting Pro/ENGINEER may vary from one system to another. On some workstations, keying in *pro* at the operating system prompt runs Pro/ENGINEER. On other systems you might select an option from a menu to start Pro/ENGINEER. Some workstations automatically run Pro/ENGINEER when you power up.

The actual command you use to start the program will vary depending on how your system is set up. Once Pro/ENGINEER is running, you will execute most commands and draw geometry using the mouse. The keyboard serves primarily for text and numerical data entry.

The left button on a three-button mouse is used for selecting commands from the menus. This button is also used for selecting geometry. Pressing the right mouse button on a menu item will provide a help window for the item. The center button can be used for the **Done** command in many menu commands. All three mouse buttons can change functions when certain commands are selected. For instance, when you are sketching feature geometry, one mouse button draws lines, another circles, and a third arcs.

> ➾ **NOTE:** *If you have a two-button mouse, you can emulate the middle mouse button by holding down the Shift key and pressing the left mouse button.*

When you are ready to quit Pro/ENGINEER, select **Exit** from the Main menu. A confirmation window will appear. Selecting **Yes** will exit you from Pro/ENGINEER, whereas **No** will abort the exit.

> ✘ **WARNING:** *Exiting Pro/ENGINEER does not automatically save your work. If you want to save your work, select* **DBMS ➥ Save** *from the MAIN menu.*

Try starting Pro/ENGINEER now, and then exit from it to familiarize yourself with how it works. Then restart Pro/ENGINEER.

Pro/ENGINEER Windows

After starting Pro/ENGINEER, you see the start-up screen consisting of the graphics window, the message area, and the menus. The following illustration shows this screen.

Figure 2-1
When you start Pro/ENGINEER and get past the copyright notice, this is the basic working window you see.

The Graphics Window

The first object you work on after beginning a session (a part, assembly, or drawing) appears in the main graphics window. In that you can work on several objects in the same session, other graphics windows may be visible in addition to the main graphics window. The main graphics window can never be deleted. It is always present, whether it contains an object or not. Any commands you select work on the object in the active window, which is the one with asterisks on either side of the window title, as shown in the following illustration.

Figure 2-2
Title bar of the active window.

The Message Window

The message window is located below the main graphics window. In fact, much of the message window is covered by the main graphics window. When you start Pro/ENGINEER, you see only two lines of text in the message window. The upper line shows you one of the following items:

- Prompts for additional information
- Your keyboard entries
- Status of an operation

> ✓ **TIP:** *Pay close attention to the message window. You may be unaware that the system is waiting for your input. Whenever your cursor crosses over a menu item, a help message for that item will be shown in the last line of the message window. This brief help message is useful in that a few of the menu item abbreviations are somewhat cryptic.*

Using your mouse, click on the thin border of the message window to make the entire message window visible, or move the cursor into the window and use the "front" key on the keyboard to toggle the window from front to back. The scroll bar on the right side of the window will allow you to see previous prompts and key-ins. The arrows on the scroll bar will remain stationary. Use successive clicks on the arrow that points in the direction you want to go. Click on the thin border of the graphics window to pop it over the message window and return to a working screen.

Figure 2-3
Pro/ENGINEER's message window.

✓ ***TIP:*** *Sometimes commands use two lines to display messages, and sometimes multiple messages are displayed faster than you can read them. To avoid missing an important message, move the entire main graphics window (click in the title bar at the top of the window and drag it up) so that you can see at least one additional line of text in the message window. If you place "visible_message_lines 5" in the configuration file* (config.pro), *the graphics window is always shifted up to allow five visible lines in the message window.*

Dialog Windows

There are several types of dialog windows you will encounter when working in the ProE environment. There are feature creation windows, model tree windows, a drawing text editor window, and so on. These windows allow you to monitor and change specific aspects of the function you are performing. They also contain commands that will be required in order to finalize the function you are performing. Most common are the **Apply** and **Okay** commands. The windows are located in either the lower right or upper left corners of the main graphics window. Pay close attention to these windows;

they will help simplify the command process. Examples of the dialog windows follow.

Command Menus

The menus on the right side of the screen contain the Pro/ENGINEER commands. Operations performed by Pro/ENGINEER execute as a result of commands you select from these menus.

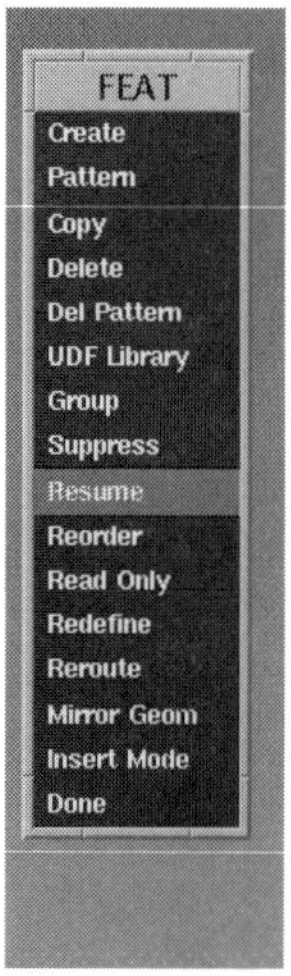

Figure 2-4 *A typical menu in Pro/ ENGINEER.*

After selecting a command, you might be required to select options from successive option menus. The selections you make define the specific action you want Pro/ENGINEER to take. Most option menus end in **Done** (or **Done-Return**, **Done/Return**, or **Done/ Accept**, and so on) and **Quit** (or **Quit/Abort**). The center mouse button will also complete a menu selection on most menus. **Done** signifies that you have finished selecting your options and would like to continue with the command. **Quit** will abort the command.

> ✓ ***TIP:*** *Pay attention to the options you select. After selecting* **Done**, *the option menu disappears, and you will not be able to return to it. If*

you think you made a mistake while selecting your options from a previous menu, **Quit** *the command and restart it.*

After a menu item is selected, it will be highlighted in red. Default menu choices are also shown in red. If you want to use the default choice, you do not need to reselect it in most cases. When your cursor crosses over a menu item, it will highlight it in pale blue. Pressing the left mouse button while a menu item is highlighted in yellow will select that menu item. If a menu item is dimmed (colored gray), you cannot choose it. The menu item may be dimmed because it is not a valid choice for your operation or because another menu item must be chosen first. Be aware, however, that these system colors can be changed by the user, so the previously mentioned color scheme applies to system default colors.

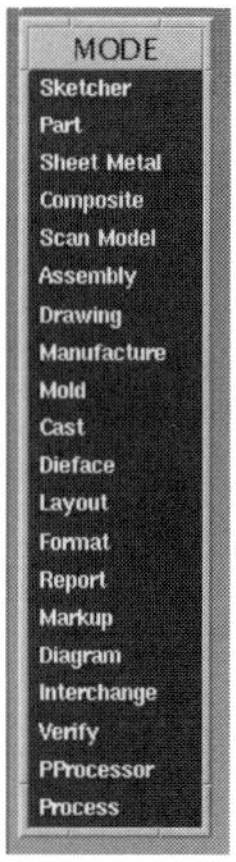

Figure 2-5 *The MODE menu.*

Understanding Modes

Pro/ENGINEER uses the term *mode* to represent various working environments. If you want to work on a drawing, use Drawing mode. For creating parts or assemblies, use the Part or the Assembly mode, respectively. Select the desired mode by choosing the **Mode** command from the MAIN menu.

All modes available within Pro/ENGINEER are listed on the MODE menu, but only the Pro/ENGINEER modules you have purchased will work. This book covers the following modes:

- **Part** — To create and modify parts
- **Assembly** — To group two or more parts into assemblies
- **Drawing** — To create dimensioned drawings of a part or an assembly

- **Format** — To create custom borders and title blocks for drawing sheets.

> ➻ **NOTE:** *If you have Pro/ENGINEER modules with floating licenses on your network, you can gain or release access to them through the* ***Misc*** ➥ ***FloatOptions*** ➥ ***Add/Delete*** *command. The FLOAT OPT menu will appear so that you can toggle the available floating modules on and off. Modes associated with these modules will not be available until you toggle them on. Ask your system administrator for assistance. Use* ***Misc*** ➥ ***List Options*** *to identify the installed modules.*

Mode Entry Options

When you select the mode you want to use, an entry menu will appear. This menu will be the same for all modes.

Figure 2-6
An example of an entry menu.

Regardless of which mode you are entering, the entry menu has the following options:

- **Create** — Creates a new object. Pro/ENGINEER prompts you for a file name.
- **Retrieve** — Retrieves an existing object. Pro/ENGINEER prompts you for a file name.
- **Search/Retr** — Lists all files in your current directory that match the mode you are entering. You can also browse through other directories until you find the desired file. Select the file from the name list to make it the object in the active graphics window.
- **List** — Lists all files within your Pro/ENGINEER search path and in memory.

- **Import** — Creates an object by importing data from standard file formats. You provide a new part file name and then select the file to import.

- **Quit** — Returns you to the MODE menu.

Working with Multiple Graphics Windows

As mentioned earlier, you can have several graphics windows open simultaneously. This is achieved by selecting the **Mode** command while a graphics window is currently in use. For example, you could work on a part in one graphics window and start Drawing mode in another graphics window.

In that only one of these graphics windows can be active, Pro/ENGINEER provides a command called **ChangeWindow** to switch between graphics windows. To use this command, select **ChangeWindow** from the MAIN menu, and then click inside the graphics window you would like to begin using.

If you want to quit your active graphics window and close it, use the **Quit Window** command on the MAIN menu. If your active graphics window is the main graphics window, it will not be closed, but its object will be removed. **Quit Window** simply removes the object from the display; however, the object will remain in memory.

> ➾ **NOTE:** *An object is* not *saved to the disk when you use* ***Quit Window****.*

> ✓ **TIP:** *If you accidentally quit the wrong graphics window, do not worry. In that the object is still in memory, just retrieve it again. The object will return as you last modified it, and you will not lose any of your work. Use* ***Search/Retr*** ➥ ***In Session*** *to retrieve the object from memory.*

Quit Window will not make another graphics window automatically active. You need to use **ChangeWindow** to select a new active window.

> ✓ ***TIP 1:*** *If you do not feel comfortable using multiple graphics windows and prefer a less cluttered screen, you can do all of your work using the main graphics window only. Just select* ***Quit Window*** *before you retrieve your next object. All of the objects will remain in memory and can be retrieved as they were last modified. Just remember to save each object prior to exiting Pro/ENGINEER.*

> ✓ ***TIP 2:*** *If you use multiple windows as your normal work environment, keep in mind that if you ever need to retrieve lost data, through the running of a trail file (trail files are explained in Chapter 15), the same number of windows will need to be displayed.*

View Orientation Commands

Looking at an object from a different direction is a common procedure in creating a part. View orientation commands are used for this purpose. There are several methods for orienting views in Pro/ENGINEER. The orientation commands are made available when you select **View ➥ Orientation** from the MAIN menu. All view orientation changes will cause Pro/ENGINEER to recalculate the view, which takes longer than the simpler view control commands described later in this chapter.

Default View

The 3D view is the default for all models in Pro/ENGINEER. To change the model orientation to the default view, select **Default** from the ORIENTATION menu. There are options that you can set in the *config.pro* file that change the orientation of the default view. These options are discussed in a later chapter.

Orientation by Two Planes

The most common method of orienting views is by selecting two orthogonal, planar surfaces. From the ORIENTATION menu, select a direction (**Front**, **Back**, **Top**, **Bottom**, **Left**, **Right**). Next, select the planar surface to face that direction. Repeat the previous steps for a different direction that is orthogonal to the first direction. The easiest way to learn this technique is to imagine that each surface has an arrow that points away from the model.

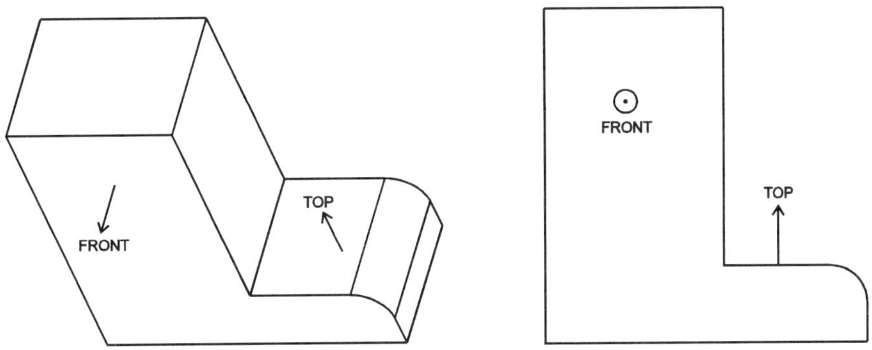

Figure 2-7
View orientation by planar surfaces:
default (trimetric) view (left) and reoriented view (right).

Usually you know which surface will face toward the front when you decide on a new orientation. As shown in the example, the L-shaped surface was selected to face the front, which means its imaginary arrow points out of the screen. Another planar surface was selected to face the top, which means its imaginary arrow points toward the top of the screen.

Orientation by Angles

Another method of orienting the view is by specifying angles of rotation. This choice is preferred by people who desire angular precision that other methods do not provide. Select **Angles** from the ORIENTATION menu. An ANGLES menu appears with the following choices:

- **Horiz** — To rotate about the horizontal screen axis, select **Horiz** and key in an angle value.
- **Vert** — To rotate about the vertical screen axis, select **Vert** and key in an angle value.
- **Norm** — To rotate about an axis normal to the screen, select **Norm** and key in an angle value.
- **Edge/Axis** —To rotate about a part edge or axis, select **Edge/Axis**, select the edge or axis to rotate about, and then key in an angle value.
- **Done/Accept** — After specifying the angles, select **Done/Accept** to reorient the view.
- **Quit/Abort** — To abort without reorienting the view, select **Quit/Abort**.

Orientation by Axis

Pro/ENGINEER provides a method for orienting models that have datum axes and no planar surfaces.

- **Axis-vert** — This command will orient the model so that a selected axis is aligned vertically. After choosing **Axis-vert** from the ORIENTATION menu, select the axis.

- **Axis-horiz** — This command will orient the model so that a selected axis is aligned horizontally. After selecting **Axis-horiz** from the ORIENTATION menu, select the axis.

These commands are ideally suited for instances wherein your model is symmetrical about the selected axis, as in a wheel. Pro/ENGINEER automatically determines the rotation about the selected axis. If you need to rotate the model to see a different side, use **Angles** ➥ **Edge/Axis** to rotate the model after using one of these commands.

Orientation by Spinning

There are several ways of "spinning" your model to create a new view. Spinning means that you rotate the model dynamically. Some workstations allow you to spin shaded models, whereas others only allow you to spin wireframe models. Spinning speed (or how fast the screen updates) is dependent on your hardware and the size of your model. Spinning will not allow you to orient your view at a precise angle. It is best used when you want to see your model quickly from a different point of view, and you do not know which combination of angles is required. After selecting **Spin** from the ORIENTATION menu, the following methods are available for spinning:

- **Mouse** — You can use the mouse to control the spinning. The *middle* mouse button must be picked in the graphics window to initiate the spinning. Move the mouse to control the direction of rotation. Press the middle mouse button again to stop spinning. The *left* mouse button will allow you to zoom in on the model, similar to the **Zoom In** command. The *right* mouse button will allow you to dynamically pan the view. The mouse method of spinning is used most frequently because of its simplicity.

- **Center** — This option provides the same functionality as mouse spinning, but you also get a control panel with sliders to control three-axis rotation, zooming, and panning.

Click on the slider with the left mouse button to begin moving the model. Click again with the left mouse button to let go of the slider. When you have finished orienting the view, click in the blank area on the control panel above the sliders to stop spinning and leave the view in its current orientation.

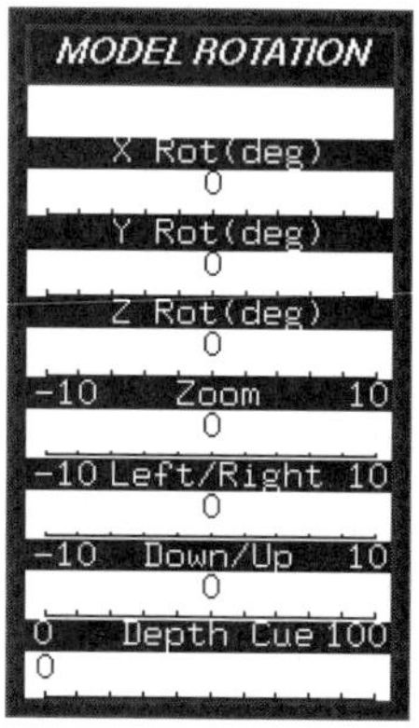

Figure 2-8
The control panel for spinning by center.

- **Edge/Axis** — You can use this option to dynamically rotate about a model's edge or datum axis. When you select this option, you are asked to choose an edge or datum axis to rotate about. After selecting this reference edge or datum axis, a control panel is provided with sliders for controlling rotation about the reference, for zooming, and for panning. When you have finished orienting the view, click in the blank area on the control panel above the sliders to stop spinning, and leave the view in its current orientation.

- **Coord Sys** — This option will allow you to spin about a datum coordinate system. When you select this option, you are asked to choose a datum coordinate system to rotate about. Once you have selected the coordinate system, you will see a control panel with sliders for controlling the rotation about each of the axes of the coordinate system, for zooming, and for panning. When you have finished orienting the view, click in the blank area on the control panel above the sliders to stop spinning, and leave the view in its current orientation.

- **Point** — This option allows you to spin about an arbitrary point you pick on the model (this is not a datum point). Once you have selected the point, you will see a control panel with sliders for controlling the rotation about each of the axes of the coordinate system, for zooming, and for panning. When you have finished orienting the view, click in the blank area on the control panel above the sliders to stop spinning, and leave the view in its current orientation.

Saving Views

Pro/ENGINEER allows you to save views and retrieve them later. It is a good idea to save a view of your model if you think you might need it again. Saved views are handy in Part mode and Assembly mode when you want to look at standard top, right, or front views of your model. They are also useful for orienting general views in Drawing mode. Keep in mind, however, that by using this technique you run the risk that views may become disoriented due to model changes.

Saved views are accessed through the **Names** choice on the MAIN VIEW menu. To use, create, or delete saved views, select commands available on the SAVE/RETR submenu.

- **Retrieve** — Selecting **Retrieve** will allow you to choose the saved view name from a name list. After selecting the name from the name list, the view will change to the saved orientation.

  NOTE: *The scale and zoom factors of this view will be the same as when it was originally saved.*

  NOTE: *These view setups are particularly useful for creating reference dimensions within the model (part or assembly).*

- **Save** — Once you have oriented your view to the desired position, select **Save** and you will be prompted for a view name.

- **Delete Name** — If you want to delete a saved view, choose **Delete Name** and select the view name you wish to delete from the name list.

View Control Commands

There are two different view menus. Part and Assembly mode use the same menu, but Drawing mode uses a menu that omits the **Refit** command. These commands allow you to adjust view scaling. Most are located on the PAN-ZOOM menu, but Repaint has been located on the MAIN VIEW menu for easier access. These commands do not alter the orientation of the model, and most do not cause Pro/ENGINEER to recalculate or regenerate the view.

- **Repaint** — The **Repaint** command located on the MAIN VIEW menu will redraw the screen with wireframe graphics. This command will dehighlight any highlighted geometry and clear any model dimensions from the display. **Repaint** will initiate a screen regeneration if you have altered the model view by turning layers on or off or by shading the model.

- **Pan** — From the PAN-ZOOM menu, the **Pan** command will scroll the screen when you select a new screen center point with the left mouse button.

- **Zoom In** — To zoom in on a section of the screen, select this command from the PAN-ZOOM menu and draw a rectangle around the area you want to view. Use the left mouse button to pick opposite corners of the rectangle.

- **Zoom Out** — Each time you select **Zoom Out** from the PAN-ZOOM menu, the screen image will be reduced by a factor of 2x.

- **Reset** — The **Reset** command on the PAN-ZOOM menu will fit your model to the window. Any dimensions or sketched entities that extend beyond the boundaries of your part are not considered when you reset the screen.

❒ **MDBO PREVIOUS MDNM** — The previous command is self-explanatory: it takes you back to your previous view orientation. This is useful for viewing objects if you are toggling between views such as top, bottom, and side.

❒ **New Window** — Another graphics window containing the object in the current active window can be created. Select **New Window** from the MAIN VIEW menu and the new graphics window is displayed. Once displayed, the new window becomes the active graphics window. The object in the window can be oriented independent of the orientation of the object in the other window. Entity selections can be made from either window. For example, the sketch plane can be selected from one window, and the horizontal or vertical orientation plane can be selected from the other. When you no longer need both windows displayed, select **Quit Window** and the currently active graphics window will disappear. Select **ChangeWindow** and then click in the remaining graphics window to make it the active graphics window.

❒ **Perspective** — If you wish to see your part or assembly in a perspective-type view, such as for product presentations, this command is well suited. A control panel is provided with sliders for controlling orientation.

❒ **Layer Display** — This command allows you to display or blank out items that reside on any given layer. This is useful for such things as text that has been added for reference but is not necessarily required, say, on detail drawings.

❒ **Alter View**—(When in assembly mode, you may want to orient the assembly or the component you are in the process of assembling. If the object you want to orient is not in the active graphics window, select **Alter View**. Next, select the graphics window of the object you want to orient. The view commands will now apply to that graphics window.

Managing Files in Pro/ENGINEER

When you need it, there is nothing as important as the ability to save a file, find a file, or, in the case of growing clutter, delete a file. The **DBMS** menu item on the MAIN menu displays a file management command menu.

Commands for saving, retrieving, and deleting your files follow:

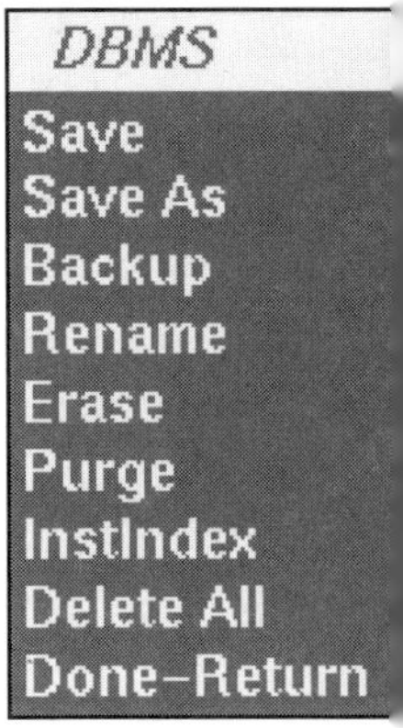

Figure 2-9
The DBMS menu.

- **Save** — This command saves an object to the disk. **Save** does not overwrite previously saved versions of the file. The object in the active graphics window will be the default choice for saving. However, you can enter the name of any object currently in memory to save it.
- **Save As** — An object can be copied to create an identical object, but with a different file name. Copying an assembly while a drawing of the same name is located in the Pro/ENGINEER search path causes the drawing file to be copied as well. Copying an assembly allows the option to copy each component and component drawing. The new files are copied to the current working directory but are not placed into memory or into a graphics window.
- **Backup** — If you want to save the object to a directory different than the current working directory, use this command.
- **Rename** — Renaming an object changes its name in memory. If the object is in the Pro/ENGINEER search path, it is also renamed on the disk.

✓ **TIP:** *Renaming a part used in an assembly or drawing can cause a failure the next time the assembly or drawing is retrieved if the renaming is done improperly. To rename a part, make sure you have all associated assemblies and drawings in memory. When all objects are in memory, rename the part. The assemblies or drawings in memory must also be saved because they now have a different reference than the previously saved versions.*

- **Erase** — Erasing an object removes it from memory (back to the last time the object was stored) but has no effect on any files on the disk. If you wish to try a new design scenario but do not wish to create a new part or save the changes, this allows you to make the changes and erase them without affecting your original design.

- **Erase NotDisp** — This command is the same as **Erase**, except that instead of erasing the current object, it will erase all objects that were active during the current session.

- **Purge** — Pro/ENGINEER has a purge utility that can be used to delete old versions of objects from the disk. Only the latest version of a file will remain.

- **InstDbms** — Pro/ENGINEER keeps track of all family table parts within a directory. If you add, delete, or rename an instance of a part, you should select this command and update the instance index (*.idx*) file.

- **Delete All** — An object and all files associated with it can be deleted from memory and from the disk with the **Delete All** command.

✗ **WARNING 1 :** *The* **Delete All** *command removes all associated files. If you delete a drawing, it will delete the object in the drawing as well. If you delete an assembly, it will delete all components in the assembly.*

✘ **WARNING 2 :** *Pro/ENGINEER provides no utilities to automatically save your work. It is your responsibility to ensure that you have saved your work before exiting Pro/ENGINEER or erasing the object from memory. Always save your work on a regular basis. Watch for the Save icon throughout the text.*

SAVE

Establishing a rhythm or habit of saving your work on a frequent basis is a small investment in the large volume of work you can quickly produce with Pro/ENGINEER.

Conclusion

By now you should have a basic familiarity with the Pro/ENGINEER environment. You know how to select commands, how to control views, and how to save files. With these basic skills, you are now ready to create a part, which will be used in an assembly in the following chapters. You will learn the philosophy of how to approach each feature creation to yield the best and most simplistic method of part creation.

Chapter 3
Building a Simple Part

An Introduction to Basic Part Building Commands

Introduction

This chapter introduces the basic concepts for designing a part. You will be shown how to begin a design and how to use Pro/ENGINEER to achieve your design intent. Commands and concepts covered in this chapter include the following:

- ❐ Deciding how to start
- ❐ Setting up and using datum planes
- ❐ Choosing and orienting a sketch plane

❒ Sketching a section

❒ Adding dimensions

❒ Adding material with extruded and revolved protrusions

❒ Removing material with extruded and revolved cuts

❒ Making rounds

You will design a wheel for a toy car. The wheel will look like the following illustration when the design is complete.

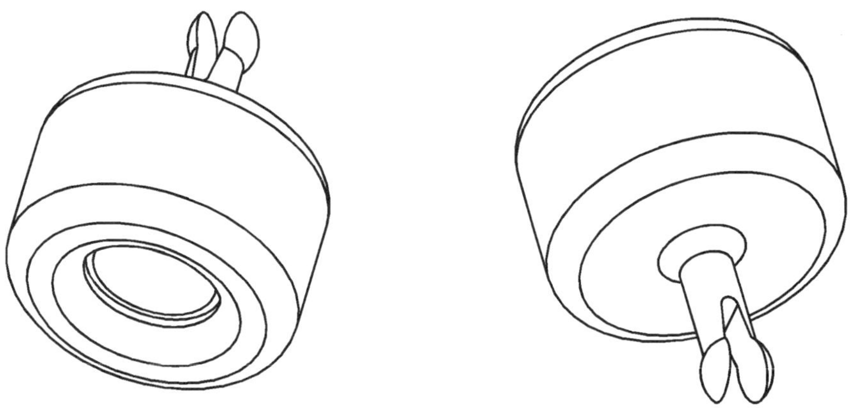

Figure 3-1
The completed wheel.

Beginning a Part

When you start building a part, you need to use Pro/ENGINEER's Part mode. Select **Part** from the MODE menu. This will place you in Part mode and display the entry menu for Part mode. In that you are creating a new part, select **Create** from the menu. A prompt for the new part's file name will appear in the message window.

```
Enter Part name[PRT0001] : wheel ↵
```

NOTE: *Pro/ENGINEER's prompts will always appear in the message window. Watch for the prompts after you have selected commands, in that they often guide your next step. You now have a part named "wheel" in memory. When you save this part, Pro/ ENGINEER will create a file called* wheel.prt.

Setting Up Datum Planes

Creating a set of default datum planes at start-up of a new part is considered good design practice. The datum planes represent three orthogonal planes to be used for sketching, view orientation, and dimensioning. To place these planes, select the following command sequence:

Feature ➡ Create ➡ Datum ➡ Plane ➡ Default

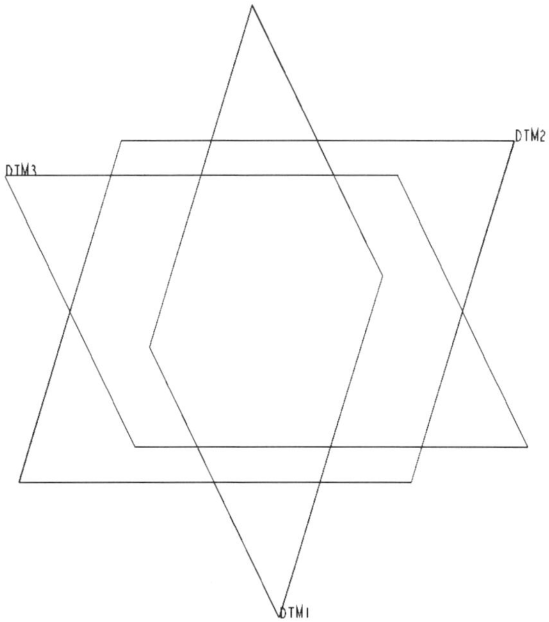

Figure 3-2
The default datum planes.

The following planes are not required, but they offer several advantages.

- **Controlling the Orientation of your First Feature**. You have the flexibility to sketch on any of the three planes for your first solid feature. When you do not have these planes to choose from, Pro/ENGINEER will assume a sketch plane for you, and you lack control over the default part orientation.

- **Aligning and Specifying Dimensions**. These datum planes provide excellent references for alignments and dimensions. Routine use of datum planes for specifying part dimensions is simply good engineering practice.

- **Minimizing Feature Dependencies**. By using these datum planes for sketching, specifying, and aligning, you will minimize the number of dependent (child) features. Features dependent on other features have a "parent/child" relationship. These relationships are often necessary and functional. Sometimes these relationships occur unintentionally, such as when you are orienting a sketch view or constraining a section. To avoid unintentional parent/child relationships, use your default datums whenever possible.

- **Creating Saved Views**. As mentioned previously, saved view orientations should be created using these default datum planes. Now that you have created your datum planes, you should also create a datum coordinate system. You can choose to create this datum coordinate system by selecting either of the following command sequences:

Create ➥ Datum ➥ Coord Sys ➥ Default ➥ Done

Create ➥ Datum ➥ Coord Sys ➥ 3 Planes ➥ Done

The first command will simply place a default coordinate system at the intersection of the three default datum planes. You do not have any control over the orientation of the axes. The second command will ask you to select three planes and to place a coordinate system at their intersection, and will allow you to orient the axes in any direction.

For this example, you will use the **Default** datum coordinate system. Placing a coordinate system is not required before you begin a part. However, you will find that certain commands will require you to select a coordinate system. It is easy to create this coordinate system now and simply have it available when your command requires it.

> ➾ **NOTE:** *You can control the display of datums with the options on the ENVIRONMENT menu. Pick* ***Environment*** *from the MAIN menu to view these options.*

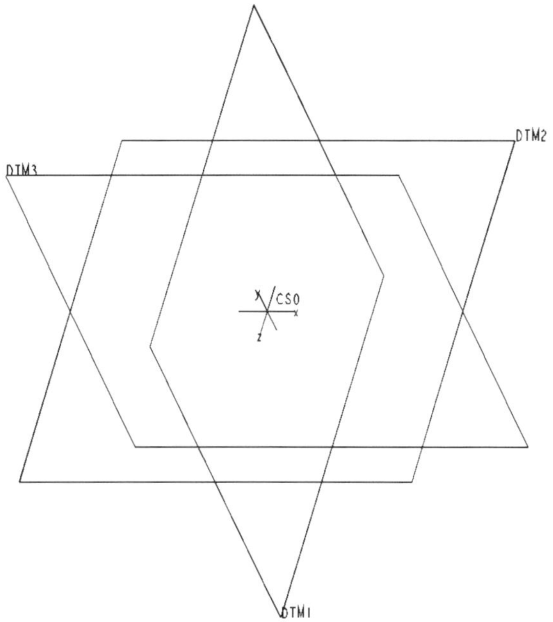

Figure 3-3
The default datum and the default coordinate system.

A Word About Selecting Geometry

There are many instances in which Pro/ENGINEER prompts you to select geometry. You will see this later in the chapter, when you are prompted to select a sketch plane. Whenever Pro/ENGINEER prompts you to select geometry, the GET SELECT menu is displayed. There are several methods that can be used to select geometry and that work in other modes.

- **Pick** —With this selection, you can use the left mouse button to pick geometry that is not hidden and is not datums nor axes. This is the default choice on the GET SELECT menu.

  NOTE: *Different types of entities are selected with the mouse in different ways. To select a surface, do not try to pick it on one of its edges. Select it by clicking the left mouse button in the middle of the surface. Because an edge is shared by two surfaces, you may not get the one you wanted. Datum planes should be selected by picking on their boundary or on their name tags. Select datum axes by picking on their dashed line. Axes cannot be selected by picking on their name tag.*

- **Query Sel** —This method of selection is used if the geometry is hidden or if several entities are in close proximity to one another. After selecting **Query Sel** from the GET SELECT menu, select the object with the left mouse button. You can either **Accept** the object that highlights, or select **Next** to highlight a different object. If you pass up the object you wanted, you can use **Previous** to go back to a previously highlighted object. Most of these commands can be accessed with different mouse buttons. This will save you the trouble of traveling back and forth from the menu to the graphics window. In that **Pick** is the default selection method, you can click the right mouse button to activate **Query Sel**. After selecting the object with the left mouse button, you can use the middle mouse button to **Accept** it or the right mouse button to go on to the **Next** object.

- **Sel By Menu** —This selection method allows entities to be selected without picking in a graphics window. The SPECIFY BY menu will display so that you can select how you will specify the desired entity. Entities that have name tags (such as datum planes, axes, or datum coordinate systems) can be selected from a name list by selecting **Name**. If you provide a name for a feature, that feature can also be selected by its name. Another method of selecting features is by typing in their feature number or feature internal ID number.

  ➾ ***NOTE:*** *If you want to select a suppressed feature, use the feature internal ID method or select it through the model trees.*

Use **Number** or **ID** from the SPECIFY BY menu to select features with the following methods:

- **Unsel Last** —When selecting multiple entities, you can pick **Unsel Last** to remove the last entity selected from the selection set.

- **Unsel Item** —When selecting multiple entities, you can pick **Unsel Item** and then pick a selected entity to remove it from the selection set.

- **Pick Many** —In Drawing mode and in the Sketcher, many entities can be simultaneously selected by using this selection method. Entities can be selected by placing a pick box around them, by specifying a chain of entities that are connected end to end, or by placing a multisided boundary around them.

- **Done Sel** —Pick this when you are finished selecting objects. This command is also initiated by pressing the middle mouse button.

- **Quit Sel** —To stop selecting objects and abort any you have selected thus far, use the **Quit Sel** command.

> ➾ **NOTE:** *By default, the message window will show only a description of the selected entity when **Query Sel** is the selection method. If you place "provide_pick_always yes" in the configuration file* (config.pro)*, the description is displayed during all selection methods.*

Choosing the First Solid Feature

Up to this point, you have created only an environment in which to design a part. Now you can add solid features to build your part. In that this is a new part without any existing solid features, Pro/ENGINEER lets you use only those solid features that ad d material. The most common command for adding material is the versatile **Protrusion** command. It has many options that will be covered as you use it throughout the following chapters.

To Start the Design

A good starting point for a new part is to determine the type of shape that will represent the majority of your part's material. In our example of the toy car wheel, the best way to start would be making a cylinder. The cylinder would be as large as the wheel and tire portion of the finished part. The easiest way to create a cylinder in Pro/ENGINEER is to project a circle to a specified length. The **Extrude** opt ion is designed for this purpose. Therefore, the first solid feature will be referred to as an extruded protrusion. To start making the feature, select the following commands, starting from the FEAT menu:

Create ➥ Solid ➥ Protrusion ➥ Extrude | Solid | Done

The ATTRIBUTES menu will appear next. The ATTRIBUTES menu lets you choose how the feature will be created with respect to the sketch plane. The **One Side** option means that the feature will be extruded to only one side of the sketch plane. The **Both Sides** option means that the feature will be extruded to both sides of the sketch plane. For this feature, select:

One Side

The SPEC TO menu allows you to choose how the extrusion will end. These options willl be explained as you use them throughout the following chapters. **Blind** means that the extrusion will stop at a specified distance from the sketch plane. For now, select **Blind** | **Done.**

Selecting the Sketch Plane

Before you start sketching your section, you need to select the plane you will sketch on. For the first solid feature, your sketch plane is your choice. You need to understand that the sketch plane you select will determine the part orientation in the default view. Having your part oriented in a particular fashion is not critical, but it is convenient if your default part orientation is one you have selected. To make your wheel with the same default view orientation as our example, select DTM3 as the sketch plane.

> **NOTE:** *Select datum planes by picking on their edges or their name tags.*

Selecting the Feature Creation Direction

The next step is to determine the direction for the feature creation (in that this is a one-sided protrusion). A red arrow will be shown, pointing from the selected sketch plane to indicate the direction of the extrusion.

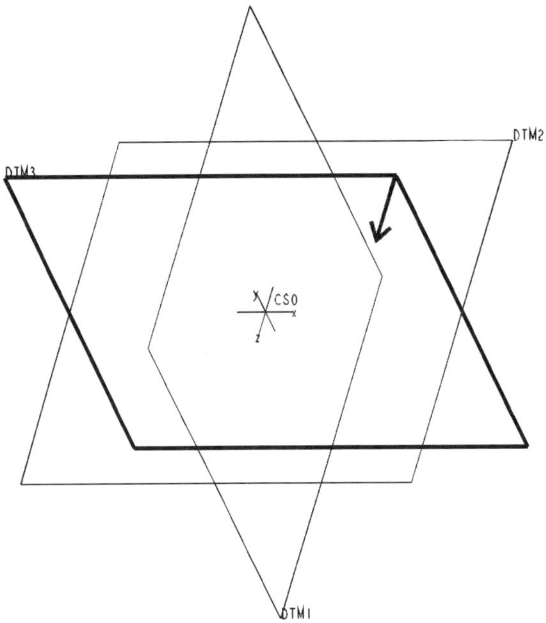

Figure 3-4
DTM3 with the feature creation direction arrow shown.

To accept the direction shown by the red arrow, go to the DIRECTION menu to select **Okay**.

Selecting a Horizontal or Vertical Reference Plane

There is one more step in the process before you can sketch. Pro/ENGINEER requires that you identify a plane to be used as a horizontal or vertical reference in the sketch view. This plane must be *perpendicular* to the sketch plane. Select **Top** from the SKET VIEW menu, and then select DTM2.

This means that the yellow side of DTM2 will face toward the top of the screen in the sketch view. This also tells Pro/ENGINEER that DTM2 will be used as the horizontal reference for this feature's section. The purpose of the horizontal or vertical reference is to fully constrain the sketch view (using two perpendicular planes) and to allow you to easily create angled sections. Most of the time you should use one of your original three datum planes for this reference.

> **NOTE:** *The sketch plane will be rotated so that it is parallel to the screen. Protrusion features orient the sketch view such that the feature is created toward you (or "out of the screen"). You should keep this in mind when selecting the horizontal or vertical reference plane so that you can anticipate how the sketch view will appear.*

Creating Sketch Geometry

Now you are ready to sketch. Your sketch view should look like the next diagram.

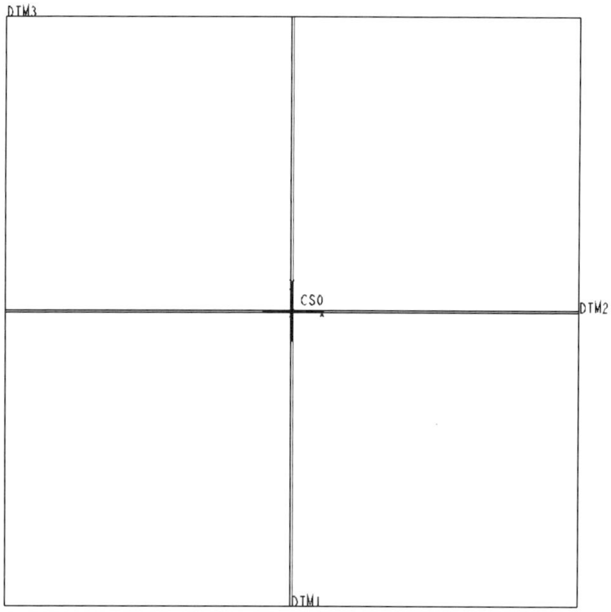

Figure 3-5
Sketch view prior to sketching the geometry.

Create a cylinder by starting with a circle. You will place a circle so that its center is located on the datum coordinate system. This can be done by moving the cursor over the datum coordinate system and then pressing the middle mouse button (for circle placement). This action establishes the center of the circle. Now move the cursor until the circle appears to fill about half the window, and then press the middle mouse button again. You have now placed a circle. You may have noticed that it is red while active and that it turns cyan when fixed.

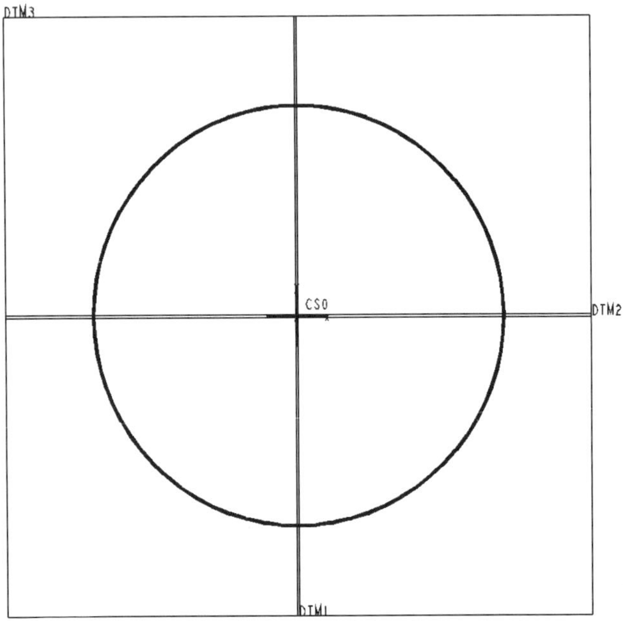

Figure 3-6
Sketch view after sketching the geometry.

Constraining the Geometry

There are two objectives for constraining the geometry: locating the section with respect to the existing part (the datums) and specifying the dimensions of the geometry. To locate the circle, align the circle to the datum coordinate system. Pick **Alignment ➡ Align** from the SKETCHER menu and select the circle. Next, select the name tag for the datum coordinate system (the "CS0" text) on the screen. This aligns the center of the circle to the origin of the coordinate system, and locates the section with respect to the part.

Check to see if the message window now reads ALIGNED. If not, repeat the prior commands and carefully select the locations for alignment. To specify the size of the circle, place a diameter dimension on the circle. Select **Dimension** from the SKETCHER menu and select the circle *twice* with the left mouse button. (Selecting only once will give you a radius dimension.) Move the cursor away from the circle and press the middle mouse button to place the dimension and then regenerate. By this time, your section should look something like the following diagram.

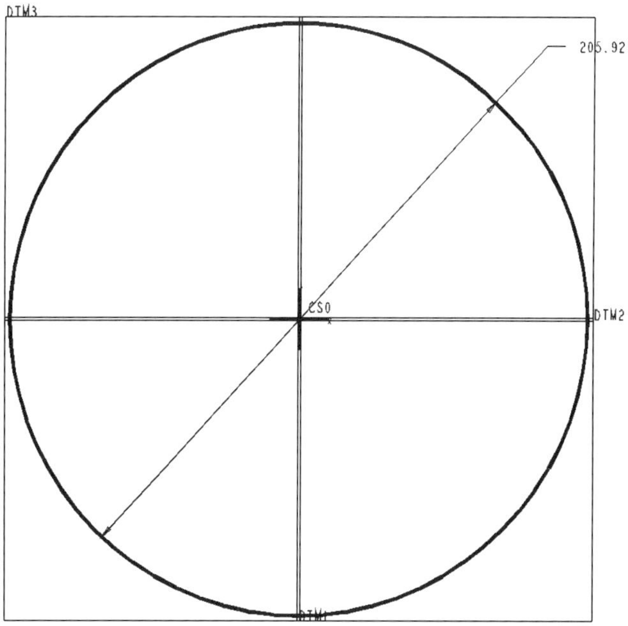

Figure 3-7
Sketch view after constraining the geometry.

Modifying the Geometry

Now that the section has been regenerated successfully, you can modify the section to the desired size. Pick **Modify** from the SKETCHER menu and select the diameter dimension text. Pro/ ENGINEER will prompt you for the new dimension value. Enter `.85`, as shown in the following diagram.

```
Enter new value [205.9200]:   .85 ↵
```

Pick **Regenerate** from the SKETCHER menu. Your section should now look like the following diagram.

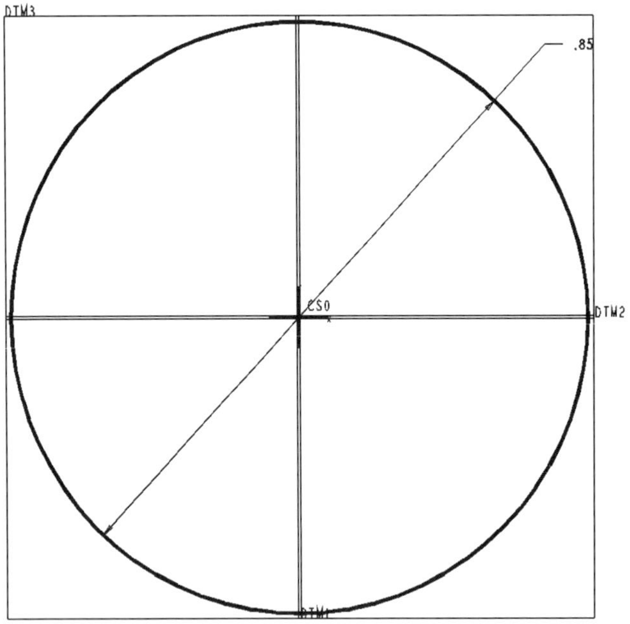

Figure 3-8
Sketch view after modifying the geometry.

Your section is now complete. Select **Done** from the SKETCHER menu to continue with the feature creation.

Completing the Feature

The last step in creating the feature is to provide Pro/ENGINEER with any other information it needs. In this example, you need to tell Pro/ENGINEER how far to extrude the section. Pro/ENGINEER will prompt you for the dimension. The SPEC TO menu will appear. These options will allow you to determine how the extrusion will be controlled. BLIND is the default, and is what you will using in the creation of this part. The **Blind** command requires that a dimension be entered for the depth of the extrusion, enter the depth of .5. The Protrusion dialog window will appear and request your final command. In this case, you want to apply all of the specifications you have entered, so the **Done** command is your last step.

The feature is now created. Congratulations! Select

View ➥ Orientation ➥Default

from the MAIN menu to see your part.

> **NOTE 1:** *The DIALOG WINDOW menu offers several options for changing any feature creation prior to finalizing it. If you oriented the viewing plane to default and selected the* ***Preview*** *command, you would see the extruded depth of the feature. If it were not satisfactory, it could be changed at this time. You would use the* ***Define*** *command and then the* ***Depth*** *command. By entering a new value, the depth of the extrusion would be changed.*

> **NOTE 2:** *Pro/ENGINEER provides several ways to display the model. The default method is to display all edges as the same solid color, such as in a traditional wireframe model. A hidden line representation is usually more desirable. To change this display characteristic, pick* ***Environment ➥ Hidden Line | Done-Return*** *from the MAIN window. Display characteristics are discussed in greater detail in Chapter 5.*

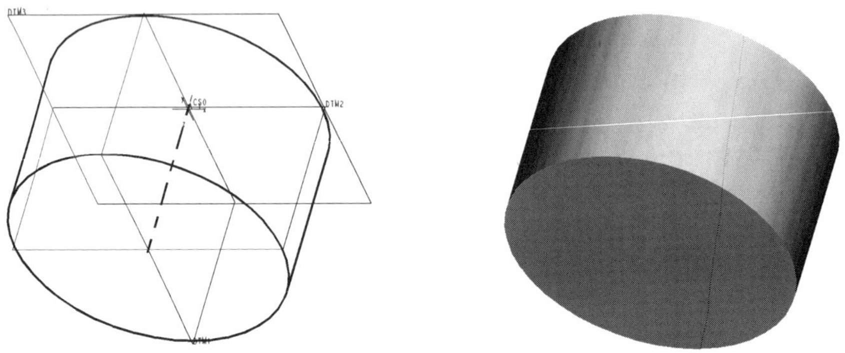

Figure 3-9
The first solid feature in a wire frame and a shaded view.

SAVE

DESIGN NOTE: *This feature references two datum planes (the sketch plane and the horizontal reference plane). This feature also references the datum coordinate system. All three of these features are "parents" of this first protrusion. Try to use your first three datum planes as references when it makes sense. This will make your new features "children" of the main part datums rather than other part features. This strategy will help you if you need to make major changes later on.*

NOTE 1: *When Pro/ENGINEER makes a cylinder by extruding a circle, it will automatically place an axis through the centerline of the cylinder.*

NOTE 2: *The default datum planes' boundaries have adjusted to match the envelope of the part. These boundaries are not fixed. In that datum planes represent infinite surfaces, their boundary size is constantly adjusted to match the overall size of your part.*

Making a Wheel with an Extruded Cut

Now that you have created some material, you may notice that the feature commands previously grayed out are now available. For the second feature, you will make an extruded cut to begin forming the wheel. The **Cut** command has the same structure as the **Protrusion** command. You will see the same menus with the same options for these two commands. Select the following commands, starting from the FEAT menu:

Create ➡ Solid ➡ Cut ➡ Extrude | Solid | Done

Pick **One Side | Done** from the ATTRIBUTES menu to tell Pro/ ENGINEER that the section will be extruded to one side of the sketch plane. Now you need to select your sketch plane. For this feature, you need to select the front planar surface of the cylinder.

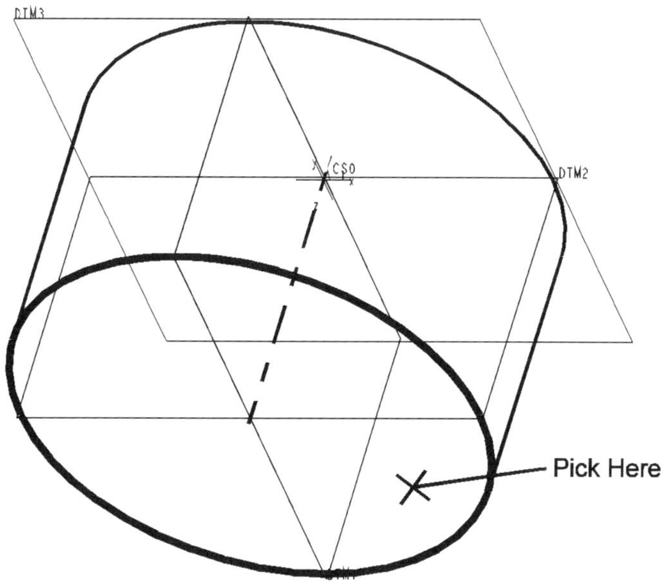

Figure 3-10
The sketch plane for this extruded cut feature.

➾ **NOTE:** *Part surfaces should be selected by picking inside the surface boundaries. Picking on the boundary of the surface is vague because that boundary represents the common edge between two different surfaces. In addition, picking near the border of a datum plane will select the datum plane rather than a part surface.*

➾ **DESIGN NOTE:** *By selecting the front surface of the cylinder as the sketch plane, you have given Pro/ENGINEER a notion of your design intent. Using this surface as the sketch plane will ensure that this cut feature maintains its extrusion depth dimension from the outside face of the part. When the wheel changes in width, this depth will not vary.*

The next step is to specify the direction for feature creation. In that you are trying to remove material from the cylinder, the direction of your cut should be toward the cylinder.

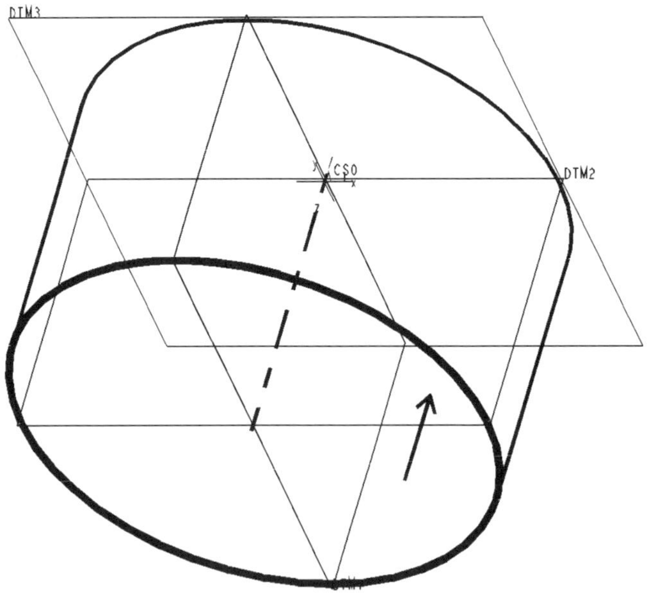

Figure 3-11
The direction for this extruded cut feature.

> ➻ **NOTE:** *Pro/ENGINEER will default to show the arrow in this direction. Pro/ENGINEER knows that it cannot cut any material from this part in the other direction.*

> ✓ **TIP:** *Sometimes the direction arrow is displayed on top of other entities in the view. This makes the direction arrow difficult to locate at times. Pick **Flip** several times to make it toggle back and forth (or flash) so that you can find it. Do not forget to leave it pointed in the desired direction before selecting **Okay**.*

From the DIRECTION menu, select the following command:

Okay

Now you need to specify the horizontal or vertical reference plane for the sketch view. You should use the same one as in the previous feature. Select

Top

from the SKET VIEW menu and pick DTM2. Your sketch view has now been oriented.

> ➻ **NOTE:** *Cut features orient the sketch view such that the feature is created away from you ("into the screen"). You should keep this in mind when selecting the horizontal or vertical reference plane so that you can anticipate how the sketch view will appear.*

Using the middle mouse button, sketch the circle, as shown in the next diagram.

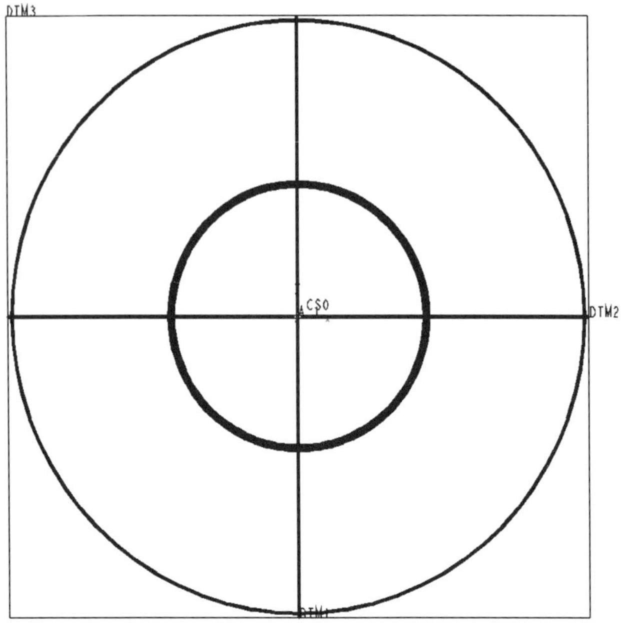

Figure 3-12
Section geometry for the extruded cut that begins to form the wheel.

Align the center of the circle to the axis of the cylinder using the following commands:

Alignment ➡ Align ➡ Query Sel

> ✓ ***TIP:*** *When there are multiple items you can align to, use* ***Query Sel*** *to select the item and read its description in the message window. This will enable you to control which item becomes aligned.*

Select the center of the circle with the left mouse button. Accept this center point with the middle mouse button. Pick **Query Sel** again, and select the center of the circle. This time, the axis (A_1) will highlight. Accept it with the middle mouse button.

➾ ***NOTE 1:*** *Axes cannot be selected by picking on their name tag. You must pick the dashed line of the axis (or its point if it is normal to the screen).*

➾ ***NOTE 2:*** *A sketched circle can be aligned by selecting its center or circumference.*

➾ ***DESIGN NOTE:*** *Aligning the center of the sketched circle to the axis of the cylinder ensures that these features will always remain concentric. If you decide to move the original cylinder and unalign its sketched circle from the datum coordinate system, this cut feature will move with the cylinder. Place a diameter dimension on the circle and regenerate the section. Modify the diameter value after you have successfully regenerated the section.*

`Enter a new value [.4500]:` **`.4`** ↵

Regenerate the section again to make the circle update to its correct size. Select **Done** from the SKETCHER menu. You will now see a new prompt asking you which side of the circle should be cut away. Point the arrow so that it indicates the inside of the circle and select **Okay** from the DIRECTION menu.

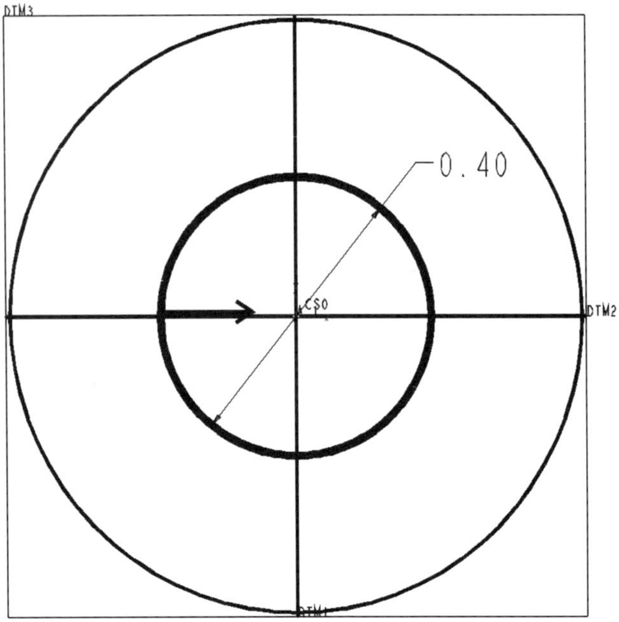

Figure 3-13
The finished section for the wheel is cut.

The selection will cut away the material to the inside of the sketched circle. Pick the following from the SPEC TO menu:

Blind ➥ Done

Enter the depth of this extruded cut feature.

`Enter depth [1.50]:` **.1** ↵

Pick **OK** from the dialog window. The cut feature should now be created. Select

View ➥ Orientation ➥ Default

to see the cut in the default view.

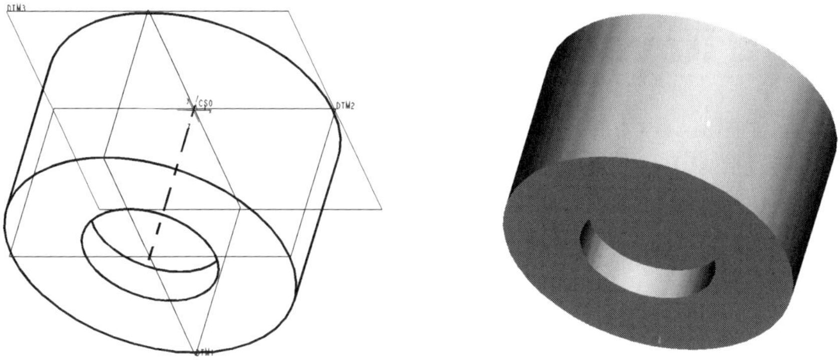

Figure 3-14
The wheel after placing the extruded cut feature.

> **DESIGN NOTE:** *You have created a parent/child relationship between the first protrusion feature (the parent) and the new cut feature (the child). Two operations you performed create this relationship: using the front surface of the cylinder as the sketch plane, and aligning the sketched circle to the axis of the cylinder.*

These relationships occur when you begin to establish your design intent. You need to pay special attention to the surfaces you use and the items you align when building features on your parts. This will help you to create parent/child relationships when you want them, and to avoid creating them when you do not.

> **DESIGN NOTE:** *This feature also used DTM2 as the horizontal reference for the sketch view. DTM2 is now a parent feature of this cut. Successful users of Pro/ENGINEER use their main datum planes whenever possible. In that these features are created very early in the part, they are stable references you should use for sketching and aligning.*

Shaping the Wheel with a Revolved Cut

The wheel has a dished shape you will create with a revolved cut. Revolved features have different options to control the feature and are driven by a sketched section. They are revolved about a centerline drawn in the section. Select the following commands, starting at the FEAT menu:

Create ➥ Solid ➥ Cut ➥ Revolve | Solid | Done

Select the following commands from the ATTRIBUTES menu:

360 | One Side | Done

Now you need to specify your sketch plane. In revolved features, the sketch plane needs to be a plane that contains the desired axis of revolution. For this feature, the axis of revolution should be the axis A_1. Therefore, DTM1 or DTM2 are valid datums for the sketch plane. Choose DTM2 as your sketch plane for this feature. Accept the default direction for viewing the sketch plane by selecting **Okay** from the DIRECTION menu. Select **Bottom** and then pick DTM3 as the horizontal reference for the sketch view.

You should be in the SKETCHER. Select

Line ➥ Centerline | 2 Points

from the GEOMETRY menu so that you can begin to draw the centerline for this section. For this feature, the centerline should be drawn on top of the axis A_1. Using the left mouse button, place the first endpoint of the centerline. Move the cursor to the second endpoint and press the left mouse button again to place the centerline.

> ✓ **TIP:** *There is another way to place centerlines on top of axes in sections. If you select* ***Geom Tools ➥ Use Edge ➥ Sel Edge***

from the SKETCHER menu and select the axis A_1, you will create a centerline that is automatically aligned to the axis A_1. This will save you the step of aligning to the axis later on.

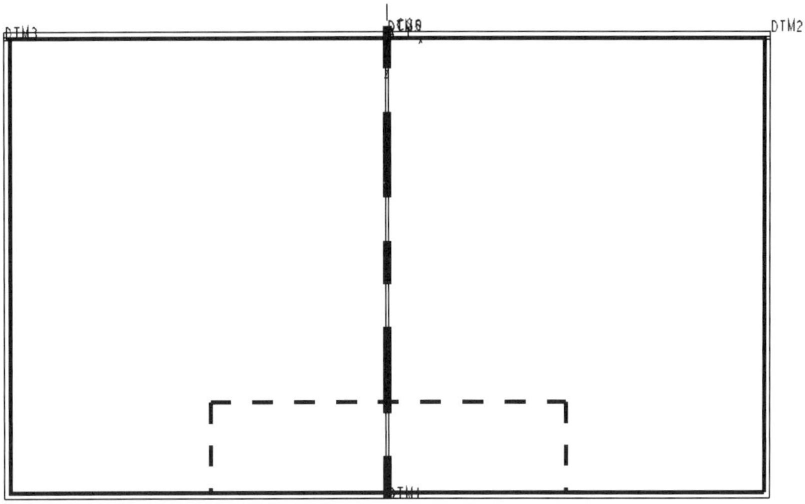

Figure 3-15
The centerline for the revolved cut.

➻ **NOTE:** *Centerlines can be used for more than just revolved features. You will frequently have more than one centerline in the same section. For revolved features with multiple centerlines, Pro/ ENGINEER assumes that the first centerline you sketched is the one you wish to revolve about.*

Next, you should constrain the position of the centerline. This step is not necessary if you used the technique shown in the previous tip to create the centerline. To constrain the position of the centerline, select **Alignment ➡ Align** from the SKETCHER menu. Pick the centerline you just sketched and then select the axis A_1.

✓ **TIP 1:** *Constrain the centerline before you sketch the other geometry in your section. Many users forget to constrain their centerlines and end up with underdimensioned sections that will not regenerate properly. Constraining the centerline immediately after you create it will help you avoid this common oversight.*

✓ **TIP 2:** *A good habit to develop is to refer to the message window when the sections you are creating will not regenerate successfully. A prompt is often displayed that will aid in the solution of the failed regeneration.*

Now you need to sketch the geometry that will define the revolved cut. The next diagram shows what you should try to sketch.

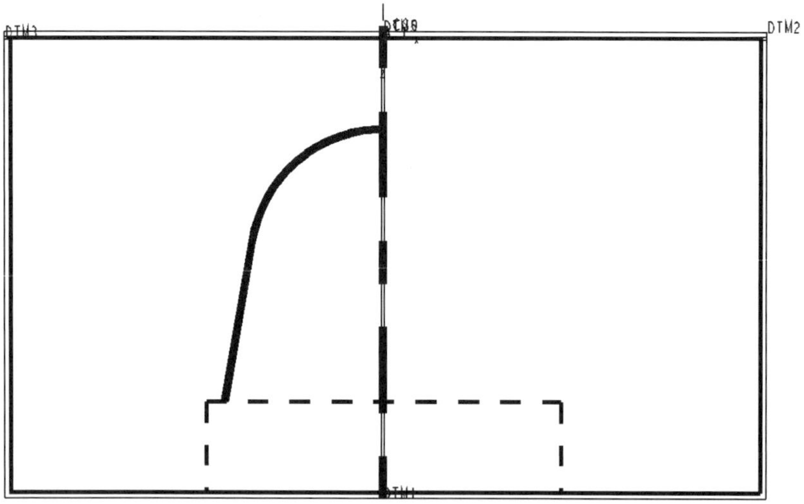

Figure 3-16

Section geometry for the revolved cut that makes the wheel's dished shape.

Start by drawing the line. Notice that the line is at a slight angle. In order to prevent the sketcher from assuming the line is drawn vertically, sketch the line at an exaggerated angle. Pick **Sketch ➥ Mouse Sketch** from the SKETCHER menu. Using the left mouse button (for line placement), place the lower endpoint. Move the cursor to the upper endpoint and press the left mouse button. Press the middle button to stop placing lines. To sketch the arc, use the right mouse button (for tangent arc placement). This method of arc placement is for drawing arcs tangent to the endpoint of previously sketched geometry. By using this method, the arc will be drawn tangent to the sketched line. The sketcher will be able to assume the tangency condition during the regeneration process.

Position the cursor at the upper endpoint of the line and press the right mouse button. Move the cursor until the center of the arc and the arc endpoint lie on the centerline. Press the right mouse button to place the tangent arc at this position. Now you should constrain your section. You need to align the free endpoint of the line to the inside planar face of the extruded cut. Pick **Alignment ➥ Align** from the SKETCHER menu and pick on the free endpoint of the line. Now pick on the hidden line that represents the edge of the inside planar face of the wheel.

> **DESIGN NOTE:** *By aligning the endpoint of the line to the edge of the wheel's planar face, you have added knowledge to this feature. When dimensions change, the sketched line will be extended to maintain this alignment criterion. This ensures that the revolved cut will always remove material up to the surface.*

Place these dimensions on the section. As you will have noticed by now, the actual number specifying size is not important at this stage, although its placement is. You will soon modify these calculations to the dimensions you want.

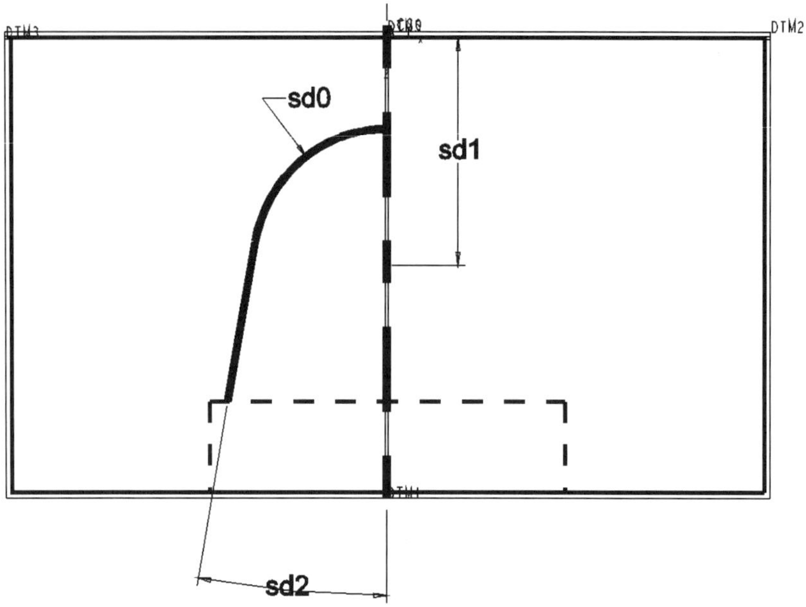

Figure 3-17
Section dimensions for the revolved cut that make the wheel's dished shape.

All of these dimensions can be placed with the **Dimension** command. When many objects (lines, centerlines, datum axes, datum planes) lie on top of one another in the sketch view, be sure to use **Query Sel** to select the desired object. Select **Dimension** from the SKETCHER menu and place the following dimensions:

❒ **The tangent arc radius** — Select the arc once with the left mouse button, and place the dimension with the middle mouse button.

❒ **A linear dimension from the arc center to DTM3** — Select the arc center with the left mouse button. Select DTM3 with the left mouse button, and place the dimension with the middle mouse button.

❐ **An angular dimension between line and centerline** — Select the line with the left mouse button. Select the centerline with the left mouse button, and place the dimension with the middle mouse button.

> **NOTE:** *If you select an entity near its end point, only the endpoint will be selected. However, if you pick an entity away from its endpoints, the entire entity is selected. When you have placed all of these dimensions, select* **Regenerate** *from the SKETCHER menu. You should see the following messages:*

```
Regeneration completed successfully.

WARNING: Not all open ends have been explicitly
aligned.
```

Do not panic. The warning is just telling you that you did not align the free end of the tangent arc. When you sketch open sections (sections that do not form closed areas), Pro/ENGINEER reminds you to align their free endpoints. Aligning the endpoints of an open section is not a requirement. In this example, you could align the endpoint of the tangent arc to the axis A_1, but you do not have to. The centerline was aligned to the axis already. In that the tangent arc endpoint lies near the centerline, the sketcher will assume that they are coincident. Select **Modify** and respond to the prompts with the following values:

❐ Tangent arc radius = .15

❐ Linear dimension = .25

❐ Angular dimension = 10

After changing these values, select **Regenerate** and pick **Done** from the SKETCHER menu. Now you need to tell Pro/ENGINEER which side of the section should be removed by this cut feature. Point the arrow toward the centerline, and select **Okay** from the ARROW FLIP menu.

Next, you need to specify the angle of revolution on the REV TO menu. This angle is the rotational distance in degrees the feature will be rotated through, starting from the sketch plane. In this example, you want to rotate a full 360 degrees. Pick the following commands from the REV TO menu:

360 ➥ Done

Pick **OK** from the dialog window. Your revolved cut feature should now be finished.

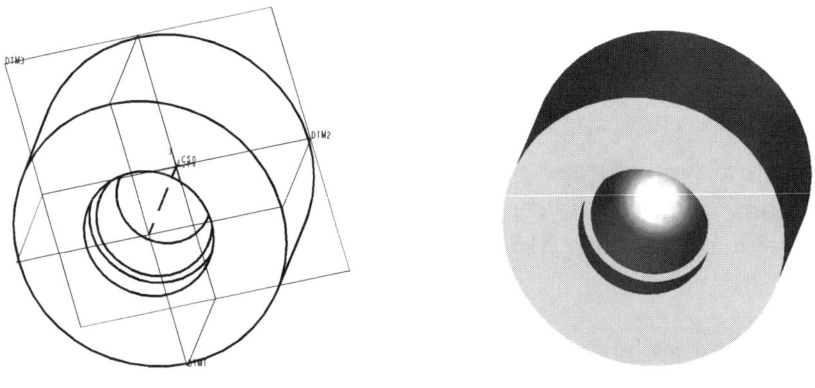

Figure 3-18
The wheel after placing the revolved cut feature.

SAVE

➾ **DESIGN NOTE:** *This feature has four parent features.*

❒ DTM2 is a parent because you used it as the sketch plane.

❒ DTM3 is a parent because you used it as the horizontal reference plane *and* because you dimensioned to it in the sketcher.

❒ The first solid feature is a parent because you aligned the sketched centerline to the axis A_1.

❒ The extruded cut for the wheel is a parent because you aligned the line's endpoint to the edge of this feature.

Adding a Wheel Hub Using a Revolved Protrusion

The hub will be created as a revolved protrusion, and it will be attached to the revolved cut you just made. Revolved protrusions have the same options as revolved cuts (the only difference being that you are adding material to the part).

This feature is very similar to the previous feature. In order to efficiently provide you with the necessary information to create the feature, this book will use a feature description box. These boxes will be used whenever you already have a step-by-step example of a similar feature.

The Wheel Hub (a revolved protrusion)

Commands:

Create ➥ Solid ➥ Protrusion ➥ Revolve | Solid | Done ➥ One Side | Done

References:

Sketch plane — DTM2.
Horiz/Vert reference —
Bottom, DTM3.

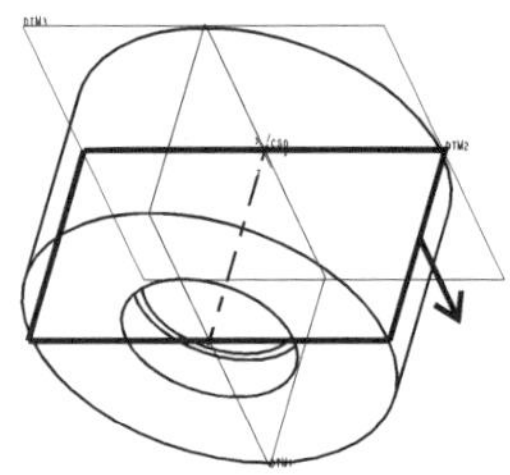

Figure 3-19

➾ ***NOTE:*** *Dimensions for this feature creation are shown in parentheses.*

1. Centerline.
2. Angled line.
3. Horizontal line.
4. Second centerline.

A. Align centerline to axis A_1.
B. Align endpoint of angled line to the arc of the wheel dish.
C. Linear dimension between horizontal line and DTM3 (.2).
D. Angular dimension between angled line and second centerline (5).
E. Diameter dimension between the centerline and the endpoint of the horizontal line (.1) To get this dimension, use the left mouse button to pick the endpoint, the centerline, and then the endpoint again. Place the dimension with the middle mouse button. You may also use the length of the horizontal line (3) for dimensioning. Note, however, that that would be viewed as a radius-type dimension and therefore represent half the finished feature diameter.

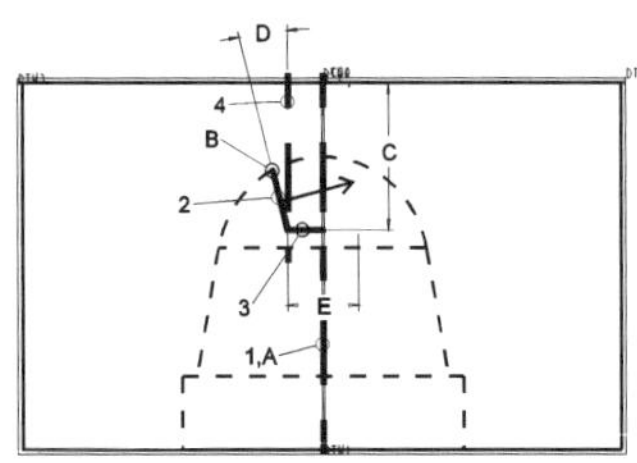

Figure 3-20

Commands:

Regenerate | Modify | Regenerate | Done | 360 | Done | One Side | OK

Parent features:

DTM2 — Sketch plane.
DTM3 — Horizontal reference and dimensioning reference.
Cylinder (First profusion) — Aligning to the axis A_1.
Wheel dish (Revolved cut) — Aligning to the arc of the wheel dish.

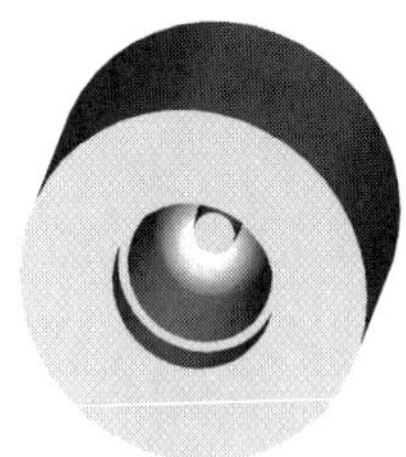

Figure 3-21

Another Revolved Protrusion

You should create the axle as another revolved protrusion. Use the following feature description box to make the axle.

The Axle (a revolved protrusion)

Commands:

Create ➡ Solid ➡ Protrusion ➡ Revolve | Solid | Done ➡ 360 One Side | Done

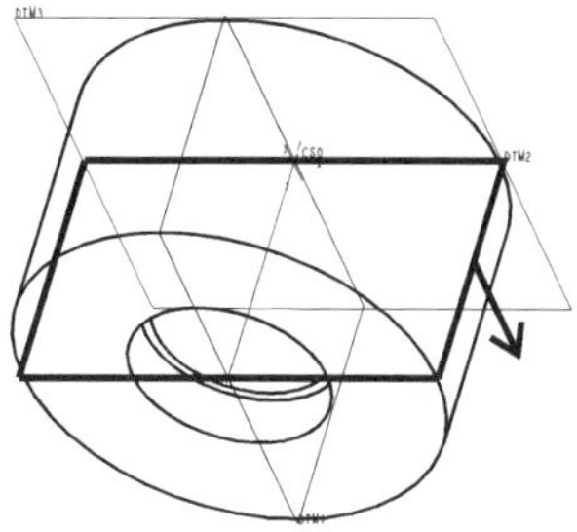

Figure 3-22

References:

Sketch plane — DTM2.
Horiz/Vert reference —
Bottom, DTM3.

NOTE: *Dimensions for this feature creation are shown in parentheses.*

1. Centerline.
2. Vertical line.
3. Three-point arc.
4. Align centerline to axis A_1.

A. Align endpoint of vertical line to DTM3.
B. Linear dimension from the center of the arc to DTM3 (.5).
C. Diameter dimension using the centerline and the vertical line (.15).
D. Radial dimension for the three-point arc (.1).

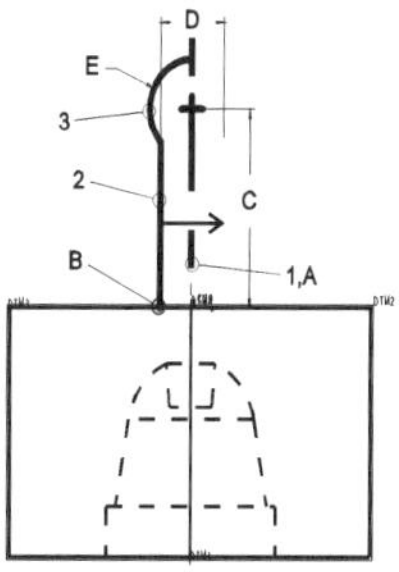

Figure 3-23

Commands:

Regenerate | Modify | Regenerate | Done | 360 | Done | One Side | OK

Parent features:

DTM2 – Sketch plane.
DTM3 – Horizontal reference, alignment reference, and diminishing reference.
Cylinder (first protrusion) – Aligning to the axis A_1.

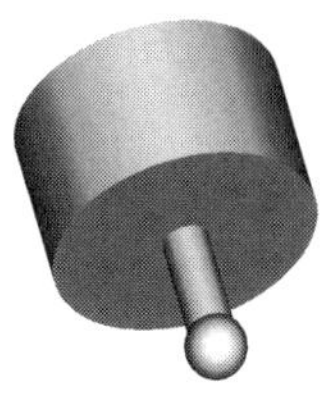

Figure 3-24

SAVE

Snap Relief for the Axle

A small cut will be made to act as a snap relief for the ball end of the axle. This will allow the wheel to snap into an axle lock in the chassis. You will use a few new options on this extruded cut. Select the following commands, starting from the FEAT menu:

Create ➥ Solid ➥ Cut ➥ Extrude | Solid | Done

Now select the **Both Sides** option from the SIDES menu. This tells Pro/ENGINEER that the cut feature will be extruded in both directions from the sketch plane. In that the sketch plane lies in the middle of the part, you can cut material in both directions.

Now you need to select the sketch plane. Use DTM2 for this feature's sketch plane. Next, accept the direction for viewing by selecting **Okay** from the DIRECTI ON menu.

> ✓ ***TIP:*** *The viewing direction is a source of confusion for many users when they are performing an extrusion using the* ***Both Sides*** *option. This direction arrow tells Pro/ENGINEER which side of the sketch plane will use the* ***Spec To*** *option. The* ***Spec From*** *option will be extruded in the opposite direction of the arrow. In this case, because the* ***Spec From*** *and* ***Spec To*** *options are the same, the viewing direction did not matter.*

To specify a horizontal reference for this sketch plane, select **Bottom** from the SKET VIEW menu and select DTM3. Sketch the following section.

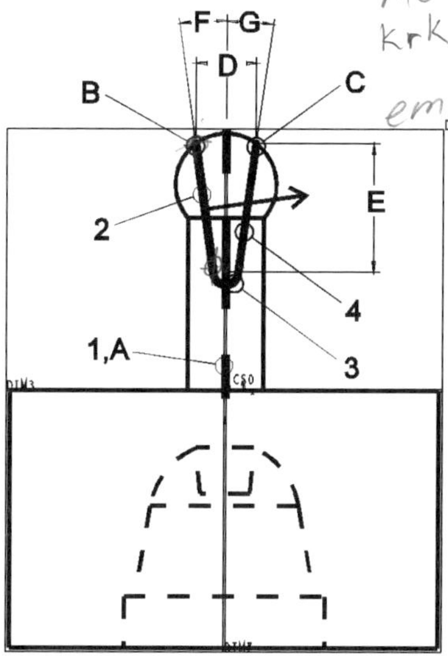

Figure 3-25
Sketch of the snap release feature.

1. Centerline
2. Angled line
3. Tangent arc (far right mouse button also performs this function)
4. Angled line (or a line tangent to an arc: **Sketch ➥ Line ➥ Geometry | Tangent**)

A. Align centerline to axis A_5

B. Align endpoint of first angled line to arc edge of part

C. Align endpoint of second angled line to arc edge of part

D. Linear dimension between the two free ends of the angled lines (.12)

E. Linear dimension from the arc center to the endpoint of the second angled line (.25)

F. Angular dimension between the centerline and the first angled line (8)

G. Angular dimension between the centerline and the second angled line (8)

Commands:

Regenerate | Modify | Regenerate | Done | Okay | Done

Then pick **OK** from the dialog window. **Note the OKAY prompt: the arrow for this prompt indicates the area in which the material will be removed.** If you choose the wrong direction and are left with nothing but the cut you just created, simply redefine the direction and your part should reappear. Choose **Thru All ➥ Done** from the SPEC FROM menu. By using the **Thru All** option, you are telling Pro/ENGINEER that the section should be extruded through the entire part. After selecting your choice from the SPEC FROM menu, you will see the SPEC TO menu. In that you are creating an extrusion to both sides of the sketch plane, you are allowed to specify different options for each direction. Now select **Thru All ➥ Done** from the SPEC TO menu. This feature is intended to be cut through the entire part in both directions.

> **NOTE:** *In this section, the centerline is used to represent an axis of symmetry between the left and right sides of the section. Once the section has been completed, the feature should be created. The cut should look like the next diagram.*

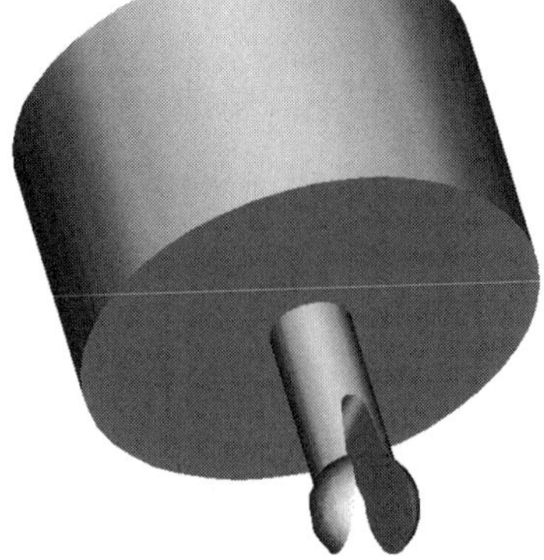

Figure 3-26
The snap relief feature in the axle.

➾ **DESIGN NOTE:** *The parent features of this cut are DTM2, DTM3, and the axle. The axle is a parent because you aligned the centerline to its axis, and you aligned the endpoints of the lines to the edge of this feature. You should be comfortable using datum planes as references by now.*

Strengthening the Axle with Rounds

Now you need to add the final touches to the wheel. To make the axle stronger, you should round the intersection between the axle and the wheel. Rounds are created to avoid sharp edges on your part.

✓ **TIP:** *You should avoid placing rounds on your part too early. It is very easy (and not desirable) to create features that are children of your round features. Rounds should be created at the point in time they are necessary.*

Round features are different from the features you have placed so far. You do not have to sketch a section. You usually create rounds by selecting the edges you wish to round. To place this feature, choose the following commands, starting from the FEAT menu.

Create ➡ Solid ➡ Round ➡ Simple | Done ➡ Edge | Constant | Edge Chain | Done | Tangent Chain | Circular | No End Srf | Done ➡ Single

Now you can begin selecting the edges you want to round. Select the edge that represents the intersection between the axle and the wheel.

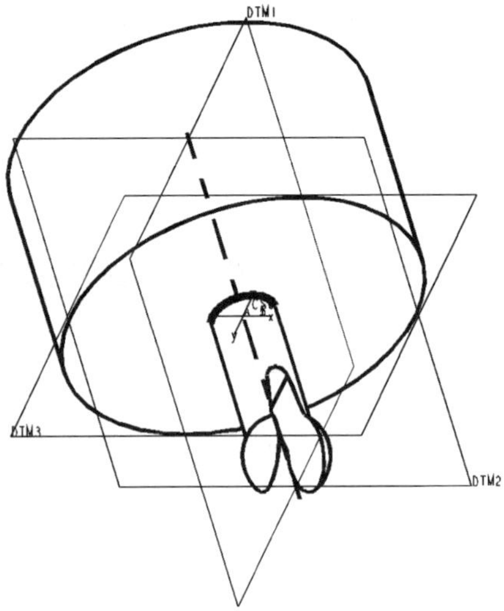

Figure 3-27
The edge to use for this round feature.

> **NOTE:** *You should notice that when you select the intersecting edge, Pro/ENGINEER highlights only half the circle. This is because Pro/ENGINEER represents complete cylinders as two equal halves. Consequently, the edges become two 180-degree arcs rather than one complete circle. When you are rounding cylinder edges, you need to select only one of these edges. The round will automatically wrap around the entire cylinder.*

After you have selected the edge, select **Done** from the CHAIN menu. Select **Enter** from the RADIUS TYPE menu so that you can enter the value for the radius.

```
Enter RADIUS [0.0038]: .05 ↵
```

Then pick **OK** from the dialog window. The round feature is now created. What are the parent features of this round? This round references a part edge; therefore, the parent features of this round are the features that created the edge. Those features are the cylinder (first protrusion) and the axle.

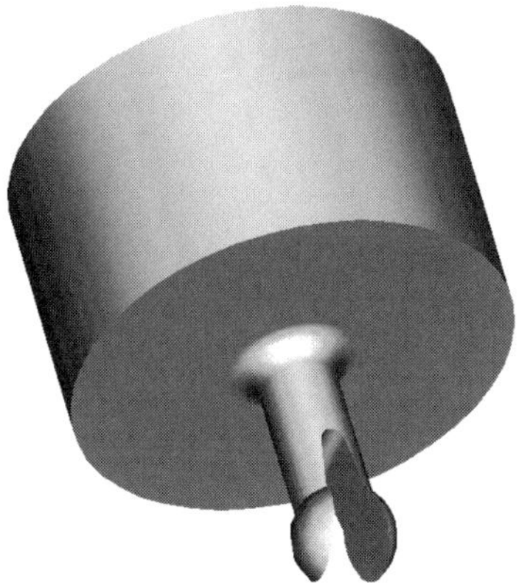

Figure 3-28
The round that adds strength to the axle.

Shaping the Tire

For the last feature of this part, you will make the tire look like a tire. You will use the round command as before, but this time you will select more than one edge to round. Select the following commands, starting at the FE AT menu:

Create ➥ Solid ➥ Round ➥ Simple | Done ➥ Edge | Constant | Edge Chain | Done | Tangent Chain | Circular | No End Srf | Done ➥ Single

Select the three highlighted edges on the part.

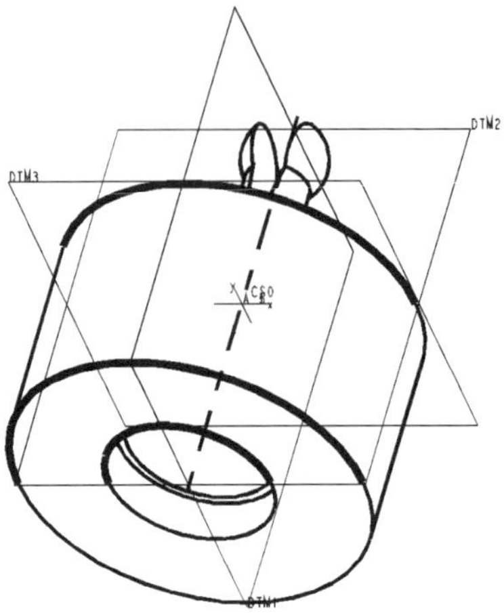

Figure 3-29
The three edges that need to be selected for this round feature.

When you have finished selecting these edges, choose **Done** from the CHAIN menu. Select **Enter** from the RADIUS TYPE menu and enter the following.

```
Enter RADIUS [0.0038]: .08 ↵
```

Then pick **OK** from the dialog window. The wheel part is now complete.

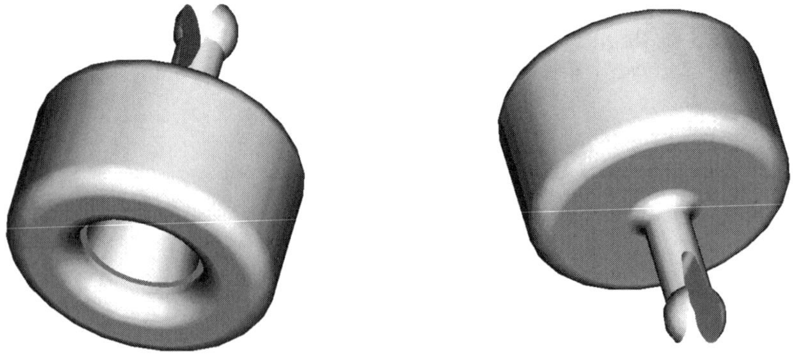

Figure 3-30
Views of the completed wheel part.

The parent features of this round are the cylinder (first protrusion) and the first extruded cut of the wheel. Two of the edges belong solely to the cylinder, and one of the edges is the intersection between the cylinder and the first extruded cut of the wheel.

Feature Numbers

You might have noticed that Pro/ENGINEER assigns a number to each feature. For example, if you pick on a surface during a **Query Sel** operation, the feature number will appear in the message window. These feature numbers are an efficient way to refer to your part's features. Rather than saying "the second revolved protrusion," we can say "feature number 9." A summary of this part's feature numbers and corresponding feature descriptions follows.

Feature No.	Description
1	Datum Plane (DTM1)
2	Datum Plane (DTM2)
3	Datum Plane (DTM3)
4	Datum Coordinate System (CS0)
5	The first extruded protrusion (cylinder)
6	The first extruded cut (wheel)
7	The first revolved cut (wheel dish)
8	The first revolved protrusion (hub)
9	The second revolved protrusion (axle)
10	The second extruded cut (snap relief)
11	The first round (axle round)
12	The second round (tire rounds)

Feature numbers are an effective method of referring to features because they are concise, and that is the way Pro/ENGINEER automatically refers to features. The feature number denotes the order in which Pro/ENGINEER creates the features. In the rest of the book, feature numbers will be used when identifying part features.

Summary of Parent/Child Relationships in the Wheel

It is important that you understand why new features become children of previous features. When Pro/ENGINEER asks you for a sketch plane or a horizontal or vertical reference, the program needs that information. The surfaces (i.e., features) you identify become parents of the feature you are creating. Also, the vertices, edges, or surfaces you reference when aligning or dimensioning will cause their corresponding features to become parents.

> **NOTE:** *An understanding of parent/child relationships is required to take advantage of the many time-saving features of Pro/ENGINEER. A practiced user will come to know which features are children of other features and why the relationship exists.*

Sometimes design changes will force you to delete a feature. This is when it is critical to know which features are children of the one you want to delete. Understanding the parent/child relationships in your parts will allow you to make modifications more easily.

If you are a new user, it can be beneficial to actually see the parent/child relationships of your parts as you build them. A table showing the parent/child relationships of the wheel would look like the following diagram.

Parent / Child	1	2	3	4	5	6	7	8	9	10	11
2											
3											
4											
5		●	●	●							
6		●			●						
7		●	●		●	●					
8		●	●		●		●				
9		●	●		●						
10		●	●						●		
11					●				●		
12					●	●					

No parents in this area.

Figure 3-31
A table showing the parent/child relationships in the wheel part.

The columns represent parents and the rows represent children. You can see that the wheel's features depend heavily on the datum planes. This is a good trait. Your datum planes should be the foundation for your parts. You never know what type of design change will be required.

> ✓ ***TIP:*** *By using your datum planes effectively, you will reduce the dependencies between your solid features. If you try to delete features that have children, you have to give the child features new references. Limiting the ties between solid features will allow you to easily restructure and remove the features when major design changes occur.*

In addition, you should notice that many features have feature 5 (the first protrusion) as a parent. This reveals the importance of your first solid feature.

✓ **TIP:** *Take your time in determining the shape of your first solid feature. If you decide to change it dramatically, the change can have a tremendous effect on the rest of your part's features.*

Conclusion

At this juncture, you should know how to build a simple part in Pro/ENGINEER. Knowing and understanding these few features are sufficient for building basic parts. Now would be a good time for you to try some of your own designs. In the next chapter you will learn how to use the Sketcher mode to its full potential.

Chapter 4
Understanding the Sketcher

Using the Sketcher to Create Sections

Introduction

Many Pro/ENGINEER commands require you to sketch a section before a feature can be created. This section is used to create a feature by extruding, revolving, sweeping, or blending. In a prior chapter, you were shown the basic sketcher methodology and how to use the various commands. The purpose of this chapter is to fully explore the sketcher. It includes a summary of sketcher assumptions and guidelines for understanding regeneration failures. You will also be shown how to select geometry that will be useful for other modes. This chapter ends with a list of sketching tips to help you become a proficient user of the sketcher without investing the time to discover these tips on your own.

The sketcher is available as a mode from the MODE menu. The Sketcher mode is for your convenience if you want to create a section and save it for use in a feature later on. However, you do not have to create your sections in advance. Pro/ENGINEER automatically opens a sketcher window for you when you create features that need sections.

What Are Sections?

Sections are sketched geometry that have been constrained. Sections are drawn in a plane commonly called the *sketch plane,* which can be viewed in 2D or 3D by setting for sketch in the ENVIRONMENT menu. Sections contain 2D geometry, dimensions, and alignment information needed to create a fully constrained feature. The sketcher can assume relationships about geometry in your section, such as equality, symmetry, and tangency. In that sketching is such an important element of using Pro/ENGINEER, you need to understand the sketcher's commands, tools, and behavior.

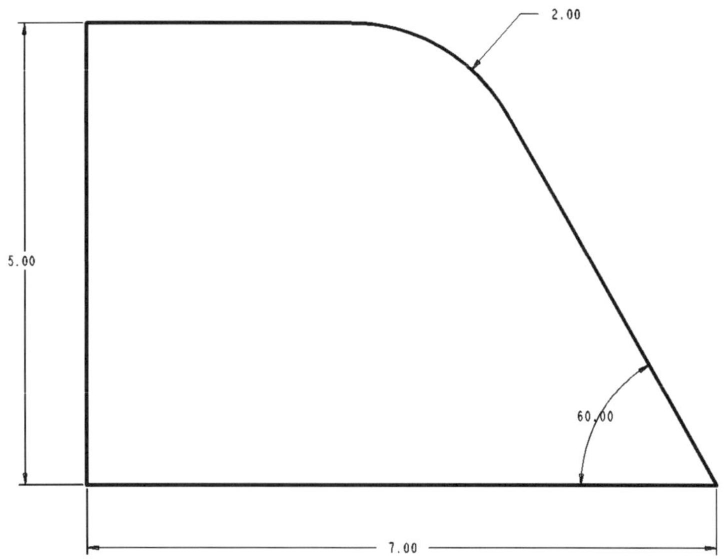

Figure 4-1
Example of a typical section.

Sketcher Methodology

There are three main steps in creating a section.

Step 1: Sketch the Geometry

Your first step is to create the geometry you need for your feature. You have the option of freehand sketching with the mouse, using edges of existing geometry, or using commands that help you sketch accurate geometry. You can also sketch construction geometry in the form of points, lines, and circles.

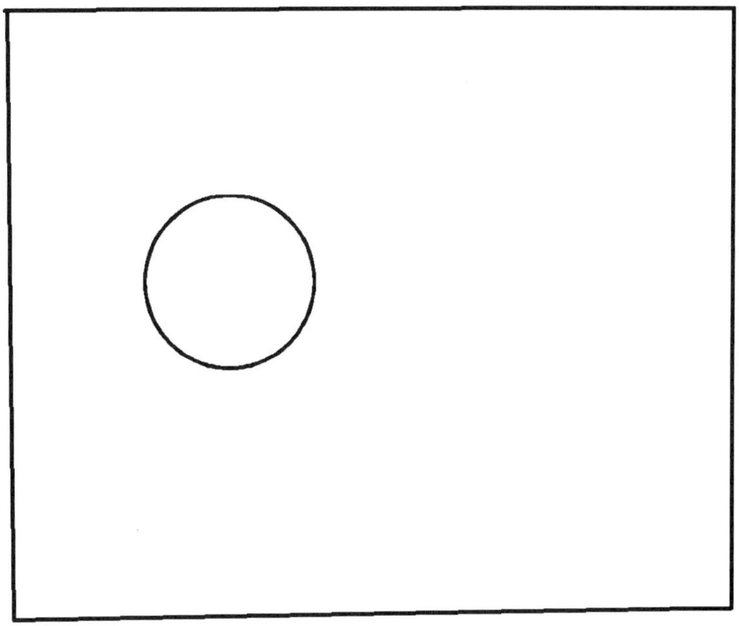

Figure 4-2
Sketched geometry.

Step 2: Constrain the Geometry

After you sketch your geometry, you need to constrain it. You do this by dimensioning and aligning the sketched geometry. Dimensions can control the size, shape, and position of the geometry. Alignments can attach the section to the vertices, edges, or surfaces of your existing part. When you have finished constraining your geometry, you need to tell Pro/ENGINEER to evaluate it. This process is called *regeneration,* and it evaluates your constraints, adds in any necessary assumptions, and cleans up the geometry.

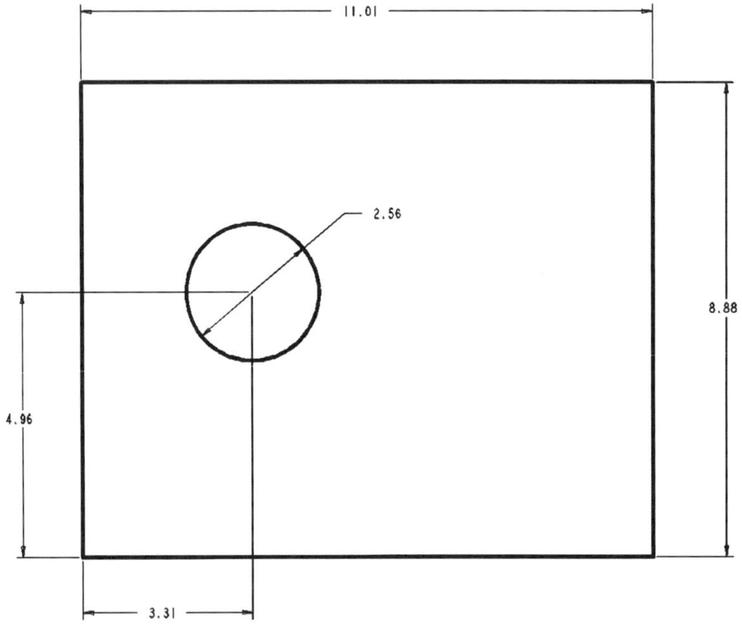

Figure 4-3
Sketched geometry with constraints.

Step 3: Modify the Geometry

At this point, your section has been constrained and successfully regenerated. (Remember that you just "sketched" it.) Now you modify the dimensions and regenerate the section so that it is the proper size and shape.

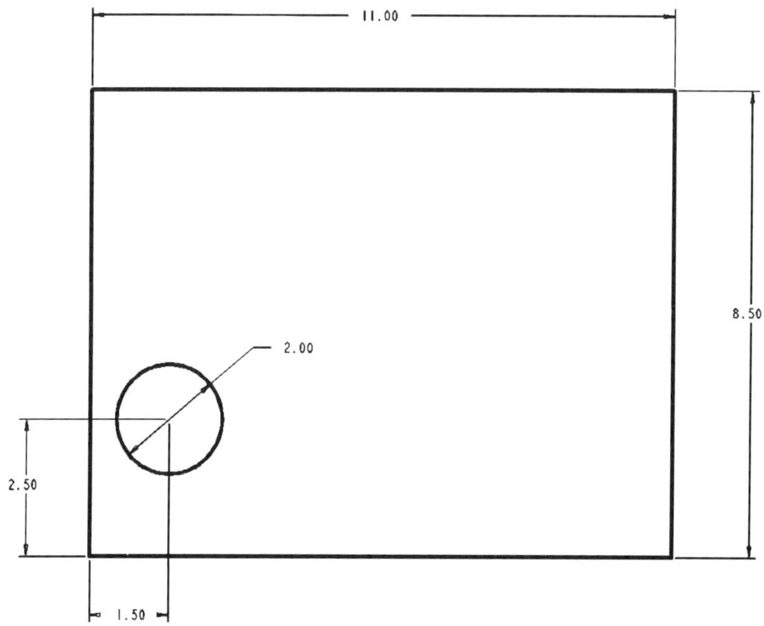

Figure 4-4
A finished section.

The sketching process is really simple. It may seem foreign to you because it is so different, but it is an effective way of creating parametric geometry. If you are familiar with 2D drafting techniques, you will notice that your sections are similar to detail views in a drawing.

Saving a Section

Sections can be saved at any time by selecting **Dbms ➥ Save** from the MAIN menu. Your section does not have to be completed or fully constrained in order to save it. Sections will be saved in a section file (e.g., as *s2d0001.sec* if you used the default section name) you can retrieve for future use. Detailed information about database management is provided in Chapter 2.

Sketcher Commands

Sketcher commands fall into four major categories:

- Basic sketcher commands control the sketcher.
- Commands for creating geometry produce new geometry.
- Commands that modify geometry change or use existing geometry.
- Commands that constrain geometry specify dimensions and align geometry.

The basic sketcher commands are usually located on the SKETCHER menu and are your major controls once you start a sketcher session.

- **Sketch** — This command activates the sketching commands associated with the mouse buttons and brings up the GEOMETRY menu.
- **Dimension** — When you want to dimensionally constrain your section, pick this command.
- **Modify** — After fully constraining the section and successfully regenerating it, select this command to change the dimensions to the correct values. Using the GET SELECT menu, select the dimension text you want to modify and enter the new value.

> ✓ **TIP:** *It is a good practice to successfully regenerate the sketch prior to modifying dimensions. Successfully regenerating the section before dimensional modification verifies that the section is fully constrained and prevents you from entering a value that makes it impossible to regenerate.*

❒ **Regenerate** — This command serves two purposes: it verifies that the section is fully constrained (using assumptions if needed), and adjusts your sketched entities to the size specified by the dimensions. Tips for regeneration are provided later in this chapter.

❒ **Unregenerate** — This command will undo a previous regeneration and return the section to its exact state prior to regeneration.

> ✓ **TIP:** *Unregenerate is useful when the sketcher makes an unwanted assumption about your section during regeneration.*

❒ **Delete** — To delete sketched entities or dimensions in your section, pick this command. Then select the objects you want to delete using the GET SELECT menu. You can undelete objects with the **Undelete Last** command on the DELETION menu.

> **NOTE:** *Only objects that have been deleted since the last regeneration can be undeleted.*

❒ **Alignment** — This command lets you associate and disassociate sketched entities to your existing part.

> **NOTE:** *You cannot align a sketched entity to another sketched entity. These types of relationships must be assumed by the sketcher.*

❒ **Geom Tools** — This command brings up the GEOM TOOLS menu, which contains commands for using part edges to create sketched geometry and for trimming, mirroring, and moving sketched entities.

- **Sec Tools ➥ Interface** — Through this command you can import a section from an IGES file or export your section to a plotter or to a computer graphics metafile (.cgm).

- **Sec Tools ➥ Restart** — Located on the SEC TOOLS menu, this command allows you to begin the sketcher session again. It will also allow the selection of a new sketch plane.

- **Sec Tools ➥ Set Up** — This command allows you to specify the number of decimal places displayed for the dimensions and to adjust the sketcher accuracy. The sketcher accuracy is a value that controls how true your sketched entities need to be drawn. For example, lines that are sketched nearly horizontal are assumed to be horizontal during regeneration. If you decrease the numeric value of the accuracy, the sketcher is not as lenient in determining that the sketched line is horizontal. Decreasing this number would force you to sketch the line more accurately. The default value of 1.0 is usually sufficient for most sections.

- **Relation** — A relation is a means of associating sketch dimensions through the use of equations. After selecting this command, the RELATIONS menu appears and provides the commands for adding equations one at a time, adding and editing equations from within an editor, or simply viewing the current equations. Sketcher relations are just like part and assembly relations, which are discussed in later chapters.

- **Integrate** — Integrate, used in conjunction with PDM (Product Data Management), allows you to merge updates to models when several users are making changes to the same model.

- **OK** — Once you have completed sketching and constraining the section, pick this command to continue creating the feature.

- **Quit** — If you decide you do not want to continue creating the feature that placed you into the sketcher, pick **Quit** from the SKETCHER menu. You will abort the creation of the feature.

Commands for Creating Geometry

Most section geometry is created with the **Sketch** command in the SKETCHER menu. When you select the **Sketch** command, the GEOMETRY menu will appear. The GEOMETRY menu contains commands used for specific sketching needs. Most common geometry can be placed by using the **Mouse Sketch** option from the GEOMETRY menu.

Lines

When you use the **Mouse Sketch** command, lines are placed with the left mouse button. Position the cursor at the location of the first endpoint and press the left mouse button. Move the cursor to the location of the second endpoint and press the left mouse button again. This will place the first line and automatically start you drawing the next line. When you want to stop placing lines, press the middle mouse button. More line commands are made available by selecting **Line** from the GEOMETRY menu. These line commands are not used very often, but they can help if you have trouble sketching accurately enough for Pro/ENGINEER to assume constraints.

- ❐ **2 Points** — Sketches a line between two points (same as **Mouse Sketch**)
- ❐ **Parallel** — Sketches a line parallel to an edge, line, axis, or centerline
- ❐ **Perpendicular** — Sketches a line perpendicular to an edge, line, axis, or centerline
- ❐ **Tangent** — Sketches a line tangent to an endpoint of an arc
- ❐ **2 Tangent** — Sketches a tangent line between two arcs or circles

> **NOTE:** *The arcs or circles will be split at the points of intersection.*

- **Pnt/Tangent** — Sketches a line through a point and tangent to an arc or circle
- **Horizontal** — Sketches a horizontal line
- **Vertical** — Sketches a vertical line

Circles

When you use the **Mouse Sketch** command, circles are placed with the middle mouse button. Position the cursor at the location for its center and press the middle mouse button. Move the cursor until the circle is the desired size and press the middle mouse button again. If you have already placed the center of the circle and wish to quit placing the circle, press the left mouse button.

When you select **Circles** from the GEOMETRY menu, you will see the CIRCLE TYPE menu appear. This menu contains more commands for drawing circles.

- **Ctr/Point** — Sketches a circle by placing its center and an edge point. (This procedure is the same as **Mouse Sketch**, except that you use the left mouse button.)
- **Concentric** — Sketches a circle concentric to a circular edge or other sketched arc or circle.
- **3 Tangent** — Sketches a circle tangent to three lines.
- **Fillet** — Sketches a circle tangent to two entities.
- **3 Point** — Sketches a circle through three points.

Arcs

When you use the **Mouse Sketch** command, arcs are placed tangent to an existing sketched entity at one of its endpoints. These tangent arcs are placed with the right mouse button. Position the cursor at an endpoint of a sketched entity and press the right mouse button. Move the cursor to the location of the second endpoint of the tangent arc and press the right mouse button again. If you have already placed the first endpoint of the tangent arc and you wish to quit placing the arc, press the middle mouse button. There are some specialty commands available that place arcs in different fashions. You need to understand the three-point arc command because it will be used most frequently.

- ❐ **Tangent End** — Sketches an arc tangent to the end of an existing line or arc. (This command is the same as **Mouse Sketch** except that you use the left mouse button.)
- ❐ **Concentric** — Sketches an arc concentric to a circular edge or other sketched arc or circle.
- ❐ **3 Tangent** — Sketches an arc tangent to three lines.
- ❐ **Fillet** — Sketches an arc tangent to two entities. (The entities will be trimmed back if they are lines. Otherwise, the entities will be split at the point of intersection.)
- ❐ **Ctr/Ends** — Sketches an arc by locating its center and the two endpoints.
- ❐ **3 Point** — Sketches an arc through three points with the first two points denoting the ends.

Splines and Text

- There are sketch commands for placing splines, conic splines (e.g., parabolas and ellipses), and for placing text in a sketch. The spline commands will not be described in great detail because of their infrequent use and complexity. These commands are located on the ADV GEOMETRY menu that appears when you select **Adv Geometry** from the GEOMETRY menu.

- **Splines** — Splines are curves that pass through a series of points. To place a spline, select **Spline** from the ADV GEOMETRY menu. Pick all the points you want the spline to pass through with the left mouse button, and then place them with the middle mouse button.

- **Conics** — Conics are special splines. Parabolas, ellipses, and hyperbolas are all types of conics. To place a conic, select **Conic** from the ADV GEOMETRY menu. Pick the first endpoint and the second endpoint, and then drag the mouse to change the shape of the conic. Place the conic with the left mouse button.

- **Text** — In the ADV GEOMETRY menu, **Text** can be used to create raised or sunken letters on a part's surface. You then enter the text you wish to place. Now you need to define a rectangular area that will represent the outer bounds of the text. To define the rectangular area, use the left mouse button and pick two locations that will identify opposite corners of the area. The text will then be placed within the rectangular area.

Construction Geometry

Construction geometry is geometry that is placed in a sketch but that does not create edges or surfaces in the feature. The most commonly used construction geometry is the centerline. Points, coordinate systems, and circles are used less often.

- **Point** — Sketched points can be placed with the **Point** command from the GEOMETRY menu. The point is placed by pressing the left mouse button with the cursor at the desired location. A centerline is simply a construction geometry line. Some features require a centerline in their section to represent an axis of revolution. However, centerlines do not always need to represent the center of your feature. They can be used to help the sketcher assume colinearity between entities in your section. The **Centerline** command is located on the LINE TYPE menu (activated by picking **Line** from the GEOMETRY menu). You can place centerlines with any method of line placement displayed on the LINE TYPE menu.

- **Construction Circles** — A construction circle can be placed when you need a circular construction element. Pick **Circle ➥ Construction** from the GEOMETRY menu, and place the circle with any of the conventional circle sketching commands displayed on the CIRCLE TYPE menu.

- **Coordinate System** — Some features require a coordinate system in their section. A coordinate system can be placed with the **Coord Sys** command from the ADV GEOMETRY menu. The coordinate system is placed by pressing the left mouse button with the cursor at the desired location.

Commands That Modify Geometry

Most geometry manipulation tools are provided on the GEOM TOOLS menu. This menu is displayed after selecting **Geom Tools** from the SKETCHER menu. Most of these tools allow you to modify existing sketched geometry. However, the **Use Edge** and **Offset Edge** commands create sketched geometry from edges of the part.

- **Intersect** — A sketched entity can be divided at the point at which it intersects another entity. In cases where the other entity is also a sketched entity and not a centerline, the second entity will also be divided at the point of intersection. Two entities can be intersected with the **Intersect** command from the GEOM TOOLS menu. The GET SELECT menu will be used to help you pick each of the two entities. If they intersect more than once, pick them near the desired intersection point.

❒ **Trim** — A sketched entity can be lengthened or shortened. Although the entity can be changed to an absolute length or incrementally lengthened or shortened, you will most often change the length to the point of intersection with another entity. To do this, pick **Trim** ➥ **Bound** from the GEOM TOOLS menu. Next, select the trimming boundary entity using the options from the GET SELECT menu. Then select the entities you want to trim to the boundary entity. For entities that will be shortened by the boundary, make sure you pick them on the portion you wish to keep. If you want to select another boundary entity, pick **Bound** from the DRAFT TRIM menu and continue with the same process. If you want two entities trimmed to their point of intersection, pick **Corner**, and then pick each of the entities on the portion you want to keep.

> ✓ ***TIP 1:*** *When trimming entities, Pro/ENGINEER makes assumptions similar to those it makes when you sketch and constrain geometry. If the portion of an entity to be removed does not appear to be very long, it will not be trimmed. In these cases, zoom in on the entity so that you can easily see that there is geometry to be removed before picking the entity you want to trim.*

> ✓ ***TIP 2:*** *If you select existing part geometry as your bound for a trimming operation, Pro/ENGINEER will automatically align your trimmed entity to that bound.*

- **Divide** — This command will allow you to take a sketched entity and slit it at any given point, as many places as you want—unlike the **Intersect** command, which required you to have another sketched entity to intersect with.

- **Use Edge** — This command will let you use edges of your part as section geometry. Pick **Use Edge** and then pick the edges using the GET SELECT menu. The edges you pick do not have to be 2D. When you select an edge to use that does not lie in your sketch plane, the edge will be projected onto your sketch plane for you. The geometry you create using this command does *not* need to be constrained like regular sketched geometry because it is already constrained. Remember, Pro/ENGINEER parts are always fully constrained. If you use an edge of a fully constrained part, you do not need to constrain it again.

- **Offset Edge** — This command allows you to select part edges and create section geometry offset from it. These part edges do not have to be 2D edges. When you select an edge to use that does not lie in your sketch plane, the edge will be projected onto the sketch plane for you. An offset dimension is automatically created.

- **Mirror** — If your section is symmetric, you can use the **Mirror** command on the GEOM TOOLS menu. You sketch half of your section and sketch a centerline to represent the axis of symmetry. Pick **Mirror**, and then, using the left mouse button, pick the centerline to mirror about. Next, the MIRROR menu will allow you to pick individual entities, pick all of the entities, or quit the command. Select **Done** to signify that you have picked all of the entities you want to mirror.

- **Move** — This command will allow you to take a sketched entity and move it. You can move single or multiple entities, with the drag commands, in the MOVE ENTITY menu. You can also rotate an entity with the **Rotate** command. The **Move** command is especially useful when redefining an already existing feature. It allows you to move existing entities that may have other features associated with them and therefore not have to delete and recreate a lost feature due to broken associations.

Commands That Constrain Geometry

After you have sketched your geometry, you need to constrain it. You can do so by dimensioning or aligning the geometry.

- **Dimension** — There is only one command for dimensions in the sketcher. The **Dimension** command allows linear, circular, and angular dimensions to be placed. The type of dimension created is determined by the type of entities picked and the method in which they are picked with the mouse. Use the GET SELECT menu to pick the geometry, and use the middle mouse button to place the dimension. If you place a dimension incorrectly, simply delete it with the **Delete ➥ Delete Item** command on the SKETCHER menu.

- **Cosm Font** — This command allows you to set the line style of cosmetic features. A dialog window displays the options. Select the desired option, then apply from the dialog window and the line style will be set.

- **Replace** — This command allows you to replace existing sketch geometry with new sketch geometry during the redefining command.

Linear Dimensions

Line length: To define the length of a line, pick the line near its midpoint and then place the dimension.

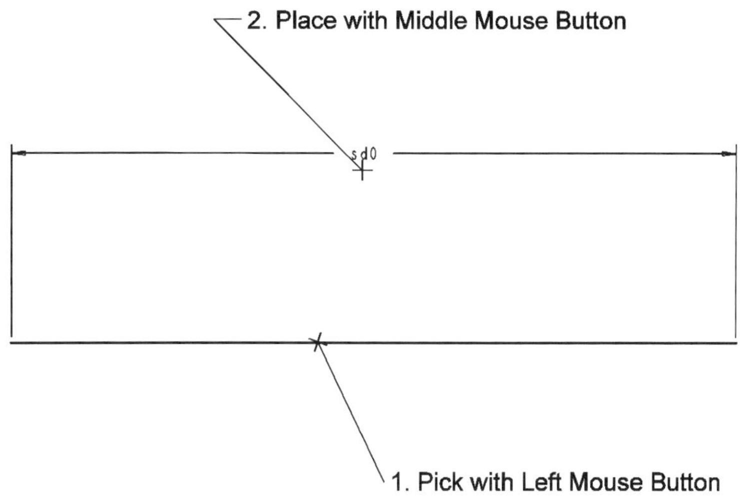

Figure 4-5
Dimensioning the length of a line.

Line to point: To define the distance between a line and a point, pick the line near its midpoint and then pick the point or vertex. Next, place the dimension.

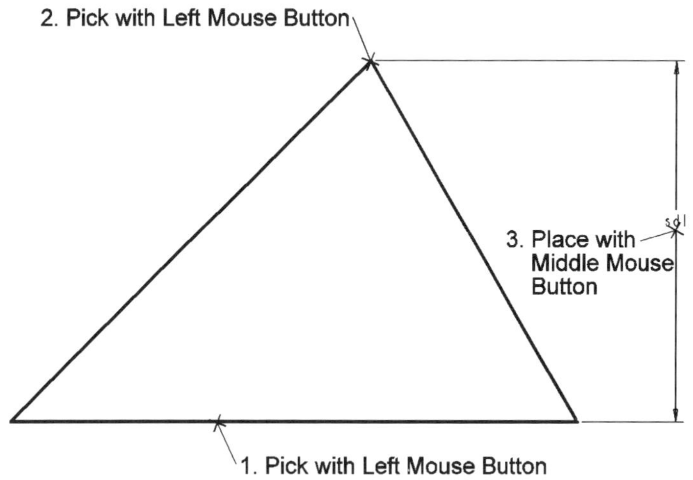

Figure 4-6
Dimensioning the distance from a line to a point.

Line to line: To define the distance between two parallel lines, pick each line near its midpoint. Then, place the dimension.

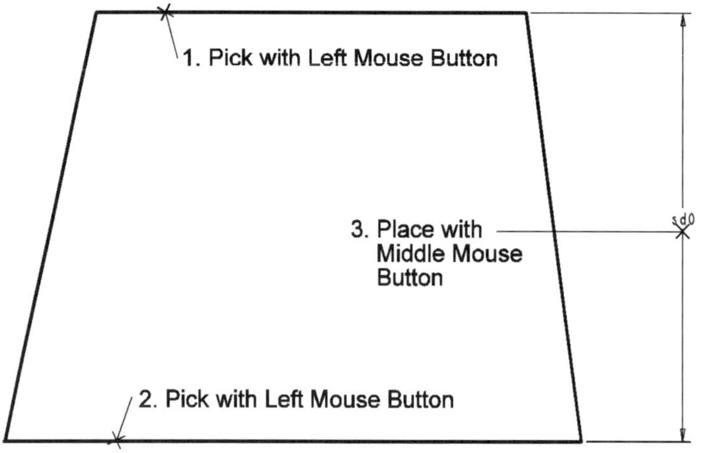

Figure 4-7
Dimensioning the distance between two parallel lines.

Point to point: To define the distance between two points, pick each point and place the dimension. If Pro/ENGINEER cannot determine which way to orient the dimension automatically, it will ask you to help by using the choices from the DIM PNT PNT menu. The choices are Horizontal, Vertical, and Slanted. After making your choice, the dimension is placed.

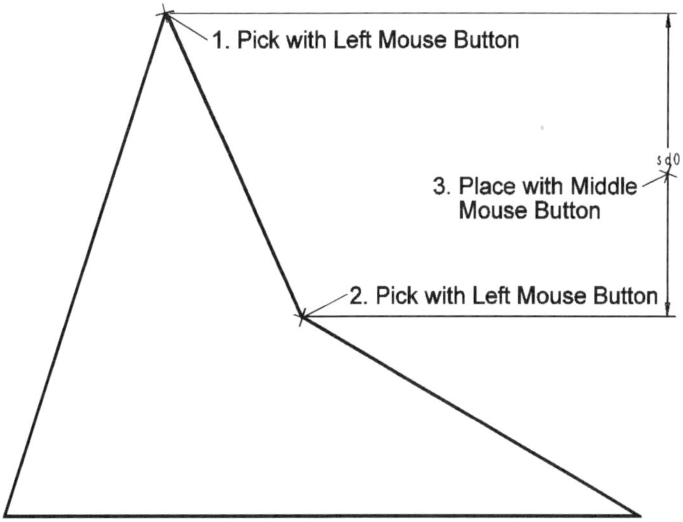

Figure 4-8
Dimensioning the distance between two points.

Circular Dimensions

Radial: To define the radius of an arc or circle, select the entity once (near its midpoint if it is an arc) and then place the dimension.

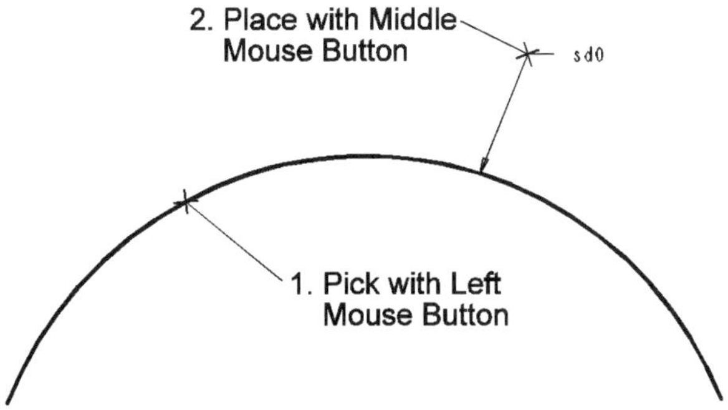

Figure 4-9
Dimensioning the radius of an arc.

Diameter: To define the diameter of an arc or circle, pick the entity twice (near its midpoint if it is an arc) and then place the dimension.

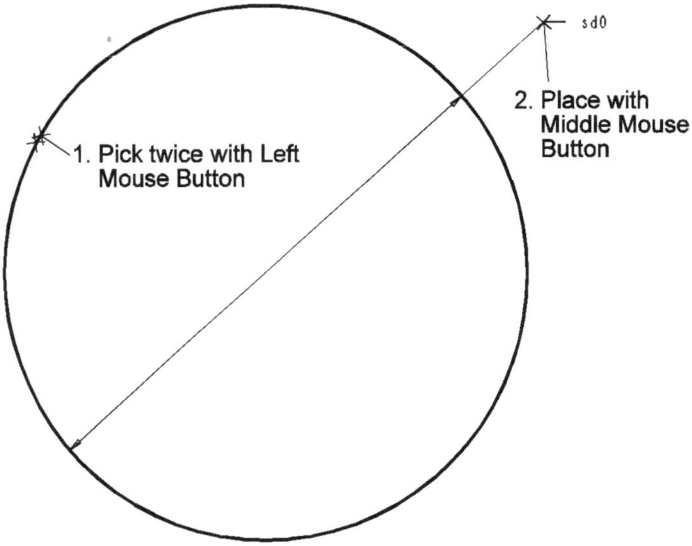

Figure 4-10
Dimensioning the diameter of a circle.

Diameter of a revolved section: To define the diameter of a revolved section, pick the entity, pick the centerline that represents the axis of revolution, and then pick the entity again. Next, place the dimension.

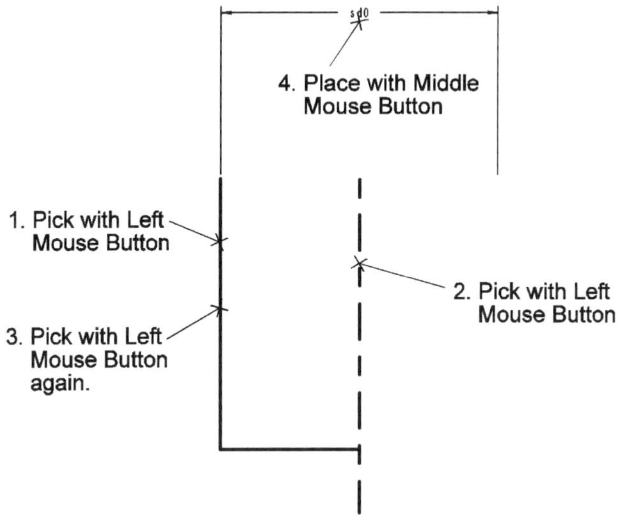

Figure 4-11
Dimensioning the diameter of a revolved section.

Angular Dimensions

Angle between two lines: To define the angle between two lines, pick near the midpoint of each line and then place the dimension. The location chosen for the dimension placement determines whether the specified angle is the acute angle or the obtuse angle.

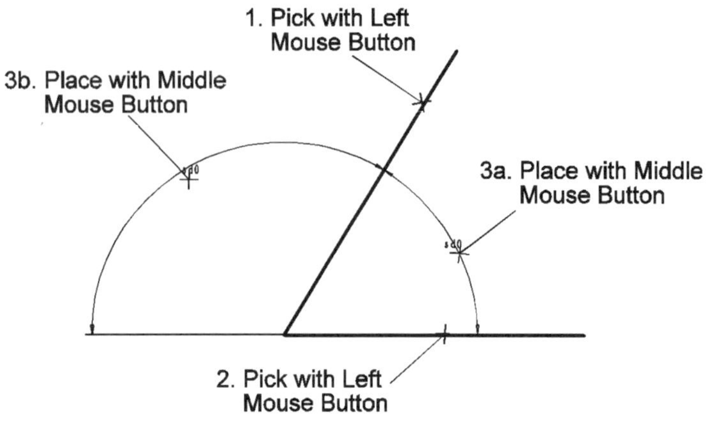

Figure 4-12
Dimensioning the angle between two lines.

Arc angle: To define an arc angle, pick each endpoint and then pick near the midpoint of the arc. Next, place the dimension.

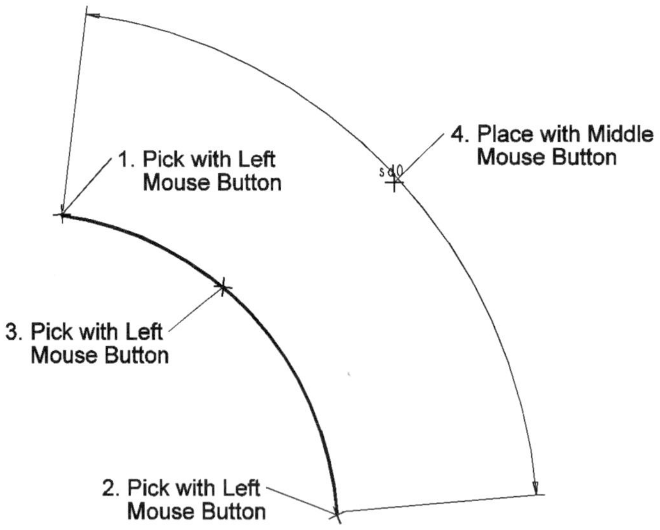

Figure 4--13
Dimensioning the angle of an arc.

- ❐ **Align** — The **Align** command is extremely important. It is a method for relating sketched geometry to existing part geometry without having to use a dimension. Pick the **Alignment ➥ Align** command. Then, using the GET SELECT menu, pick the sketch geometry and pick the part geometry you want to align to. You can align points, endpoints, and centerpoints to part vertices, part edges, axes, datum planes, and planar part surfaces. You can align lines and centerlines to part vertices, linear part edges, axes, datum planes, and planar part surfaces. You can align arcs and circles to circular part edges. If your sketched entity is not sketched close enough to the part geometry, the **Align** command will fail.

- ❐ **Unalign** — **Unalign** removes an alignment constraint. To unalign, select **Alignment ➥ Unalign**: All of the geometry that has been used for aligning will highlight in green. Pick the part geometry you no longer want to be aligned with. **Unalign Many** and **Unalign All** are additional options available on the ALIGNMENT menu for removing alignment constraints.

Practice Sections

Now you can practice creating sections and sketching entities. You will create a few example sections on one face of a rectangular box. Because the goal is to practice sketching and not feature creation, you will begin creating a feature, but you will never actually complete it. You will use the **Cancel** command (from the dialog window) between each of these example sections in order to clear the previous section.

Before you start the examples, you need to make the box. Pick the following commands, starting from the MODE menu:

Part ➥ Create

```
Enter Part name [PRT0001]:  ↵
```

Feature ➥ Create ➥ Solid ➥ Protrusion ➥ Extrude | Solid | Done

You should see the sketcher window appear. Using the line command built into the left mouse button, sketch the four sides of a rectangular box. Notice the "rubberbanding" between the first endpoint and the cursor as you move the cursor to the location of the second endpoint of the line segment. The line segments do not have to be perfectly straight: they just need to be close. Use the gridlines as a guide.

When you have completed a rectangular shape, press the middle mouse button to stop placing lines. Pick **Dimension**, pick the upper horizontal line with the left mouse button, and place the dimension over the line with the middle mouse button. Pick the rightmost vertical line with the left mouse button and place the dimension to the right of the line with the middle mouse button. Your box should now be constrained with a width and height dimension.

Pick **Regenerate** to tell Pro/ENGINEER to evaluate the section and to determine that the section is fully constrained. Now, use **Modify** to specify the dimensions such that the width of the box is 6 and the height of the box is 3. Select **Regenerate** again to let Pro/ENGINEER resize the box. Pick **Done** to dismiss the sketcher window and continue with the feature creation. For this type of feature, you will be prompted for the depth dimension.

```
Enter extrusion DEPTH [1.12]:  1 ↵
```

Pick **OK** from the dialog window. The rectangular box is completed.

Sketching Planes

In the box previously created, you did not have to identify a sketching plane. That was because there were no planes to choose from. Notice that the default orientation has been determined by Pro/ENGINEER. If you want to control your default orientation, you need to sketch your first protrusion on a datum plane of your choice. Now that you have created a solid box, you have a choice of sketching planes.

The next time you need to create a section, the SETUP PLANE menu will appear and give you a few options for choosing or creating a sketch plane. To get to the SETUP PLANE menu for these example sections, select the following commands:

Create ➡ Solid ➡ Cut ➡ Extrude | Solid | Done ➡ One Side | Done

The first option on the SETUP PLANE menu, **Plane**, is used in conjunction with the GET SELECT menu to select an existing plane to sketch on. The second option, **Make Datum**, will be discussed in a later chapter. You should choose the **Plane** option and pick the larger front face of the box as the sketching plane.

Figure 4-14
The sketching plane.

Next, Pro/ENGINEER will prompt you for the direction of feature creation. This direction indicates the direction of the cut feature from the sketching plane. Accept the default direction by choosing **Okay** from the DIRECTION menu. The last step is to identify another surface. This surface will control the orientation of the sketching view. To orient a sketch plane, you need to find a reference plane perpendicular to the sketch plane. You use the options on the SKET VIEW menu to tell Pro/ENGINEER which side of the sketch window the reference plane should face. For example, to orient your sketch view like the following diagram, you could select **Top** from the SKET VIEW menu and then pick this surface.

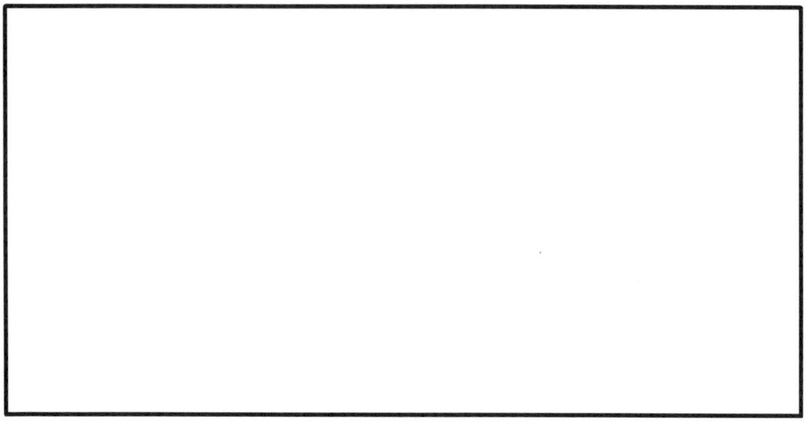

Figure 4-15
Desired sketch view orientation.

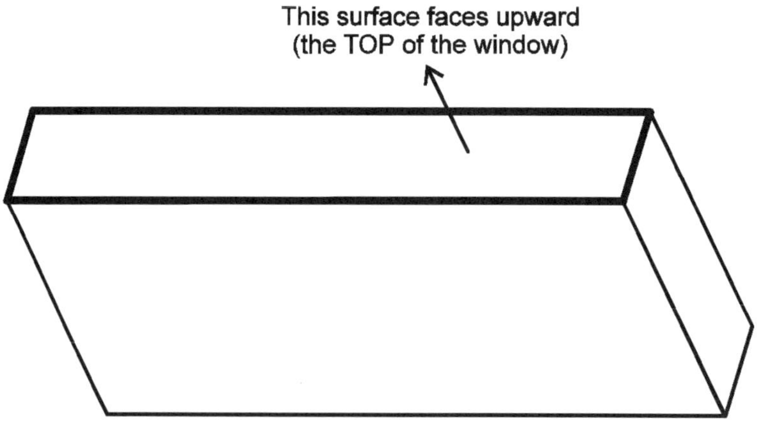

Figure 4-16
Picking the top surface of the box.

Another option would be to select **Right** from the SKET VIEW menu and pick this surface.

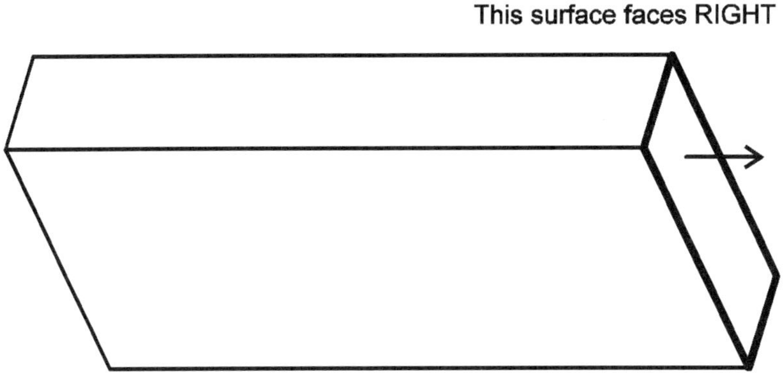

Figure 4-17
Picking the side surface of the box.

As you can see, there are a number of choices that would give you the same result. It is up to you to pick a reasonable reference plane for orientation.

> ✓ **TIP:** *You do not want to pick just any planar surface for orienting the sketch plane because the new feature will be a child of the feature that contains the reference plane. The horizontal or vertical reference establishes the base for the horizontal and vertical assumptions in the sketcher. Picking an angled surface is rarely required, but if you do pick one, your horizontal and vertical assumptions will be based on the orientation of this surface. If you understand how this reference works, you can predict what the sketch view will look like, and you will not waste time and energy trying to determine how Pro/ENGINEER has reoriented your view.*

When you have identified the reference plane, Pro/ENGINEER reorients your view so that you can sketch on the plane you choose.

Example Practice Sections

The following example sections are provided to show you how the various sketcher commands work together. Because the examples are intended to help you practice sketching, pick **Quit** between examples to clear out the previous section before you begin the next one. Using the **Quit** command will also provide you with practice in selecting and orienting sketching planes.

In that the **Quit** command leaves your view oriented in the sketch view, you should choose **View ➥ Orientation ➥ Default** from the MAIN menu to place you back in the default view. This will allow you to pick the sketching plane and the reference plane more easily. Start with the first example and work your way through all of them. The examples will show you many of the assumptions Pro/ENGINEER can make about your sections. Refer back to the sketcher command descriptions if you need extra help.

Plan your sketching prior to drawing the entities in these examples. Use your grid to help you make the entities of equal size and equal length. Minor variations in entity size can prevent the sketcher from making all of the necessary assumptions.

Example 1

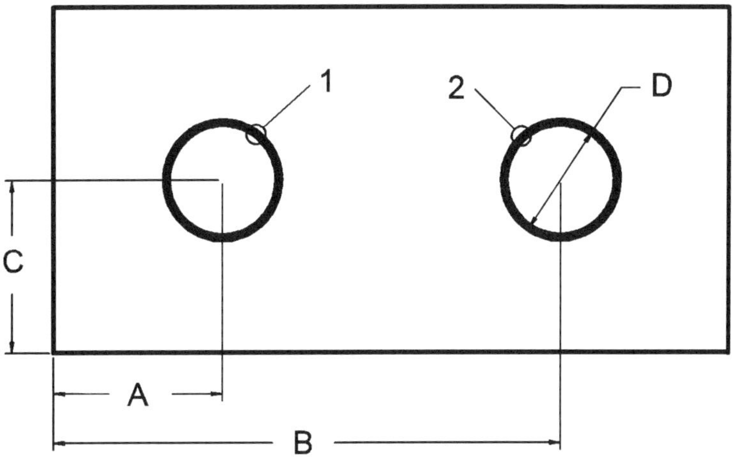

Description

1. Circle.
2. Circle.

A. Linear dimension from left edge to circle center. (1.5)
B. Linear dimension from left edge to circle center. (4.5)
C. Linear dimension from bottom edge to circle center. (1.5)
D. Diameter dimension for circle. (1.0)

Assumptions

1. Vertical alignment of circle centers.
2. Equal radius circles.

Figure 4-18

Example 2

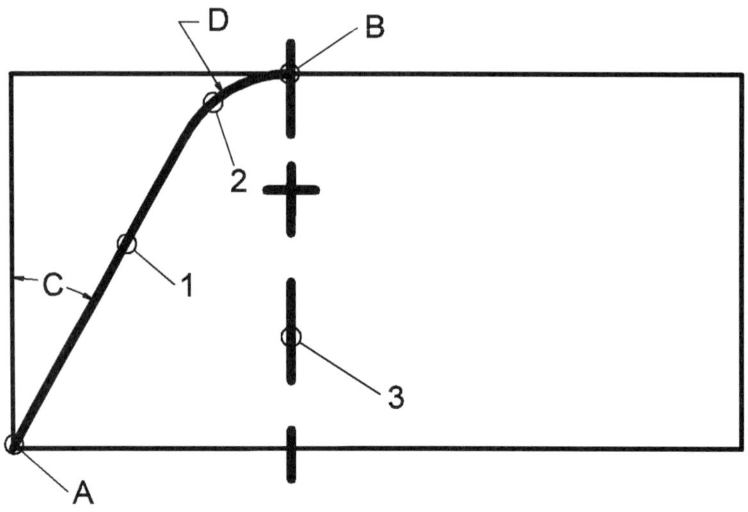

Description

1. Angled line.
2. Tangent arc.
3. Vertical centerline.

A. Align endpoint of line to vertex.
B. Align endpoint of arc to edge.
C. Angular dimension between left edge and angled line. (30)
D. Radial dimension for tangent arc. (1.0)

Assumptions

1. Tangency between line and arc.
2. Centerline is vertical.
3. Arc endpoint and arc center lie on centerline.

Figure 4-19

Example 3

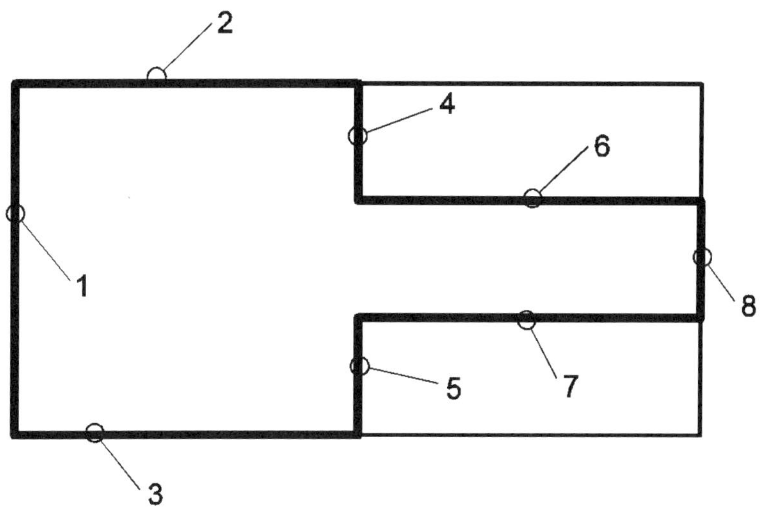

Description

1. Use edge of part.
2. Use edge of part and trim back to vertical line.
3. Use edge of part and trim back to vertical line.
4. Vertical line.
5. Vertical line.
6. Horizontal line.
7. Horizontal line.
8. Use edge of part and trim back to both horizontal lines.

Assumptions

1. Lines are vertical and horizontal.
2. Equal length lines.
   (Lines 1, 2, 3, 6, and 7 are equal, and lines 4, 5, and 8 are equal.)

Figure 4-20

Example 4

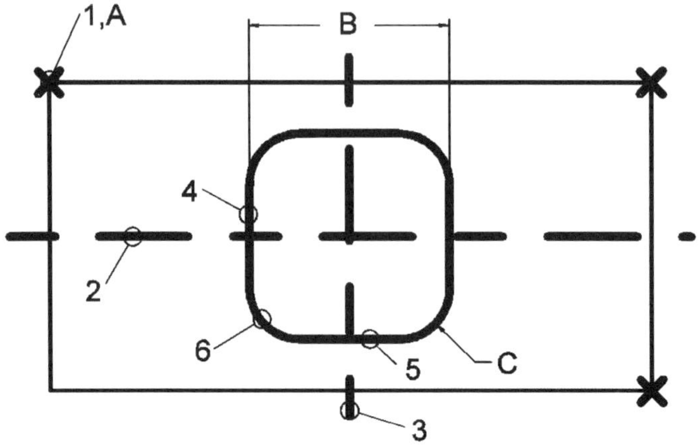

Description

1. Sketched point. (3 total)
2. Horizontal centerline.
3. Vertical centerline.
4. Vertical line. (2 total)
5. Horizontal line. (2 total)
6. Fillet. (4 total)

A. Align point to vertex. (3 places)
B. Linear dimension between vertical lines. (2.0)
C. Radial dimension for fillet. (0.5)

Assumptions

1. Tangency between all lines and arcs.
2. Centerlines are vertical and horizontal.
3. Lines are vertical and horizontal.
4. Horizontal and vertical symmetry. (Geometry is centered on the centerlines; centerlines are centered between points.)
5. Equal length lines and equal radius arcs.

Figure 4-21

Sketcher Assumptions

The following list will explain all of the assumptions that can be applied when you regenerate a section. Not all of these assumptions will apply to every section, but they will be used as necessary by the sketcher.

- **Horizontal or vertical lines** — Lines close to being horizontal or vertical will become horizontal or vertical.
- **Parallel or perpendicular lines** — Lines close to being parallel or perpendicular to another sketched entity will be constrained that way.
- **Centerpoint alignment** — Arc and circle centerpoints close to being aligned horizontally or vertically will be aligned.
- **Arc angles** — Arcs created close to an increment of 90 degrees (90, 180, or 270 degrees) with an endpoint being nearly tangent to either horizontal or vertical will be adjusted to an angle that is an increment of 90 degrees, and its endpoints will become tangent to horizontal or vertical.
- **Tangent entities** — Entities sketched nearly tangent to a sketched arc will become tangent to the arc.
- **Point location** — A point sketched near a sketched entity will be assumed to lie on the entity.

The following assumptions are evaluated for similar entities after one of the entities has been properly constrained.

- **Collinear lines** — Lines sketched nearly collinear will be adjusted to collinear.
- **Equal line lengths** — Lines of nearly equal length will be adjusted to equal length.

- **Arc or circle radius** — Arcs or circles created with approximately the same radius will be adjusted so that they have the same radius.

- **Centerline symmetry** — Similar entities with a centerline placed between them will be assumed to be equal distance from the centerline.

Most of Pro/ENGINEER's assumptions can be overridden by providing the appropriate dimensions. However, once successfully regenerated, lines assumed to be vertical or horizontal cannot have the assumption overridden. The same is true about lines assumed to be parallel or perpendicular.

> ✓ **TIP:** *Use the* ***Unregenerate*** *command to undo a regeneration that makes unwanted assumptions.*

Understanding Regeneration Failures

A list of common regeneration errors and their associated corrective actions follows. The list may help you in determining what to fix if you cannot get a section to regenerate properly.

```
Underdimensioned section. Please align to part or add
dimensions.
```

This error means you have not fully constrained your section. Look at the red highlights to see which entities need more dimensions. If an entire entity is shown in red, its size needs to be constrained by a dimension. If all of the endpoints of your sketched entities are highlighted, the program tells you to locate the section with respect to your part. This means that your sketched geometry is constrained properly, but you need to reference it to the existing part geometry by either aligning or specifying dimensions from the section to the part.

```
Extra dimensions found.
```

This error indicates that you have over-constrained the section. You need to delete extra dimensions. Pro/ENGINEER will highlight any dimensions it thinks are extra, but you can delete whichever ones you like.

```
Regen failed. Highlighted segment is too small.
```

When your section is too small for Pro/ENGINEER to interpret, you will get this error message. Try zooming in on the section and see if it successfully regenerates.

> **NOTE 1:** *Imagine showing your sketch to another person. If you are zoomed out too far, the details of small entities will not be clear. If you zoom in on those details to make them clear to the other person, Pro/ENGINEER will also understand them. However, there is a drawback if you have gaps between endpoints of your sketched entities. When zoomed in, they may appear so large that Pro/ENGINEER may assume you wanted the gaps. In these cases, you will need to trim the entities to their intersection points.*

> **NOTE 2:** *Your entire section does not need to be visible in order to be regenerated. You can zoom in to a portion of the section if necessary.*

```
Regen failed. No entities to regenerate.
```

If you have not sketched anything yet, and you try to regenerate, you will get this message.

```
Multiple loops must all be closed in this section.
```

This error message tells you that you have more than one "open" loop in your section. The following illustration shows the difference between open loops and closed loops.

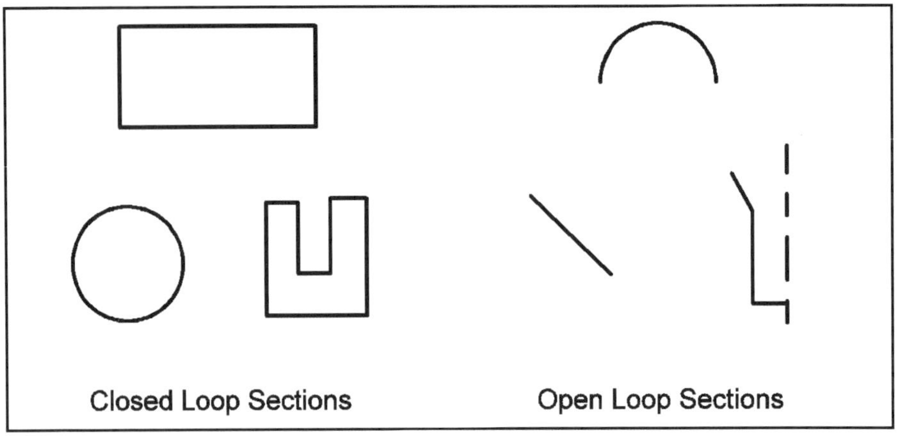

Figure 4-22
Examples of open and closed loops.

You may have any number of closed loops in a section. If you want to sketch an open loop section, you are allowed to have only one loop. If you get this message, you must either close all of the loops or reduce the section to one open loop.

```
Warning: Not all open ends have been explicitly
aligned.
```

This warning can appear if you have an open loop section. This is only a warning, not an error or failure message. The message is simply telling you that you have endpoints of an open loop that have not been aligned to your existing part. If you needed to align the endpoints, align them and regenerate again.

Regeneration failures can often be caused by failing to meet the section requirements of the particular feature you are trying to create. For example, revolved sections require a centerline to represent the axis of revolution. If you do not have a centerline in your section, the regeneration will fail.

Conclusion (or Top Ten Sketcher Hints)

Sketching will require a little bit of practice. These tips will help you become a better user of the sketcher without all of the usual trial and error. If you read and understand this section, you will find that you can make the sketcher do what you want it to do, rather than believing that it has a mind of its own.

1. **Exaggerate sizes** — If two entities are not intended to be the same size, do not sketch them so that they appear to be the same size. Sketch them differently so that Pro/ENGINEER will not assume them to be equal in length or in radius.

2. **Exaggerate angles** — If a line is intended to be at a 1-degree angle, sketch it at 10 degrees, regenerate the section, and modify it to a 1-degree angle. You do not want Pro/ENGINEER to assume that the line is vertical or horizontal when it should not be.

3. **Keep it simple** — Do not try to put everything into a single section. Use multiple features to break your design down into simple steps with simple sections. If you have to make a large, complex section, try sketching and regenerating a little bit at a time to approach the final section.

4. **Capture your design intent** — Do not just dimension the section in order to constrain your geometry—capture your design intent. If a feature is meant to be 4 inches from a surface of your part, specify it to be 4 inches from the surface. By dimensioning this way, your feature will move with the surface if the part changes size.

5. **Exaggerate nonaligned entities** — Do not sketch on or near a part edge if you do not want to associate the

sketched geometry to that edge. Pro/ENGINEER will ask you if you want to align these items and, most of the time, you do not want to. Exaggerate the placement of your sketched geometry to avoid this problem.

6. **Use the tangent arc commands** — If you want to place a tangent arc in your section, use the fillet command or the tangent arc button. Do not try to approximate tangency with three-point arcs unless you are very accurate with placement. Let the commands work for you, not against you.

7. **Avoid fillets galore** — Do not place unnecessary fillets in your sections. It is a better practice to create these as round features when they are required. You may think that you are saving a feature, but in reality you are adding needless complexity to a section and making your part more difficult to modify.

8. **Regenerate before you modify** — It is possible to modify dimensions before you have regenerated your section. However, it is better to regenerate your section before you modify the dimensions so that you can ensure the geometry is properly constrained. You can induce regeneration failures by inputting an invalid value for a dimension. You may then think that the geometry is under-constrained, when the only problem may be that an invalid value was entered for a dimension.

9. **Use the grid** — Use the grid lines to help you draw horizontal and vertical lines. Use them to help you make sketched geometry of equal length.

10. **Practice, practice, practice** — Nobody becomes an expert overnight. Use these methods to help you become more and more successful as you design bigger and better parts.

Chapter 5
Construction and Work Management Tools

Datum Features and More About Views

Introduction

This chapter describes in greater detail some basic tools associated with two things that help you in your design work: datum features and views of your work in progress. You have already been introduced to the use of datum planes in construction. You will learn specific information about datum planes, the most frequently used datum feature, and the basics of other special datum types. The use of views is the next best thing to having the model in your hands. This chapter will teach you to become more competent with using a greater variety of view commands. It concludes the first section of the book, which provides information to learn basic skills for using Pro/ENGINEER.

Datum Features

Datums are a category of features available in Pro/ENGINEER. Datums are used as references while you design. For example, a hole can have a datum axis to represent its center. You can dimension to this axis and create more features relative to the hole (so that if the hole moves, the other features will move). Datums are also the way you create 3D point and 3D wireframe geometry in Pro/ENGINEER.

Uses of Datum Planes

Datum planes are the most common and most important of the various types of datums. Experienced Pro/ENGINEER users know how and when to use datum planes in their designs. Datum planes are planar references you can place in your design. The following is a partial list of some of the uses of datum planes.

- **Dimensioning References** — Creating three datum planes as the first features of your part is a good practice (use the Default Datum Plane menu selection for this). During the design process, whenever you have a choice between dimensioning to one of these original datum planes or a part feature, select the datum plane. This will prevent unnecessary parent/child relationships between your features.

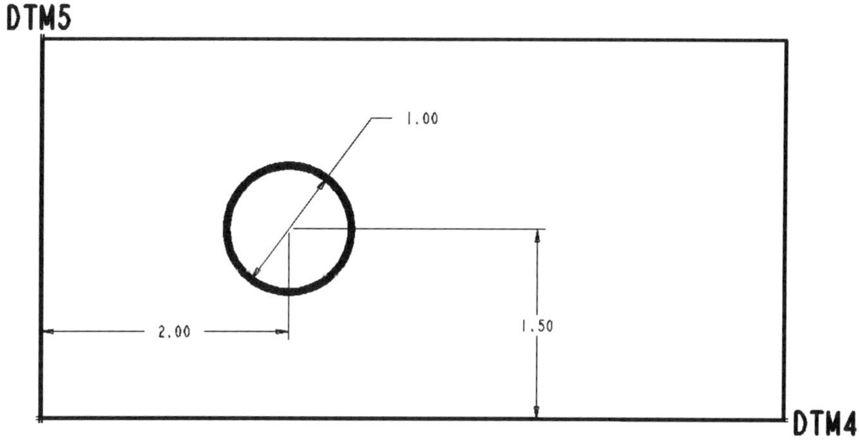

Figure 5-1
Using datum planes to dimension features.

- ❒ **Orienting Views** — Imagine trying to orient a pyramid side view. Without two planar surfaces normal to each other, there is no way to specify an orientation. However, after placing a datum plane through the apex and normal to the base, orienting the view with the two orthogonal planes is easy.

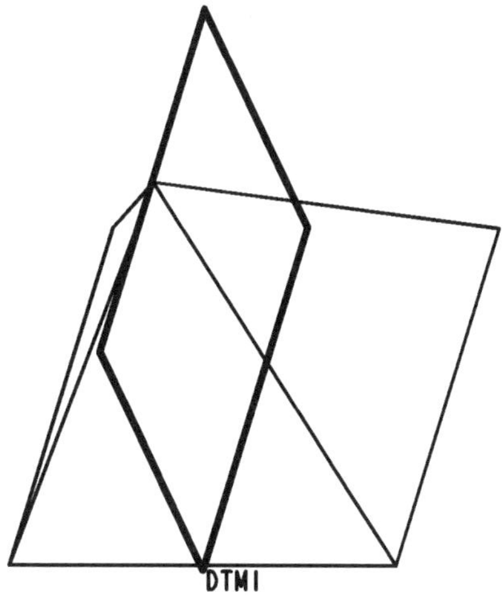

Figure 5-2
Using datum planes for view orientation.

- **Sketching Planes** — Some features require that you sketch a section. If a planar surface is not available or is not located in the correct position, a datum plane can be created to sketch on.

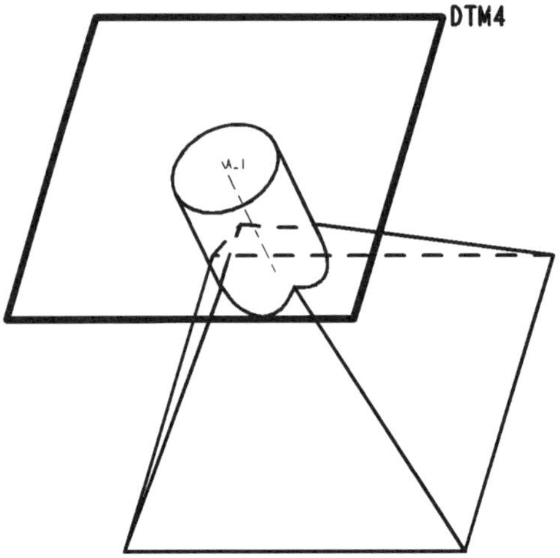

Figure 5-3
Using datum planes for a sketch plane.

- **Assembling Parts** — Sometimes it is best to assemble two mating parts by aligning their datum planes. If you have created each part with common datum planes, these planes should line up when the parts are assembled.

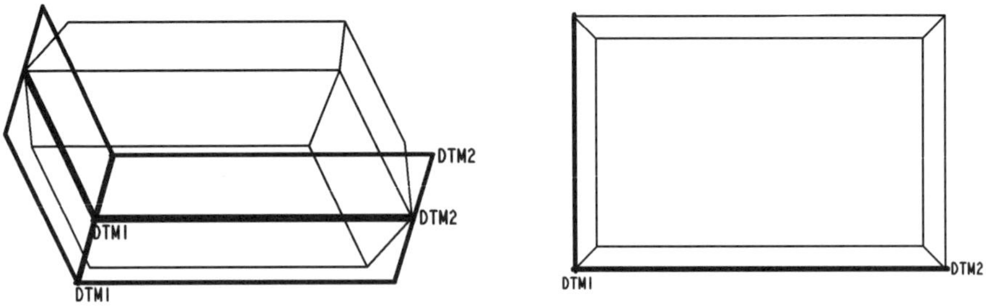

Figure 5-4
Using datum planes for assembly.

❐ **Locating Cross Sections** — A datum plane can be placed in a part in order to display and use a planar cross section in a drawing.

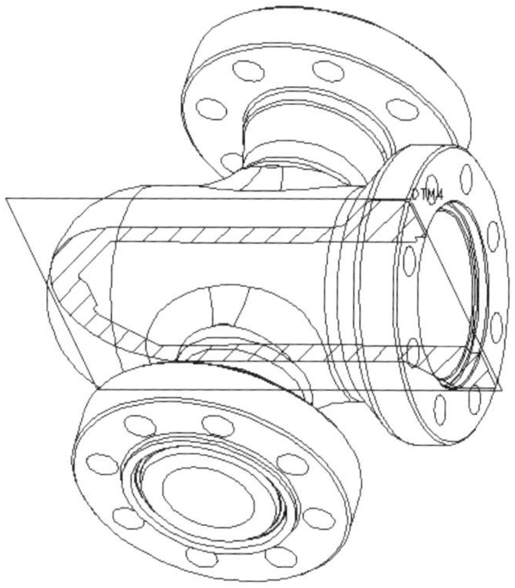

Figure 5-5
Using datum planes for a cross section.

Yellow and Red Sides of Datum Planes

Datum planes have two sides, just as a part surface has an inside and an outside. These sides are displayed in yellow and red. The yellow side of a datum plane can be compared with the outside of a part surface. When a datum plane is used for view orientation, you are specifying the direction the yellow side should face. Using the previous pyramid part example, select the base to face toward the bottom, and the datum plane to face toward the right.

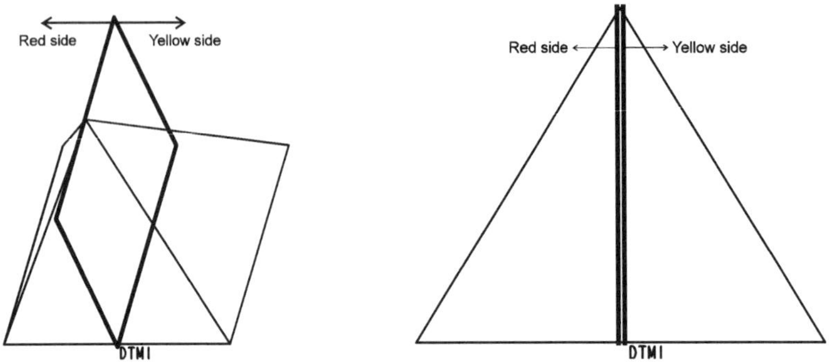

Figure 5-6
The yellow side of the datum plane faces to the right.

Size Control of Datum Planes

Although datum planes represent infinite planes, they are displayed with boundaries. You have control over the size of the displayed boundary. To change the size of a datum plane, pick **Redefine** from the FEAT menu and then select the datum plane. Next, select **Attributes | Done** from the REDEFINE menu. Select one of the options described in the material that follows from the OPTIONS menu, and then pick the appropriate geometry as required by the option.

> **NOTE:** *Datum planes will always be created with the default size. You must use the previously described method to select a different-size control option.*

- **Default** — The datum plane boundary will always encompass the part or assembly. As the latter increases in size, the boundary will grow.

- **Fit Part** — In assembly mode, the datum plane boundary can be fitted to one of the component parts.

- **Fit Feature** — The datum plane boundary will always encompass the chosen feature. After selecting this option, you will need to select the feature around which to fit the datum plane boundary.

- **Fit Surface** — The datum plane boundary will always encompass the chosen surface. After selecting this option, you will need to select the surface around which to fit the datum plane boundary.

- **Fit Edge** — The datum plane boundary will always encompass the chosen edge. After selecting this option, you will need to select the edge around which to fit the datum plane boundary.

- **Fit Axis** — The datum plane boundary will always encompass the chosen axis. After selecting this option, you will need to select the axis around which to fit the datum plane boundary.

- **Fit Radius** — The datum plane boundary size will be determined by the keyed-in radius value.

Methods of Placing Datum Planes

Datum planes are placed by providing constraints. The number of constraints you need to specify will vary with the type of constraints used. For example, a datum plane can be placed through three points. This example makes use of the **Through** constraint three times. An example of when you need only one constraint would be when you wanted a datum plane offset from an existing planar surface. This would require the use of only the **Offset** constraint. A list of the available constraints follows.

- **Through** — You can place a datum plane through an axis, edge, planar curve, point, vertex, planar surface, or cylinder.

- **Normal** — You can place a datum plane normal to an axis, edge, planar curve, or planar surface.
- **Parallel** — You can place a datum plane parallel to a planar surface.
- **Offset** — You can place a datum plane offset from a planar surface or datum coordinate system. This constraint requires you to key in a distance and cannot be used in conjunction with another constraint.
- **Angle** — A datum plane can be placed at an angle to a planar surface. This constraint is selected after using the **Through** constraint to provide the pivot line.
- **Tangent** — A datum plane can be placed tangent to a cylinder.
- **BlendSection** — A datum plane can be placed through one of the sketch sections of a blend feature.

Datum Coordinate Systems

Pro/ENGINEER does not require that you design within a specific coordinate system. Because you are never asked to supply values such as *xy=0,2,3* or *3,4,5,* you do not need a coordinate system that facilitates these key-ins. In that features can be placed relative to other features (parametrically), you are not required to create a coordinate system during your design process.

Uses of Datum Coordinate Systems

There are only a few occasions when you need a coordinate system. Most of these situations occur when you import or export geometry from or to other CAD file systems that require the geometry to be mapped to a coordinate system. A few examples of situations that require datum coordinate systems follow.

- **Graphical Import and Export** — IGES, FEA, and STL (stereolithography) files require that the geometry be supplied with respect to a coordinate system. When you import an IGES file, you need to place it relative to a datum coordinate system.

- **Manufacturing** — Toolpaths created by Pro/MANUFACTURE are created with a datum coordinate system as a reference.

- **Mass Properties** — Pro/ENGINEER calculates the center of gravity with respect to a datum coordinate system when you choose **Mass Props** from the INFO menu.

- **Measure ➥ Distance ➥ Increment** — Provides a distance between two selected entities along with the x, y, and z incremental distances in the directions defined by the selected coordinate system.

- **Copying Features** — The **Copy ➥ Translate** command allows you to copy a feature using a datum coordinate system as its reference.

Because of these different functions, you can have multiple datum coordinate systems within the same part. Datum coordinate systems have names assigned to them automatically. The default name for the first coordinate system is CS0. The number is incremented by one for each subsequent datum coordinate system.

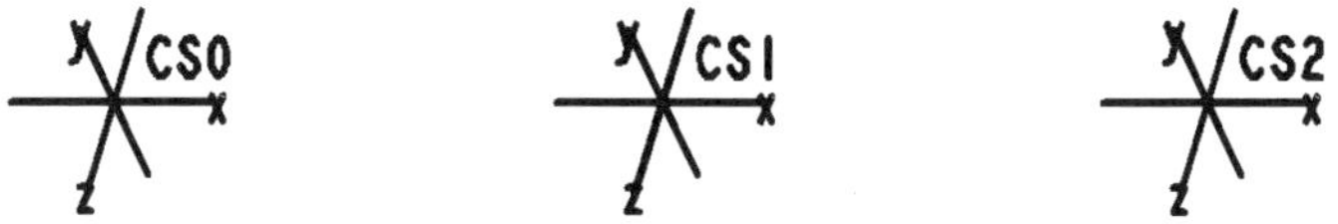

Figure 5-7
Examples of datum coordinate systems.

Datum coordinate systems are always shown with x, y, and z axes (Cartesian coordinates), but they can represent the origin of cylindrical and spherical coordinate systems as well. The following diagram shows how the various coordinate system types are mapped onto datum coordinate systems.

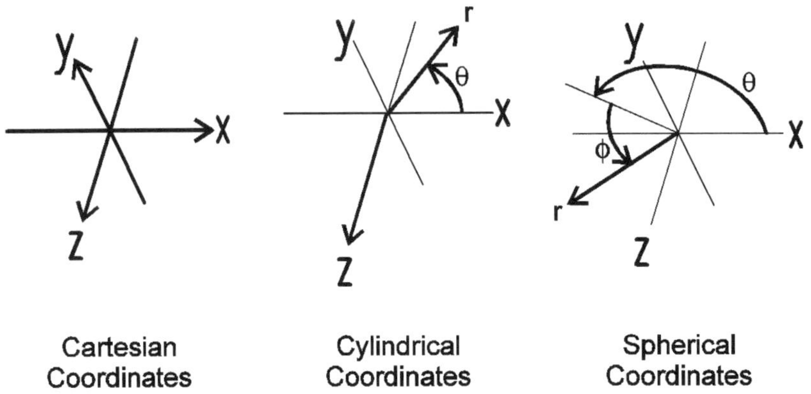

Figure 5-8
How coordinate system types relate to datum coordinate systems.

Uses of Datum Points

Datum points are reference points that you can identify and place in your design. They can help you identify points of interest in your design. Datum points can be used in a variety of ways. A list of common uses of datum points follows.

- Not many features require datum points, but some can use them to permit you more control over the feature's shape. For example, **curve X srf** datum points can be placed on an edge to create control points in variable radius rounds. This allows you to enter a different radius value at each of the datum point locations. The same technique can be applied to variable draft features using datum points. Blend features that have a profile (trajectory) can use datum points to establish locations for sketched sections along the profile.

❒ Datum points can identify locations for applied loads in an FEA mesh.

❒ For calculating geometric tolerance, datum points provide a location for attaching datum targets.

A datum point is represented by a yellow x, which is assigned to the point automatically. The default name for the first point is PNT0. The number is incremented by one for each subsequent datum point.

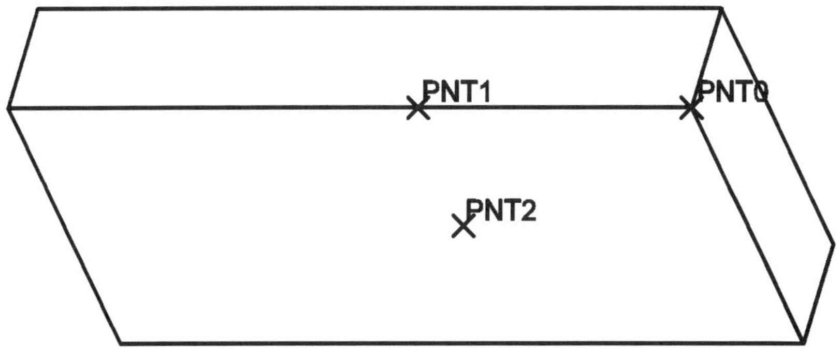

Figure 5-9
Examples of datum points.

Uses of Datum Axes

Datum axes are linear references that often represent the centers of revolved features. They have no defined endpoints and are infinite in length. Two uses of datum axes follow:

❒ **Centerlines** — Datum axes represent centerlines of cylinders, holes, and revolved features. Datum axes are automatically created when you extrude a circle or revolve a section.

❒ **Concentric features** — When you need to create concentric features, you can align them to a common axis to maintain their concentricity.

Datum axes are represented by a yellow centerline and names are assigned to them automatically. The default name for the first axis is A_1. The number is incremented by one for each subsequent datum axis.

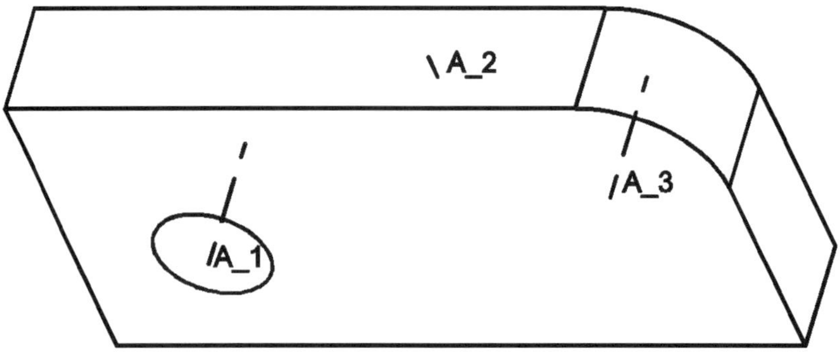

Figure 5-10
Examples of datum axes.

Uses of Datum Curves

Wireframe geometry in Pro/ENGINEER is created through datum curves, which are displayed in orange. Datum curves help you create many advanced features in Pro/ENGINEER, including the following:

- **Sweep and Blend Trajectories** — Datum curves can be used as the path in sweep and blend trajectories.
- **Surface by Boundaries** — Datum curves can define boundaries of surface features.
- **Manufacturing Boundaries** — Datum curves can define the boundaries and tool paths for manufacturing operations.
- **Curve Driven Draft** — Datum curves can be used to define the parting line of a draft feature.

Summary of Datum Features

There are many options for various types of datums. In the real world, you will not use all datum features. Ninety percent of your datums will be datum planes, and they will be used at the default size. If you concentrate on learning just a few of the datum plane placement constraints—such as **Offset**, **Parallel**, **Angle**, and **Through**—you will be in good shape. Do not get bogged down trying to understand all options and datum features. These commands are best learned at a gradual pace and only when you need to use them.

View Enhancements and Cosmetic Views

Sometimes you need to see your model in an enhanced form to help clarify its shape. Commands that provide cosmetic view attributes are located on the COSM VIEW menu, which is displayed when you select **Cosmetic** from the MAIN VIEW menu. To create output from a cosmetic view, use the PostScript output capabilities from the SHADE menu.

Shading the View

The most frequently used cosmetic attribute is shading. The shading commands are made available when you select **Shade** from the COSM VIEW menu.

- ❒ **Display** — Pick this command to shade the model in the active graphics window.

Figure 5-11
A shade view of a part.

- **Save** — Shaded views can be saved to a file in a proprietary raster format. To save the shaded view, you must shade the view with the **Display** command immediately prior to selecting **Save**. Pro/ENGINEER will prompt you for a file name.

- **Restore** — Pick this command to redisplay a previously saved view. Pro/ENGINEER will then prompt you for the file name.

  ✓ **TIP:** *There are a few drawbacks to using the saved shaded view capabilities. If the size of the active window does not exactly match the size of the window when the image was saved, it will not be displayed. In addition, these saved views are not transportable across various workstation platforms, nor are they supported between major releases of Pro/ENGINEER.*

- **Quality** — You can adjust the quality of the shaded view. If your view is zoomed in closely to a curved surface, and you shade the view, the surface might look rough because planar facets are used to approximate the curved surfaces during shading.

This command will allow you to increase the accuracy (decreasing the facet size), which makes the surface look much smoother. Any increase in quality will increase the time required to shade the display.

- **PostScript** — Shaded views can also be saved to a PostScript file. To save the shaded view to a PostScript file, you must shade the view with the **Display** command immediately prior to selecting **Postscript**. Pro/ENGINEER will then allow you to choose the type of printer, resolution, and paper size. You will be prompted for a file name.

Surface Meshing

Individual surfaces can be meshed with lines in order to show their contours. Pick the **Mesh Surface** command from the COSM VIEW menu. You can then pick the surface you want to mesh. If the default values for mesh lines are not appropriate, you can adjust them with the **Change Mesh** command.

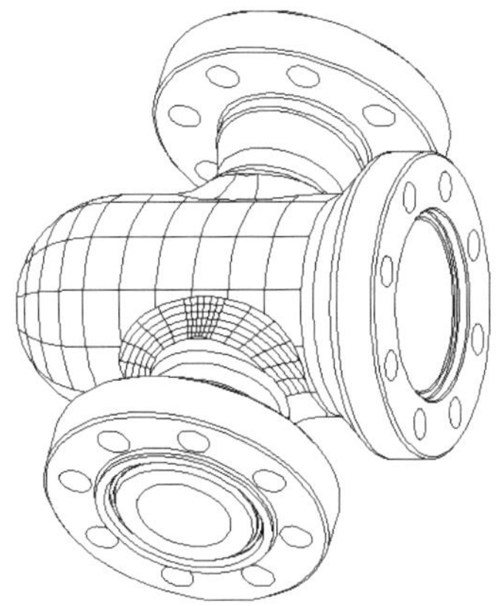

Figure 5-12
A curved surface shown after meshing.

Perspective Views

Perspective views can be displayed with the **Perspective** command located on the MAIN VIEW menu. Perspective views can be shaded and spun, but you cannot work in a perspective view.

> **NOTE:** *As soon as you dismiss the VIEW menu, the perspective view will be replaced by the standard view. After selecting* ***Perspective*** *from the MAIN VIEW menu, you are presented with the PERSP TYPE menu.*

You have the following three options for perspective views:

- **From/To** — Pick two points that define a linear path for viewing the model.
- **Follow Path** — Pick an edge or curve to follow for the perspective view path. This option is followed by the PATH DIR menu, which allows you to define the direction of the path.
- **Default** — Uses the center of the model as the start point and defines the path as the vector normal to the screen through this point.

All options are followed by the DYNAMIC PERSPECTIVE window, which is similar to the spin control window. Sliders control the following view characteristics:

- **Eye Dist** — Imaginary distance from the camera eye to the model (default perspective views only)
- **Walk Thru** — Normalized (0 to 1) distance along the defined path (From/To and Follow Path perspective views only)
- **X Rot** — Rotates the view around the x axis of the screen
- **Y Rot** — Rotates the view around the y axis of the screen

- **Z Rot** — Rotates the view around the z axis of the screen
- **Zoom** — Zooms out of the current view
- **Left/Right** — Pans the view left and right
- **Up/Down** — Pans the view up and down
- **View Angle** — Controls the field of view

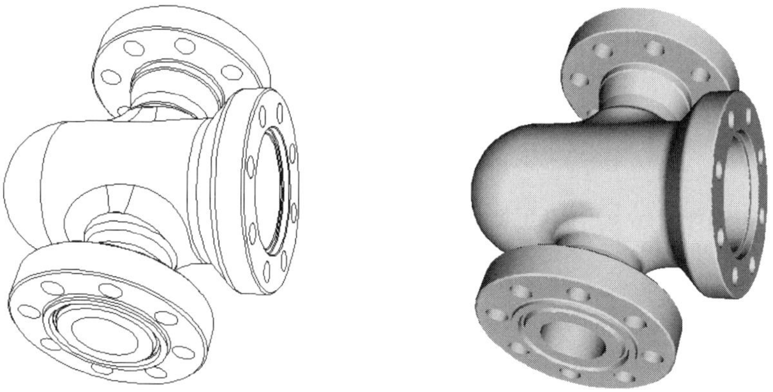

Figure 5-13
A perspective wireframe view and a perspective shaded view.

Colors

Colors can be created, and surfaces, parts, and assemblies can be assigned these colors. Color commands are accessed by picking **Appearances** on the COSM VIEW menu.

- **Show** — Displays the current color palette.
- **Define** — Creates new colors through the Appearance Editor dialog window. Color is controlled by setting its red-green-blue (RGB) values. Each color can contain different settings for special attributes, such as ambient light effectivity, reflectivity, shininess, and transparency. Threshold values for colors can also be

defined. Colors are very useful when dealing with large parts or assemblies. Experiment with the options offered in the dialog window, but avoid the use of red because Pro/ENGINEER highlights in red during geometry selection.

- **Set**—Applies a predefined color to a surface or part. If you set a color to a surface, you will see only that color when you shade the model. If you set a color to a part, the wireframe and shaded display of the part will be in the set color, instead of the system geometry color.
- **Unset** — Removes colors assigned to surfaces or parts. When colors are unset from a model, the model will return to the default system geometry color.
- **Change** — Modifies the settings of a model color or a palette color.
- **Store Map** — Saves newly defined colors. A file named *color.map* will be created. Because color palettes are not saved with individual parts or assemblies, the best way to share colors between parts and assemblies is to use a common *color.map* file.

Lights

Pro/ENGINEER has commands for setting multiple light sources. These light sources affect the shading of your model. Because these commands are rarely used, they will not be covered in detail. Available light types are directional, point, and spot. You can set the lights around your model and adjust their colors and positions. All light commands are shown when you pick **Lights** from the COSM VIEW menu.

Environment Settings for Views

The Pro/ENGINEER ENVIRONMENT menu has a few options that affect the display characteristics of your view. These options can be toggled at any time during your design session to help clarify your model geometry.

Model Display

There are five options for displaying your parts and assemblies in Pro/ENGINEER.

- ❒ **Wireframe** — Shows the entire model with solid white (object) lines.
- ❒ **Hidden line** — Shows all hidden lines in gray.
- ❒ **No hidden** — Displays no hidden lines.
- ❒ **Shading** — Displays the model as shaded. It differs from the **Cosmetic ➡ Shade ➡ Display** method of shading the model. The model will always remain shaded, even in Sketcher and after repainting. Datum planes, points, curves, coordinate systems, and axes will also be visible.
- ❒ **LOD** — Level of detail for shaded images during view spsin and pan.

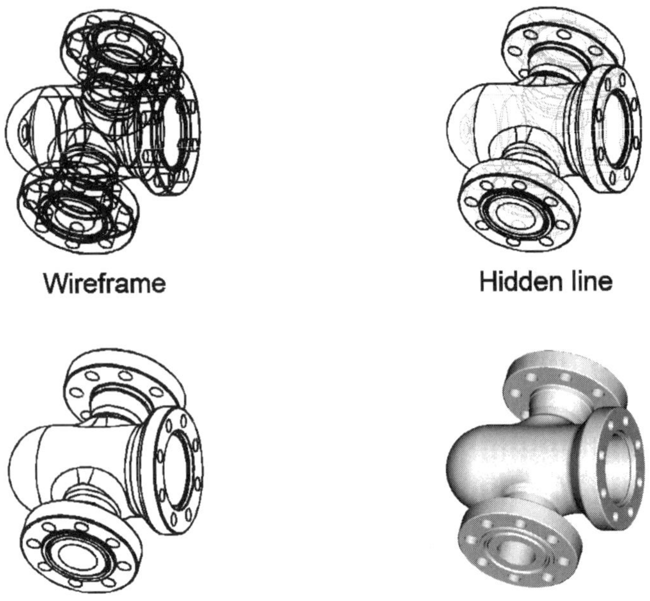

Wireframe Hidden line

No hidden line Shaded

Figure 5-14
Pro/ENGINEER's hidden line display options.

The **Hidden line** and **No hidden** options are unique to Pro/ENGINEER, and they are extremely useful. Most people use the **Hidden line** option when they are designing with Pro/ENGINEER. The **Shading** option may not be appropriate if shading parts and assemblies is a time-consuming process on your workstation. If your model is extremely large, you can switch to the Wireframe option and shorten the time required to regenerate the display.

Tangent Edge Display

A tangent edge exists between two tangent surfaces of a part. The five options for displaying tangent edges follow.

- **Solid Tan** — Displays tangent edges as solid white (object) lines
- **No Disp Tan** — Turns off the display of all tangent edges
- **Tan Ctrln** — Displays tangent edges as dashed centerlines
- **Tan Phantom** — Displays tangent edges as phantom lines
- **Tan Dimmed** — Displays tangent edges slightly dimmer than major object lines

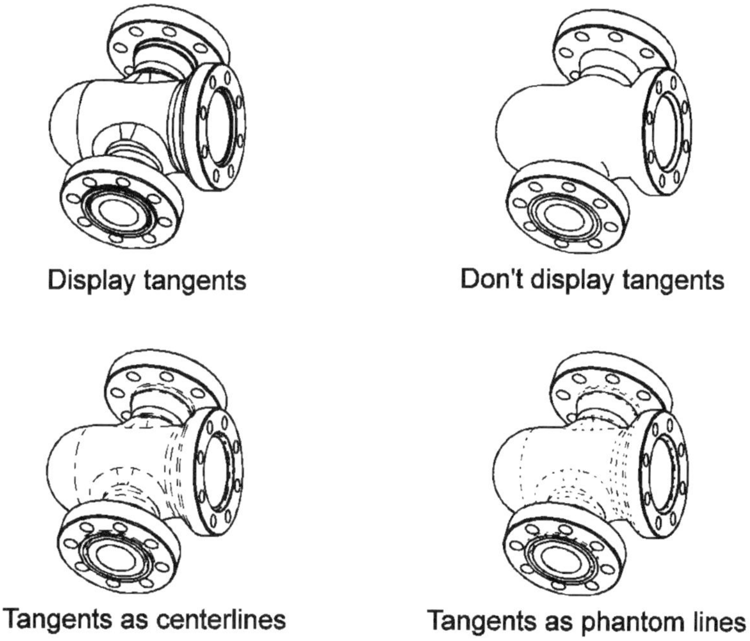

Figure 5-15
Pro/Engineer's tangent edge display options.

NOTE: *Most people use* **Tan Solid** *when designing in Pro/ ENGINEER.*

✓ ***TIP:*** *After you create a feature you will occasionally want to determine whether two surfaces are tangent to each other. The easiest way to accomplish this is to set the* ***No Disp Tan*** *option, and watch to see if the common edge between the two surfaces disappears.*

Default Orientation

The **Isometric** and **Trimetric** options control the orientation of the default view. The difference between these orientations is shown in the following illustration.

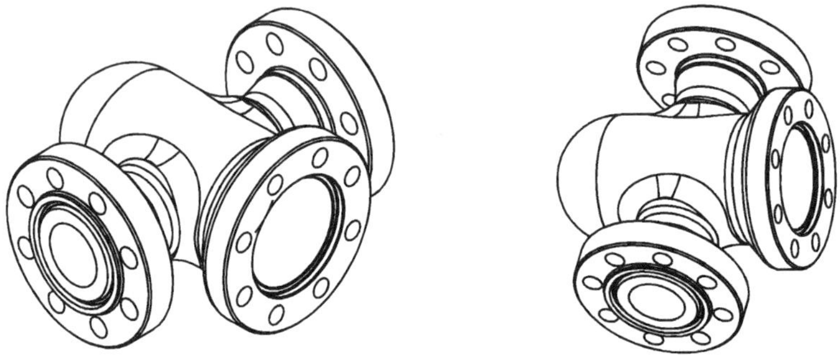

Figure 5-16
Pro'ENGINEER's isometric (left) and trimetric (right) orientation options.

Parametric View Behavior

As seen in previous chapters, Pro/ENGINEER is parametric in nature. In Chapter 1, you saw how changing one feature can make other dependent features change automatically. Pro/ENGINEER views can (and do) exhibit the same behavior.

Why Are Views Parametric?

The best way to explain why views should be parametric is to show an example. A part will be used to illustrate Pro/ENGINEER's view capabilities.

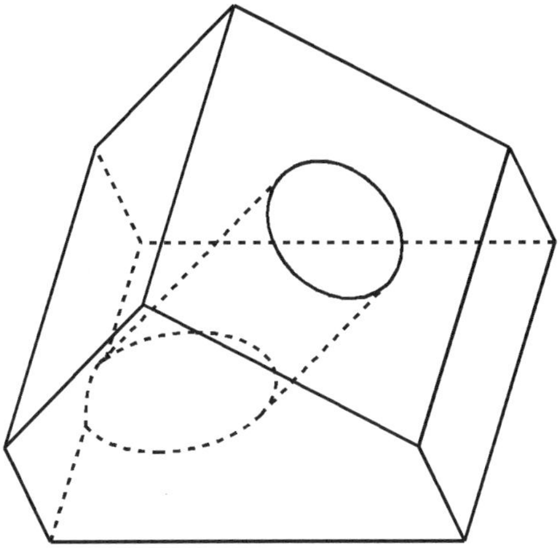

Figure 5-17
Example part for parametric views.

The part was constructed with the three features shown in the following illustration.

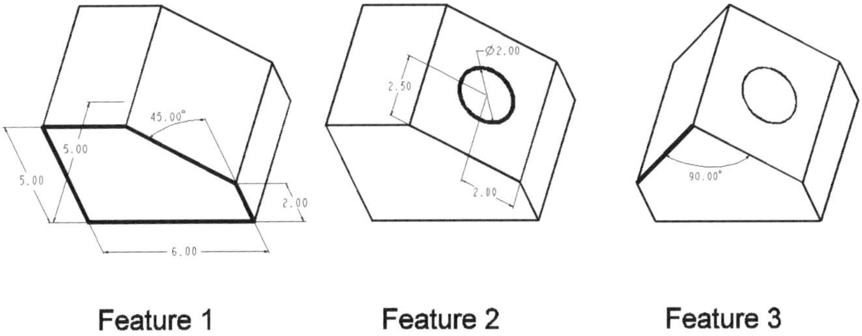

Figure 5-18
Features for the example part.

The goal is to create a view that lets you view along the axis of the hole.

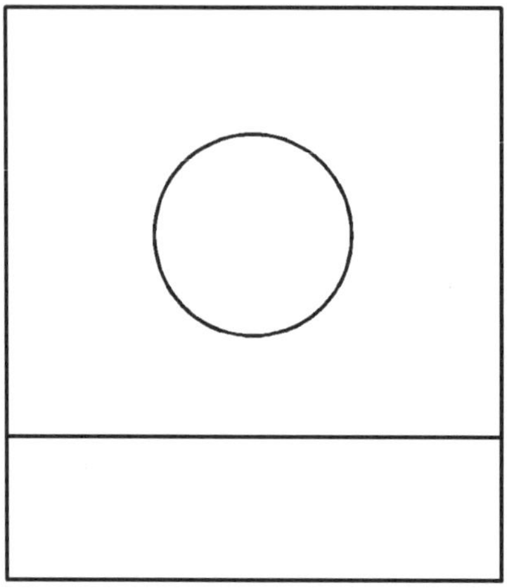

Figure 5-19
View orientation intent.

Use the Orientation by Two Planes method to create this view. As described earlier, you need two orthogonal planes to orient the view. Surface 1 should be the front plane, and surface 2 should face to the left.

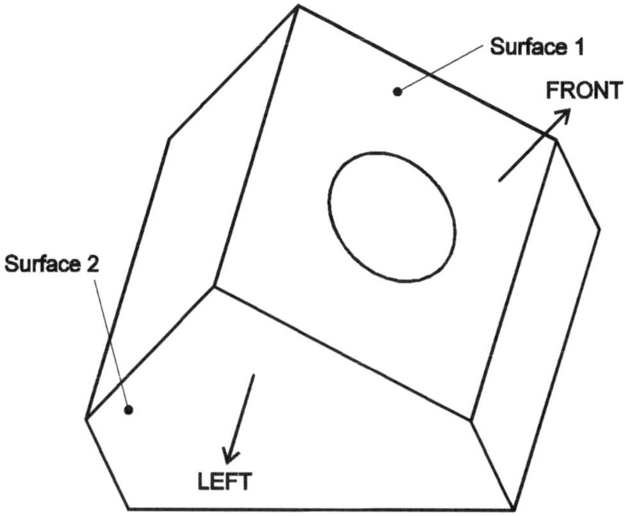

Figure 5-20
Reference surfaces for the view orientation.

These two surfaces were picked for view orientation because they have a unique property. Not only are they orthogonal, but they will *always* be orthogonal. Any dimensional change you make will not affect the orthogonal relationship between these two surfaces. Now would be a good time to save the view.

View ➥ Names ➥ Save

`Enter NAME:` **`downhole`** ↵

If you modify the angle of surface 1, you can see how the view updates properly. Change the angle dimension in feature 1 to 30 degrees and retrieve the saved view. Select the following commands:

View ➥ Names ➥ Retrieve ➥ Downhole

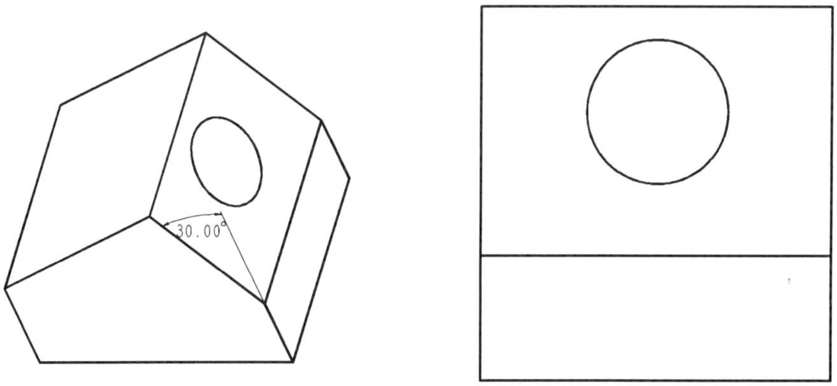

Figure 5-21
The example part after modification and the updated view.

Views in Pro/ENGINEER are not simple, static pictures of your model. Views need to be parametric because your parts are parametric. When you modify a dimension that affects the way a view is set up, you need the view to update its orientation along with the part. If the view were static, the previous modification would have resulted in the view shown in the following illustration, which does not meet the requirements of the original goal.

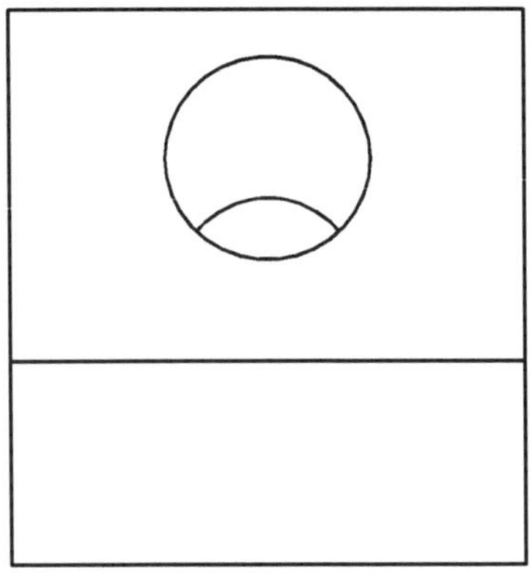

Figure 5-22
Improper view orientation caused by a static view.

Parametric View Failures

There are instances when Pro/ENGINEER will not be able to display a saved view after the model has been changed. To illustrate this, a similar view will be created using different reference surfaces. This time, use surface 1 as the front plane, and surface 2 as the plane that faces to the top.

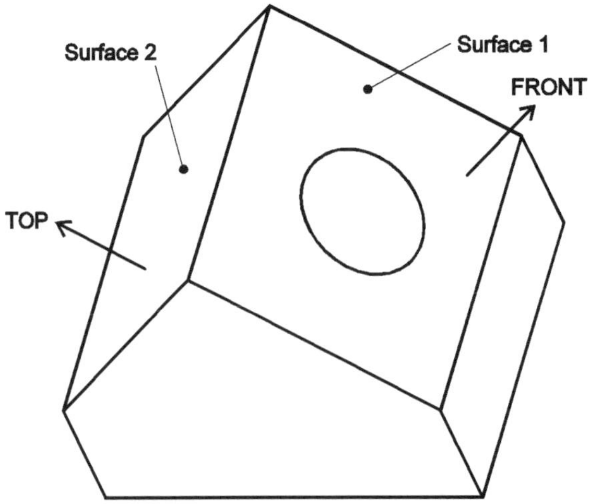

Figure 5-23
Reference surfaces for the view orientation.

Save this view. Now modify the angle in feature 3 to 80 degrees. When you try to retrieve the previously saved view, Pro/ENGINEER will give you the following message:

```
The selected view may contain invalid instructions.
```

In this case, the view will not display because the two planes are no longer orthogonal. Now, modify the angle in feature 3 to 90 degrees. The view will display when it is retrieved.

Another way to make a view retrieval fail is to delete geometry referenced by the view orientation instructions. In the example part, delete feature 3 and try to retrieve the view. Pro/ENGINEER will not be able to display the view and will provide the same error message.

Review of View Management

Pro/ENGINEER's views are parametric (just like the features). When you are creating your views for drawings or sections, you need to be careful about which objects you reference. You should always know which features are parents of the views you create and save. Here are a few tips to help you create retrievable saved views.

- When creating a view, try to reference major part geometry—or better yet, the default datum planes—rather than surfaces or edges of small details. Because details are often changed or removed, they can make a view unretrievable.

- When selecting orthogonal planes for view orientation, verify that the selected planes will remain orthogonal. If you can modify a dimension and cause the planes to no longer be orthogonal, they are probably bad choices for defining a saved view orientation.

- Create the basic views of your model (top view, front view, right view, and so on) using your first feature or a set of default datum planes. Then you can change to these views quickly during a design session.

Conclusion

After the design experience exercises and information overview of the first five chapters, you should feel familiar with the capabilities and potential of Pro/ENGINEER. You have developed an acquaintance with the basic mode functions, the window configuration, and design protocols—such as using datum features—that maximize the efficiency of the program. Now that you have created a basic wheel, you will move on to a more complex design session. You are ready to develop a new level of skill using Pro/ENGINEER. In Part 2 of the book you will have an opportunity to fine tune the base skills you have acquired in Part 1. You will participate in a real-time design process, using advanced techniques to design parts and create assemblies.

Part 2

Working with Parts and Assemblies

Chapter 6
Designing a Larger Part

Emphasis on Feature Creation Technique

Introduction

This chapter introduces intermediate-level concepts used to design parts. You will be shown how to build a part using many copied or patterned features. More complex sections will be shown, as well as a few more options used for extruded cuts and protrusions. Commands and concepts covered in this chapter include the following:

- Sketching complex sections
- Using text in sections
- Using part edges in sections

- ❒ Aligning to part geometry in sections
- ❒ Setting relations
- ❒ Making datum planes "on the fly"
- ❒ Copying features
- ❒ Mirroring features
- ❒ Patterning features
- ❒ Using the **UpTo Surface** and **Thru Next** extrude options
- ❒ Creating simple, blended protrusions
- ❒ Creating holes
- ❒ Making more complex rounds

You should now feel comfortable enough to move on to a bigger part. We will develop the chassis for the toy car.

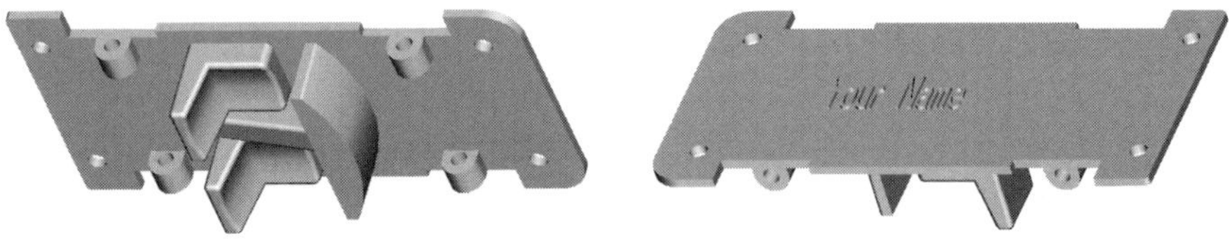

Figure 6-1
The finished chassis.

Starting a Large Part

Start this session by selecting **Part** from the MODE menu. Pick **Create** from the ENTERPART menu.

```
Enter Part name [PRT0001]:  chassis ↵
```

You have now created the chassis part in memory.

Creating Datums

Like the wheel, you should start the chassis by creating the default datum planes and the default coordinate system. Pick the following commands:

Feature ➥ Create ➥ Datum ➥ Plane ➥ Default ➥ Create ➥ Datum ➥ Coord Sys ➥ Default | Done

You should now see the default datum planes and coordinate system.

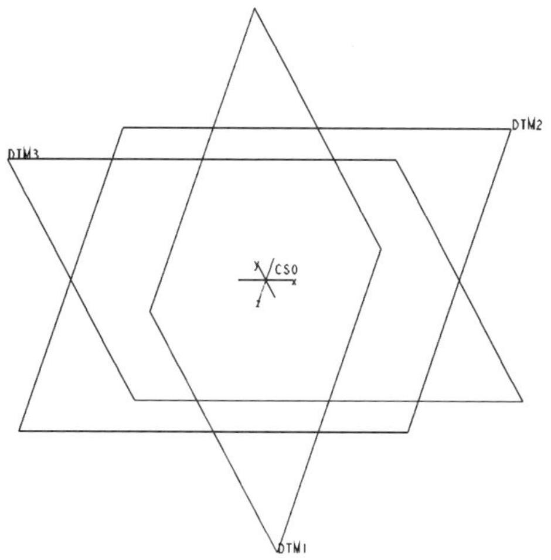

Figure 6-2
The basic datums.

These datums are the first four features of the part.

- ❐ Feature 1 — Datum Plane DTM1
- ❐ Feature 2 — Datum Plane DTM2

- ❒ Feature 3 — Datum Plane DTM3
- ❒ Feature 4 — Datum Coordinate System CS0

> **DESIGN NOTE:** *These datums will be used as references to help minimize unwanted parent/child relationships between your features.*

The Base Feature of a Large Part

A good place to start on the chassis is building a protrusion that makes a platform for the rest of the features. This platform will be a basic rectangle that is extruded to a specified thickness. Use the following information to create this feature.

Feature 5, the Base Protrusion

Commands:

Create ➡ Solid ➡ Protrusion ➡ Extrude | Solid | Done ➡ One Side | Done

References:

Sketch plane — DTM3.
Horiz / Vert reference — **Top**, DTM2.

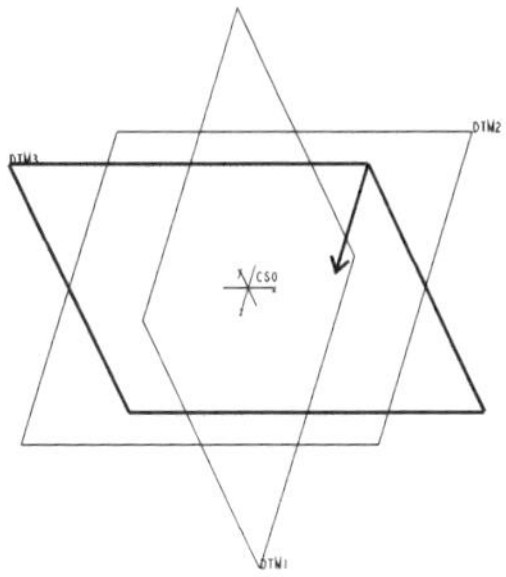

Figure 6-3

➡ ***NOTE:*** *Feature dimensions are shown in parentheses.*

1. Horizontal line
2. Vertical line
3. Horizontal line
4. Vertical line

A. Align lower horizontal line to DTM2
B. Align leftmost vertical line to DTM1
C. Linear line length dimension of the rightmost vertical line (5.6)
D. Linear line length dimension of the upper horizontal line (2.7)

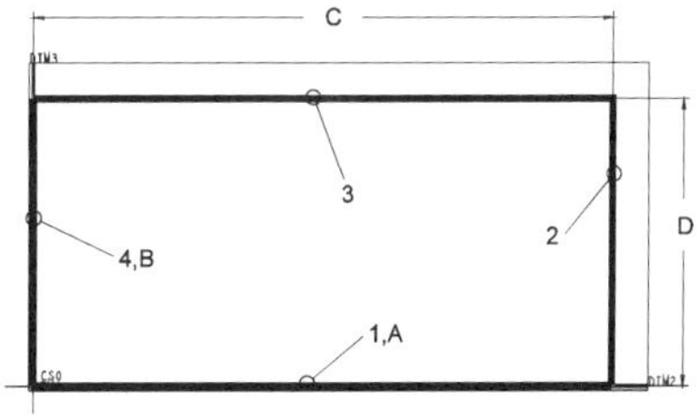

Figure 6-4

Extrusion depth is .1

Commands:

Regenerate ➡ Modify ➡ Regenerate | Done ➡ Blind | Done

Parent features:

DTM3 — Sketch plane.
DTM2 — Horizontal reference and alignment reference.
DTM1 — Alignment reference.

Figure 6-5

➾ **DESIGN NOTE:** *This feature will also be used as a reference for many of the other solid features.*

SAVE

Adding a Midpoint Datum Plane

Many of the features on the left side of the chassis will be mirror images of the features on the right side of the chassis. To create the left-side features, you will mirror the right-side features about a datum plane in the center of the chassis. In order to create this mid-plane datum, you should use a two-step process.

Making the Midpoint Datum Plane

The midpoint datum plane is simply another datum plane created as an offset of one of the original three. Pick the following commands to make this datum plane:

Create ➥ Datum ➥ Plane ➥ Offset

Now pick the datum plane DTM2. Pick **Enter Value** from the OFFSET menu. Pro/ENGINEER will ask you for the offset value and provide an arrow showing the direction of positive offset. Entering a negative value will offset a datum in the opposite direction.

Figure 6-6

The offset datum arrow showing the direction of the positive offset.

In that the arrow points in the direction you want to offset, enter a positive value.

```
Enter offset in the indicated direction, <ESC> to
quit [0.16]: 1 ↵
```

Select **Done** from the DATUM PLANE menu to create the datum plane.

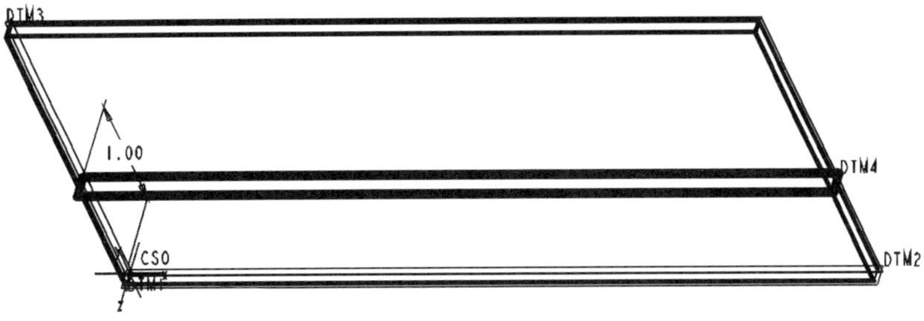

Figure 6-7
The new datum plane, DTM4, offset from DTM2 by 1 inch.

Adding the Design Intent to the Midplane Datum

By using the offset dimension of 1, you can see that the datum plane is not quite in the middle. The next step in this process is to add a relation (an equation) that drives the position of the middle datum plane using the width dimension of the chassis. Pick **Done** from the FEAT menu to go back to the PART menu. Pick **Relations** from the PART menu.

When you pick features, you will see their dimensions appear as variables. By creating simple equations using these variables, you can make dimensional relationships between features. Pick the chassis solid to show its dimensions. Pick DTM4 to show its offset dimension.

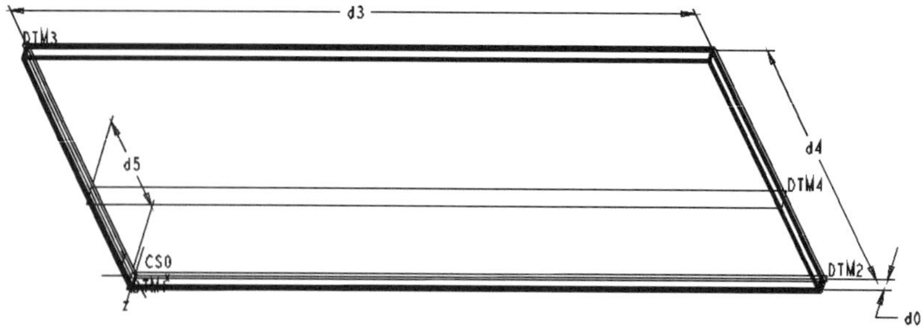

Figure 6-8
The dimensions shown in variable form.

> **NOTE:** *Your dimension variables will be different if you did not complete the steps in exactly the same order. The dimension variables are assigned in the order created. If you place your section dimensions in a different order, or place some dimensions and then delete them, your dimension variables will be different. Do not panic: use the variables that correspond to the dimensions shown in the illustration when you write your equation.*

Now pick **Add** from the RELATIONS menu. Pro/ENGINEER is ready for you to enter the equation. This equation is very simple. If you want the datum plane to always be in the middle of the chassis, its offset dimension needs to be half the width of the chassis.

```
Enter RELATION [QUIT]: d5=d4/2 ↵
Enter RELATION [QUIT]: ↵
```

Pick **Switch Dim** from the RELATIONS menu to make Pro/ENGINEER show the dimensions in numeric form. Now select **Done** from the MODEL REL menu to go back to the PART menu. Pick **Regenerate** from the PART menu to tell Pro/ENGINEER to update the chassis. Now DTM4 lies in the middle of the chassis.

Figure 6-9
The fully generated chassis with a midplane datum.

DESIGN NOTE: *You might think this was a lot of extra work. It would have been easier to create the offset datum plane at 1.35 inches rather than 1 inch. Although this is true, it would cause problems if you ever needed to change the width of the chassis. This relation adds design intent to the midplane datum. When the width changes, this datum will move to maintain its midpoint position.*

How Relations Behave

Pick **Modify** from the PART menu and select DTM4. You should see its dimension of `1.35` appear. Now pick on the dimension value and try to modify it. You will get the following message:

```
This dimension is driven in CHASSIS by relation
d5=d4/2.
```

You cannot modify dimensions that are driven by relations. This prevents you from accidentally moving this datum plane and losing its design intent. Relations do not cause parent/child relationships among features.

DTM4 is a child of DTM2 because it was offset from it. DTM4 is not a child of the base protrusion, feature 5. Relations can be written such that a child feature can drive the parent feature. Because of this usage, relations cannot create or affect parent/child relationships in any way.

Shaping the Part

The next step will be to shape the sides of the chassis. You need to remove material to form the wheel wells and to clear the sides of the body. You should perform the cut on one side of the chassis and mirror a copy of the cut to the other side.

The First Side Cut

This first cut will be created as an extruded cut. Use the following instructions to create this feature.

Feature 7, the Side Cut

Commands:

Create ➡ Solid ➡ Cut ➡ Extrude | Solid | Done ➡ One Side | Done

References:

Sketch plane — DTM3.
Horiz/Vert reference — **Top**, DTM2.

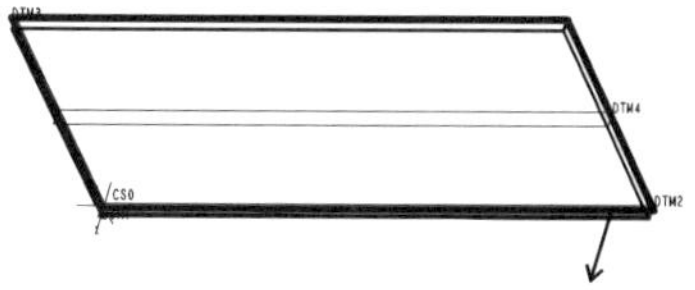

Figure 6-10

➾ ***NOTE:*** *Feature dimensions are shown in parentheses.*

1. Vertical centerline
2. Vertical centerline
3. Vertical line
4. Horizontal line
5. Vertical line
6. Horizontal line
7. Vertical line
8. Horizontal line
9. Vertical line

H. Align endpoint of vertical line to DTM2
I. Align endpoint of vertical line to DTM2
J. Linear dimension between centerline and DTM1 (1.1)
K. Linear dimension between centerline and DTM1 (4.35)
L. Linear dimension between centerline and leftmost vertical line (.5)
M. Linear dimension between centerline and rightmost vertical line (.5)
N. Linear dimension between lower horizontal line and DTM2 (.2)
O. Linear dimension between upper horizontal line and DTM2 (.3)
P. Linear line length dimension of upper horizontal line (1)

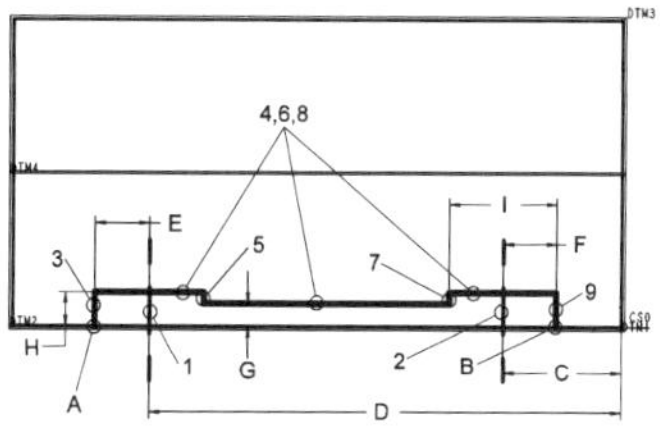

Figure 6-11

Commands:

Regenerate ➡ Modify ➡ Regenerate | Done | Okay ➡ Thru All | Done | Okay

Figure 6-12

Parent features:

DTM3 — Sketch plane.
DTM2 — Horizontal reference, alignment and dimensioning reference.
TM1 — Dimensioning reference.

> **NOTE:** *This section may appear under-constrained at first glance. The sketcher is assuming that some of the lines are of equal length. Three pairs of lines should meet this criterion. If you have trouble getting this section to regenerate, it probably means that the line lengths were not sketched accurately enough. Use the grid to help you sketch equal-length lines.*

> ✓ **TIP:** *If Pro/ENGINEER does not seem to recognize that the lines are of equal length, you can dimension the line lengths, regenerate the section, modify the dimension values to make them the same, and then delete one of the dimensions.*

Next, regenerate the section. Now that the lines are truly equal in length, you can delete the extra dimensions you placed and regenerate the section. Pro/ENGINEER will now assume that the lines are of equal length.

> **DESIGN NOTE 1:** *The centerlines in the section are used as dimensioning references. Centerlines do not only represent the axes of revolution of revolved features.*

> **DESIGN NOTE 2:** *The centerlines in the section actually represent the centerline of the wheel. By basing the dimensioning scheme on these centerlines, you control the dimensions that are most important to you: the position of the wheels relative to the chassis wheel wells.*

Mirroring Features

Now you will mirror the last feature about the midplane datum to create the cut on the other side. Mirroring features is very quick and easy. When features are mirrored, the new feature's dimensions can be controlled by the parent feature. This characteristic is determined by the options you select when you make the copy. The **Dependent** option tells Pro/ENGINEER that the new feature's dimensions should be controlled by the original feature. The **Independent** option will make the new feature without creating the "invisible" link between their dimensions. To mirror the cut, pick the following commands, starting at the FEAT menu:

Copy ➡ Mirror | Select | Dependent | Done

Select the features you want to mirror. In this case, pick feature 7, the side cut. Pick **Done** from the SELECT FEAT menu. Now you need to tell Pro/ENGINEER which plane you will mirror about. You can use actual part surfaces or datum planes to perform this mirror function. Pick DTM4 as the mirror plane. When you do this, all features you selected will be mirrored.

Figure 6-13
The mirrored side cut.

➻ ***DESIGN NOTE:*** *The parent features of this mirrored cut, feature 8, are the mirroring plane (DTM4) and the parents of feature 7 (DTM1, DTM2, and DTM3). Any dimensional modifications to the parent cut will be reflected in the mirrored copy. Likewise, if you change the width of the chassis, the mirror plane will move to the center, and this mirrored feature will be created in the proper place. When mirroring a feature, keep in mind what will happen further on in the drawing mode when dimensions are shown. They may not appear where you would like them to, in that they will reside with the original feature.*

SAVE

More Protrusions

You need to add features to hold the wheels in place. One axle lock feature will be created and then copied and mirrored to the other three wheel locations.

A Protrusion with Multiple Closed Loops

The axle lock will be a protrusion with a hole in it that retains the axle of the wheel. Pick the following commands, starting at the FEAT menu:

Create ➥ Solid ➥ Protrusion ➥ Extrude | Solid | Done ➥ One Side | Done

For the sketch plane, use the surface of the original side cut shown in the illustration. The feature will be extruded in the positive Y direction, as indicated by the arrow.

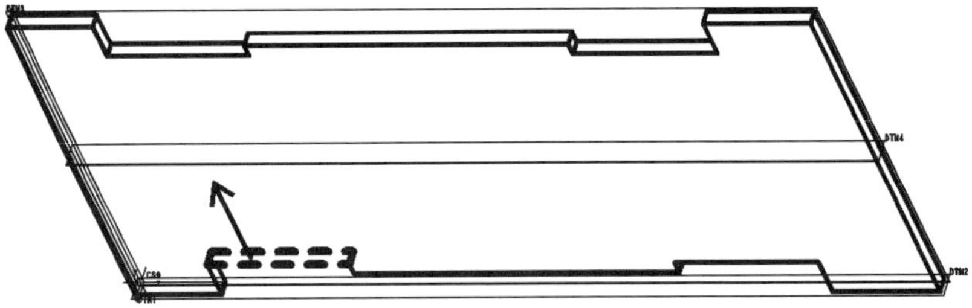

Figure 6-14

Surface to use as the sketch plane and feature creaion direction for feature 9.

Pick **Top** and **DTM3** for the horizontal reference plane. For one of the lines in this section, you will use an existing edge of the part. You should notice that in the sketch view you see many edges of the part. Picking an edge of your part in the sketch view can be ambiguous because different part edges can appear collinear in the sketch view. You do not really know which edge you are picking.

✓ **TIP:** *When you need to use an edge of your solid to create a section entity, change to the default view before you pick the edge. This way, you know which feature's edge you are picking, and therefore know which feature will be a parent of the feature you are currently sketching.*

Pick **View ➥ Orientation ➥ Default** from the MAIN menu to return to the default view. Pick **Geom Tools ➥ Use Edge** from the SKETCHER menu. Pick the top edge of the sketching plane surface. You will see that it turns cyan.

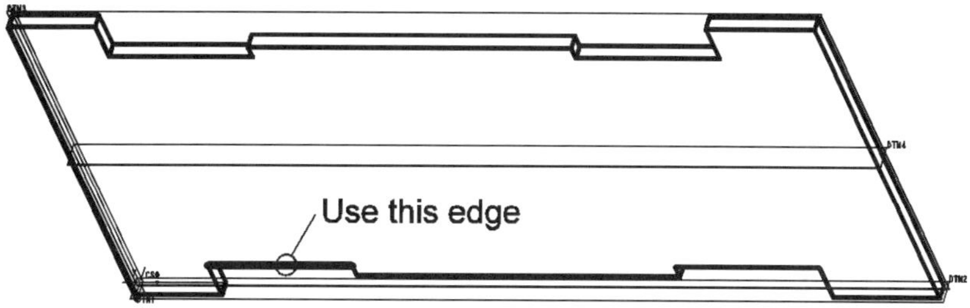

Figure 6-15
Edge to pick with the Use Edge *command.*

Now pick **Sketch** from the SKETCHER menu to return to the sketch view. Place a sketched point at each endpoint of the line. To create these points, you need to pick **Point** from the GEOMETRY menu and then place the point with the left mouse button. Your section should look like the following figure.

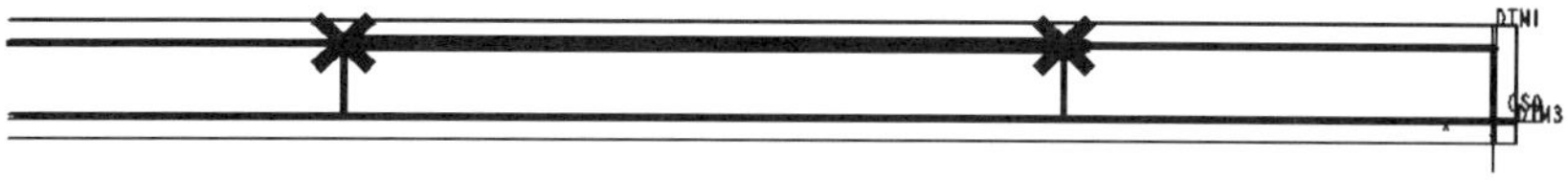

Figure 6-16
The section after placing the two sketched points.

Sketch a centerline, two lines, a tangent arc, and a circle, as shown in the next figure. Locate the circle center so that the circle is concentric to the arc. Use the **Geom Tools ➥ Trim** command to trim back the line created with the **Use Edge** command.

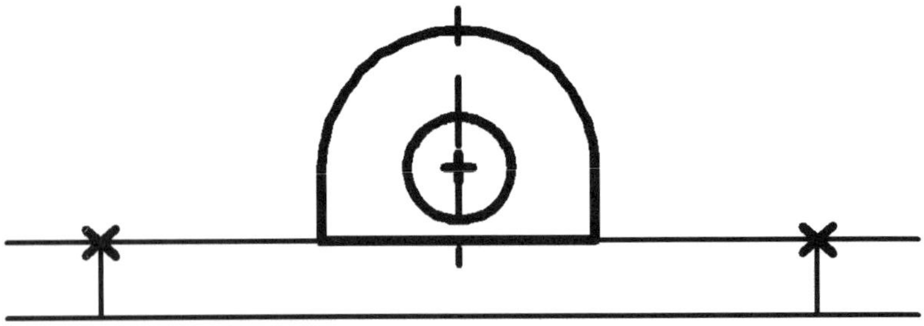

Figure 6-17
The section after sketching and trimming.

Now you need to align the sketched points to vertices of the part. You should perform this alignment while in the default view.

> ✓ **TIP:** *When you need to align to part surfaces, edges, or vertices, you should always change to the default view. This way, you know which features you are aligning to, and therefore which feature will be a parent of the feature you are currently sketching.*

Select **View ➡ Orientation ➡ Default** from the MAIN menu. Pick **Alignment ➡ Align** from the SKETCHER menu. Pick the sketched point. Now pick on the part vertex that is coincident with the sketched point. Pro/ENGINEER will align the sketched point to the part vertex. Perform the alignment on both points.

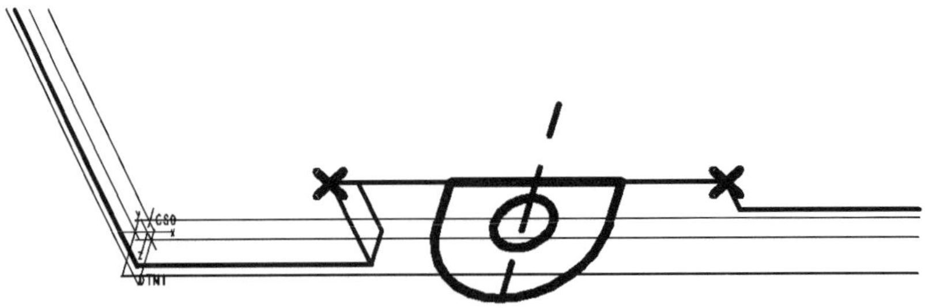

Figure 6-18
Aliging the sketched points from the default view.

Change back to the sketch view by selecting **Sketch** from the SKETCHER menu. Add the following dimensions and regenerate the section.

A. Linear dimension between the center of the circle and DTM3 (.2)

B. Linear dimension between the two vertical lines (.38)

C. Diameter dimension on the circle (.18)

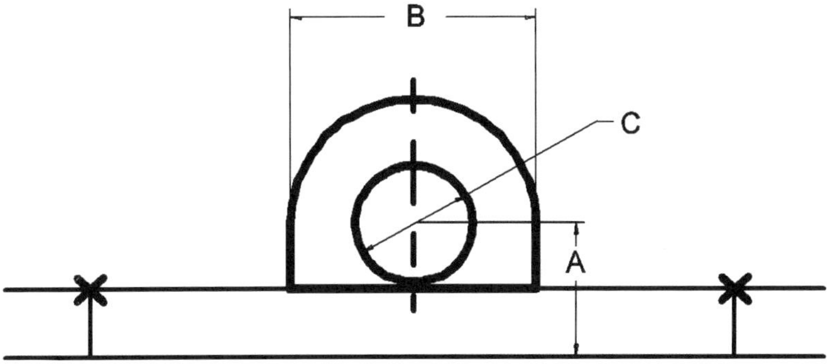

Figure 6-19
The completed section for feature 9.

➾ ***NOTE:*** *If you cannot get the section to regenerate using these dimensions, it is probably because the sketched geometry is not centered between the two sketched points. If this is the case, add a linear dimension from the center of the circle to one of the sketched points and regenerate. If this solves the problem, change the dimension to .5 and regenerate the section.*

This will move the section to the center of the points. Now, delete the .5-inch linear dimension and regenerate the section to let Pro/ENGINEER assume the symmetry with respect to the centerline.

> ➾ **DESIGN NOTE:** *The Sketcher is assuming symmetry about the centerline. When you anchored the sketched points at the part vertices, you constrained the section from moving left and right. This forces the center of the circle to be at the center of the wheel well.*

Select **Done** from the SKETCHER menu to proceed with the feature. Select **Blind** from the SPEC TO menu, then **Done**.

```
Enter depth [1.55]: .3 ↵
```

Then pick **OK** from the dialog window. The feature should be created. See the next figure.

Figure 6-20
The completed feature.

> ➾**DESIGN NOTE:** *The parents of feature 9 are DTM3, feature 5 (the base), and feature 7 (the side cut). DTM3 was used as the horizontal reference and as a dimensioning reference. A surface of feature 7 was used for the sketch plane. Both feature 5 and feature 7 are parents because of the part edge that was used in the section and the part vertices that were aligned to.*

Copying Features Requiring New References

You need to place another axle lock on the same side of the chassis. Pro/ENGINEER has a couple of copy commands that let you copy old features to create new ones. You have already used one of these commands to create feature 8 (the mirrored cut). The axle lock feature is constrained in the x direction by the alignments to the wheel well. Ideally, you would want the same type of constraints for the copied axle locks. By copying the axle lock with the **Copy ➡ New Refs** command, you can provide different references to the copied axle lock. These references include any surfaces, edges, or vertices used in the creation of the original feature. By swapping a few old references with a few new ones, you can actually create a copy of the axle lock at a different location.

Pick the commands **Copy ➡ New Refs | Select | Dependent | Done**, starting from the FEAT menu. Now select the axle lock feature you just created. Pick **Done** from the SELECT FEAT menu.

Pro/ENGINEER will show all of the dimensions for the axle lock. At this time, you have the opportunity to make the dimensions on the copied feature different from the original feature. In this case, we want the axle locks to be identical in size. In that you do not need to change any of these dimensions, pick **Done** from the GP VAR DIMS menu.

Now Pro/ENGINEER will highlight each reference of the original axle lock. If you want to keep the same reference for the copied feature, choose **Same** from the WHICH REF menu. If you want to pick a new reference, pick **Alternate** from the WHICH REF menu and then select the new reference geometry.

The first reference is the sketch plane. You need to specify a new sketch plane for the copied feature. Pick **Alternate** and then pick the other wheel well surface on the same side of the chassis.

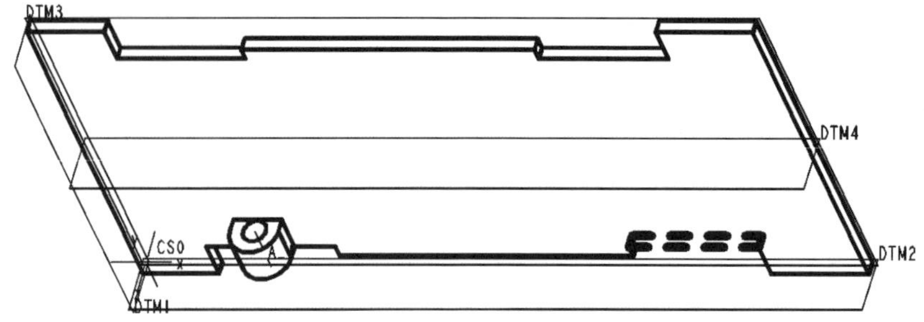

Figure 6-21
Pick a new sketch plane for the copied feature.

✓ ***TIP:*** *If you accidentally pick the wrong option for a particular reference, you can go back to the previous reference by picking* ***Prev Prompt*** *from the MODIFY PREV menu.*

The second reference is the horizontal reference plane, DTM3. DTM3 will still be a valid horizontal reference for the copied feature, so pick **Same** to keep this reference for the copied feature. The third reference is the part edge that was used in the section. You need to change this reference to the equivalent part edge in the other wheel well. Pick **Alternate** and select the other edge.

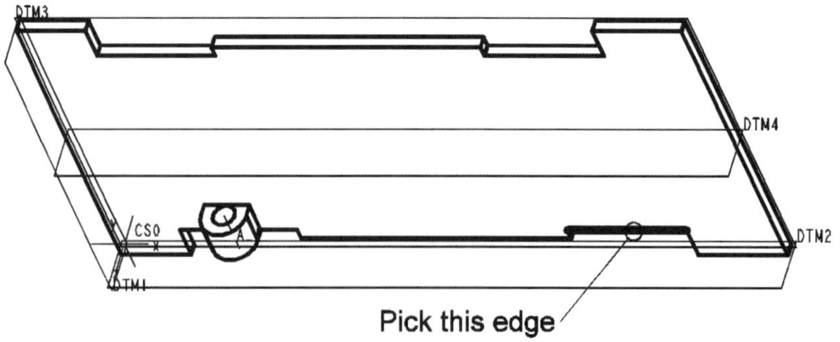

Figure 6-22
Pick a new edge for the copied feature.

The fourth and fifth references are the part vertices that the sketched points are aligned to. You need to select the equivalent part vertices in the other wheel well. Pick **Alternate** and select the other part vertices for both of these references. Pick **Done** from the MODIFY PREV menu.

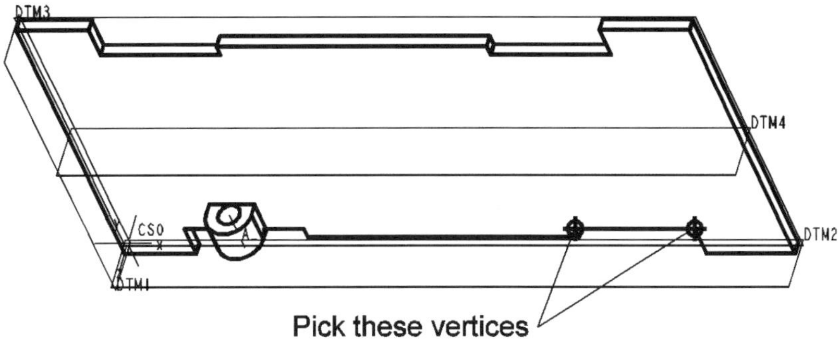

Figure 6-23
Pick new vertices for the copied feature.

The last question that Pro/ENGINEER asks is somewhat confusing to most users.

```
Select upward direction of the horizontal plane for
PROTRUSION section 1.
```

Pro/ENGINEER is trying to determine the proper orientation of the copied feature's section. If you think about the original axle lock, you selected **Top** and DTM3 as the horizontal reference for the section. Pro/ENGINEER wants to know which side of the horizontal reference faces the top in the sketch view of the copied feature. In this case, you want the upward direction to be the same for both the original feature and the copied feature. Make sure that the arrow points in the positive z direction, and pick **Okay** from the DIRECTION menu, and **Done** from the GRP PLACE menu. The copied feature should be created in the other wheel well.

Figure 6-24
The copied axle lock at its new location.

➾ ***DESIGN NOTE:*** *The copied axle lock has the same parents as the original axle lock. Although you picked new references for the copied feature, the new sketch plane, the new edge, and the new vertices have the same parent features. The original axle lock is also a parent because the copied feature's dimensions are shared with those of the original. If you change one of the axle lock's dimensions, you will modify both axle locks.*

Mirroring Multiple Features

Just as you did with the side cut, you will mirror the two axle locks to the other side. Pick the commands **Copy** ➡ **Mirror** | **Select** | **Dependent** | **Done** from the FEAT menu. Pick the two axle lock features (features 9 and 10). Pick **Done** from the SELECT FEAT menu. Pick DTM4 as the plane to mirror about. The two axle locks on the other side will be created.

Figure 6-25
The mirrored axle locks.

NOTE: *Although you have performed only one operation, you have created two more features. Copied features are assigned feature numbers as if you had made them one at a time. These two new features are also grouped together. If you ever need to delete one of them, you must first ungroup them with the* ***Group ➥ Ungroup*** *command.*

DESIGN NOTE: *The same parent/child relationships pertain to these mirrored copies. In addition, DTM4 is a parent feature to both of these mirrored copies. If you change the dimension of any of the axle locks, they will all update accordingly.*

SAVE

Creating the Mounting Holes

You need to cut some holes in the chassis. These holes will be used to mount the body to the chassis when these parts are completed. You should first place one hole in the corner nearest the coordinate system. This hole will be dimensioned to the datum planes. Once this hole is placed, you can pattern it and put a hole at each corner of the chassis.

Placing "Thru All" Holes

To place the first hole, pick the following commands, starting from the FEAT menu:

Create ➡ Solid ➡ Hole ➡ Linear | Done ➡ Straight | Done | Linear | Done ➡ One Side ➡ Thru All | Done

Now you need to identify a placement plane for the hole. The placement plane can be a datum or a planar surface. Because the hole will go through the entire chassis, it can be cut from either side. Pick the top surface of the chassis for the placement plane.

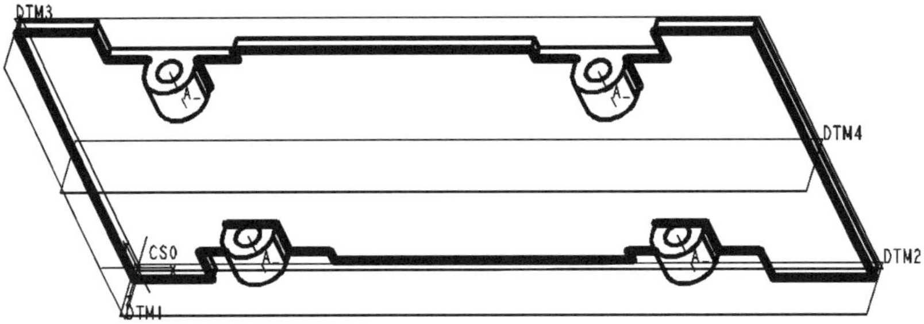

Figure 6-26
The placement plane for the hole.

After you have selected the placement plane, Pro/ENGINEER will prompt you for two dimensioning references. The location of the hole will be dimensioned to whatever geometry you pick. This hole should be dimensioned to your datum planes. Pick DTM1. Pro/ENGINEER will prompt you for the dimension from this reference.

```
Distance from reference [2.91]: .25 ↵
```

Now pick DTM2. Pro/ENGINEER will prompt you for this dimension also.

```
Distance from reference [1.97]: .45 ↵
```

Now enter the following commands:

One Side | Done ➡ Thru All | Done

Pro/ENGINEER will prompt you for the hole diameter.

`Enter DIAMETER [0.06]:` **.2** ↵

Now pick **OK** from the dialog window. The hole will now be created.

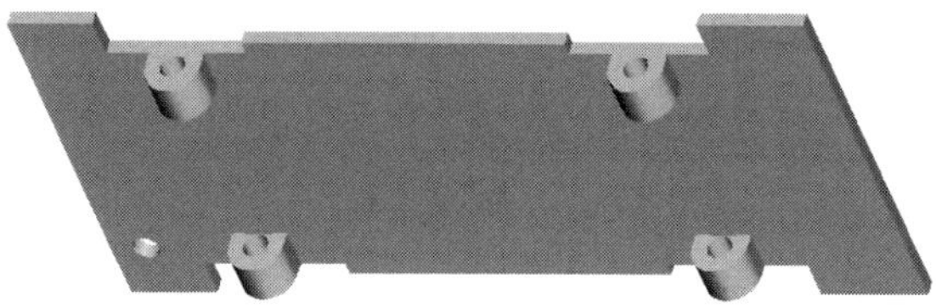

Figure 6-27
The first mounting hole in the chassis.

➡ **DESIGN NOTE:** *The hole has three parent features. The base (feature 5) is a parent because its surface was chosen as the placement surface for the hole. DTM1 and DTM2 are parents because they were used as the dimensioning references.*

✓ **TIP:** *You should notice that this hole could have been placed with a **thru all,** extruded cut using the top surface of the chassis as the sketch plane. It is a little easier to place straight holes with the hole command because there are fewer prompts, and there is no sketching involved.*

Making Identical Patterns

The **Pattern** command can be used to create copies of features. This command will create an array of features that have equal spacing. Patterns can be made in one direction or two. The controlling dimensions for a pattern are the spacing between instances and the total number of instances for each pattern direction.

To create a pattern of holes, you need to pick **Pattern** from the FEAT menu. After selecting **Pattern** and picking the hole feature, the PAT OPTIONS menu will appear. There are three types of dimensional patterns.

- ❒ **Identical** — All pattern instances are identical. All instances intersect the same part surfaces and cannot intersect each other.
- ❒ **Varying** — Instances can vary in size and intersect different surfaces, but they cannot intersect each other.
- ❒ **General** — Instances can vary in size, intersect different surfaces, and intersect each other.

In this example, the hole pattern can be created using an identical pattern. Pick **Identical | Done** from the PAT OPTIONS menu.

> ✓ ***TIP:*** *Using an identical pattern whenever applicable will speed up the regeneration of the pattern features. If you have a very large pattern, you will realize a tremendous time savings.*

Pro/ENGINEER will prompt you to pick the dimensions to pattern in the first direction. Pick the linear dimension between DTM1 and the hole axis (the `.25` dimension). Pro/ENGINEER will prompt you for the dimension increment in this direction (the distance between the holes).

```
Enter dimension increment:  4.9 ↵
```

Pick **Done** from the EXIT menu to tell Pro/ENGINEER that you have finished picking pattern dimensions for the first direction. You will be prompted for the total number of instances in this direction.

```
Enter TOTAL number of instances in this direction
(including original):  2 ↵
```

Now you need to pick the dimension to pattern in the second direction. Pick the linear dimension between DTM2 and the hole axis (the `.45` dimension). Pro/ENGINEER will prompt you for the dimension increment in this direction.

```
Enter dimension increment:  1.8 ↵
```

Pick **Done** from the EXIT menu to tell Pro/ENGINEER that you have finished picking pattern dimensions for the second direction. You will be prompted for the total number of instances in this direction.

```
Enter TOTAL number of instances in this direction
(including original):  2 ↵
```

After entering this number, the pattern will be created. Your chassis should look like the following figure.

Figure 6-28
The chassis with the pattern of the mounting holes.

➾ ***DESIGN NOTE:*** *The parent features of these three patterned holes are the first hole and the three parents of the first hole (DTM1, DTM2, and feature 5).*

➾ ***NOTE:*** *Each instance of the pattern is assigned a feature number. This operation added three features to your part (features 14, 15, and 16).*

Combining Features

The toy car will need seats for the driver and passenger (driver and passenger are sold separately). A seat will be made with a protrusion that undergoes two cuts. This seat will then be copied to make the second seat.

Making a Datum "On the Fly"

The protrusion that makes the basic seat shape will be extruded to a blind depth. The main difference between this extruded protrusion and all the other ones you have made is that this one will be sketched on a plane that does not yet exist. You will use the **Make Datum** command just before selecting the sketch plane. This is commonly referred to as making a datum "on the fly." Pick the following commands:

Create ➡ Solid ➡ Protrusion ➡ Extrude | Solid | Done ➡ One Side | Done

Now you need to select **Make Datum** from the SETUP PLANE menu. The next menu that appears is the same menu shown when you create a datum plane as a feature. To make the sketch plane for the seat, pick **Offset** and pick DTM2. The offset plane needs to be created in the positive y direction. Pick **Enter Value** from the OFFSET menu and look for the green direction arrow displayed on DTM2. Pro/ENGINEER will prompt you for the offset value.

> **NOTE:** *If you have your environment set with the datum planes off (not shown), the option to create a datum will be grayed out.*

```
Enter offset in the indicated direction, <ESC> to
quit [0.16]: .4 ↵
```

Pick **Done** from the DATUM PLANE menu. The datum plane will appear and you are asked for the direction of feature creation. The direction of feature creation will also be in the positive y direction. Pick **Okay** from the DIRECTION menu when the arrow is pointing in the desired direction.

Pick **Top** and **DTM3** for the horizontal reference plane. Sketch the following geometry.

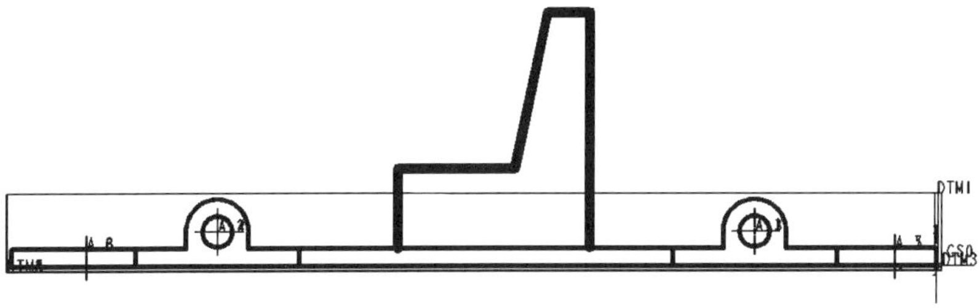

Figure 6-29
The section geometry for the seat protrusion.

You need to align the open ends of the section to the top surface of the chassis. If you tried to do this from the sketch view, you would not be able to pick the top surface: instead, you would only be able to pick the edge between the top surface and the side cut. This is another example of the need to change to the default view before you perform this alignment.

> ➾ ***DESIGN NOTE:*** *Aligning the endpoints to the adjacent surface edge in the sketch view will make the seat protrusion a child of the side cut. There is no need for this relationship to exist between these features. Take the time to perform the alignment to the top surface in the default view to avoid this problem.*

Pick **View ➥ Orientation | Default** to change to the default view. Pick **Alignment ➥ Align** from the SKETCHER menu. Align both open endpoints to the top surface of the chassis base.

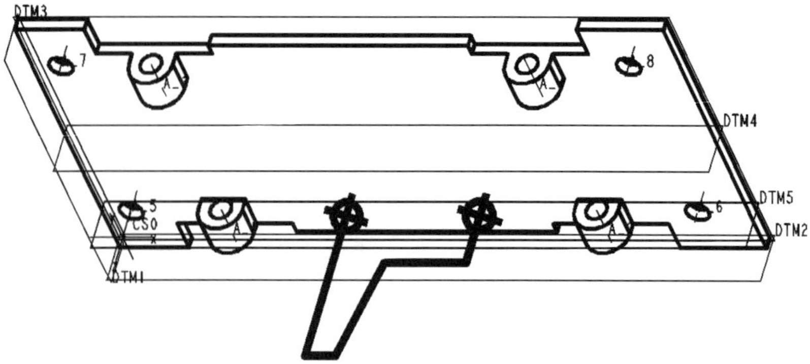

Figure 6-30
Aligning the open endpoints in the default view.

Pick **Sketch** from the SKETCHER menu to return to the sketch view. Dimension the section as seen in the following figure.

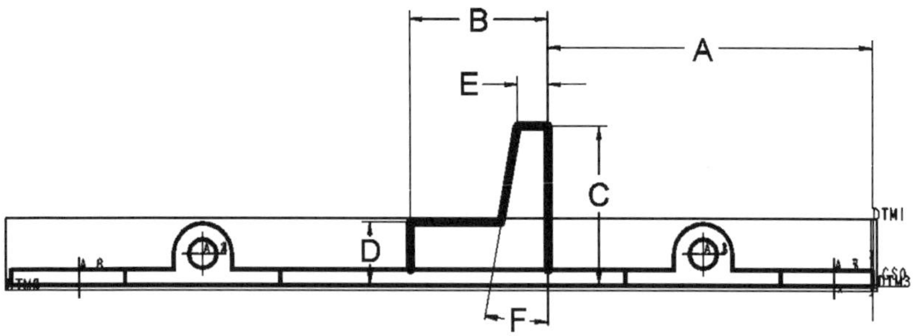

Figure 6-31
Dimensions for the seat protrusion.

A. Linear dimension between the rightmost vertical line and DTM1 (1.9)

B. Linear dimension between the two vertical lines (.9)

C. Linear dimension between the upper horizontal line and DTM3 (1)

D. Linear dimension between the lower horizontal line and DTM3 (.4)

E. Linear line length dimension of the upper horizontal line (.2)

F. Angular dimension between the rightmost vertical line and the angled line (10)

Regenerate the section and pick **Done** from the SKETCHER menu. Now you need to tell Pro/ENGINEER which side of the section will be added. Because you need to add the material to the inside of the section, orient the arrow in that direction and pick **Okay** from the DIRECTION menu. Pro/ENGINEER will prompt you for the depth of the blind protrusion, from the SPEC TO menu.

`Enter depth [1.56]:` **`.75`** ↵

The seat will be created and should look like the next figure.

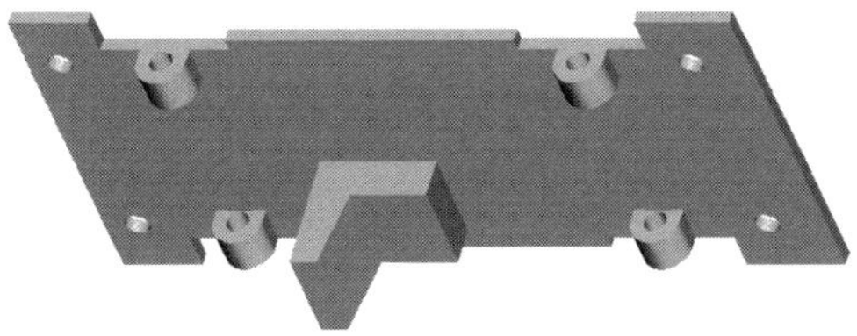

Figure 6-32
The chassis after placing the seat protrusion.

➾ ***DESIGN NOTE:*** *The parent features of the seat protrusion are DTM1, DTM2, DTM3, and the base protrusion (feature 5). DTM2 is a parent because it was used as the origin for the offset datum plane (the sketch plane). Feature 5 is a parent because the section endpoints are aligned to the top surface of the chassis base. DTM1 and DTM3 are dimensional references. DTM3 is also the horizontal reference plane.*

➾ ***NOTE:*** *The datum plane you made "on the fly" is not visible. This is a characteristic of datum planes made in this fashion. Large parts with many on-the-fly datum planes would look very confusing if the datum planes were always displayed.*

"UpTo Surface" Cuts

You need to add some curvature to the seat bottom. To cut the curve shape into the seat, you need to make an extruded cut. This time you will use the **UpTo Surface** extrude option. Pick the following commands, starting from the FEAT menu:

Create ➡ Solid ➡ Cut ➡ Extrude | Solid | Done ➡ One Side | Done

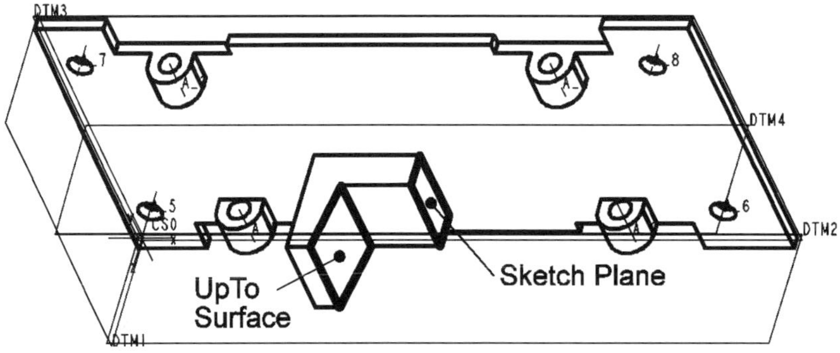

Figure 6-33
The sketch plane and the surface to extrude up to.

Next, select the front surface of the seat as the sketch plane. Accept the default direction for the sketch view by picking **Okay** from the DIRECTION menu. Pick **Top** and DTM3 for the horizontal reference plane.

You need to use an existing edge of the seat in this section. Change to the default view, pick **Geom Tools ➡ Use Edge** from the SKETCHER menu, and pick the upper, front edge of the seat.

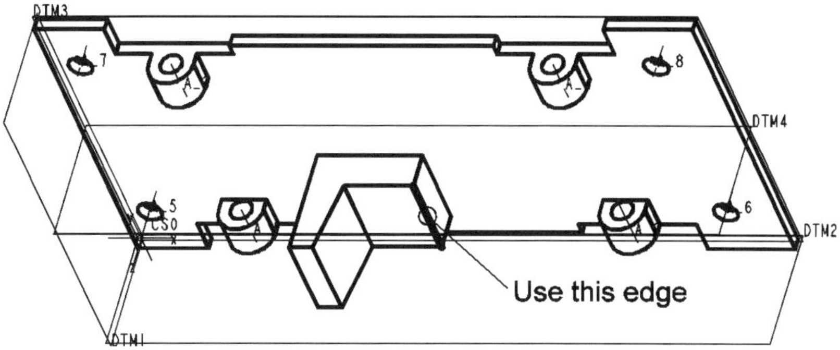

Figure 6-34
The edge to use in this feature's section.

Now change back to the sketch view by picking **Sketch** from the SKETCHER menu. Sketch a three-point arc, as shown in the following illustration, and give it a radius dimension.

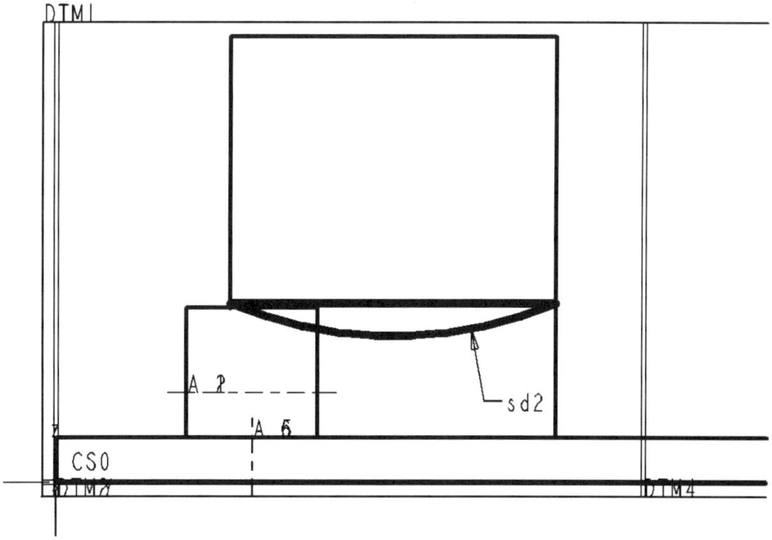

Figure 6-35
This feature's section.

Regenerate the section. Modify the radius value to 1 and regenerate it. Pick **Done** from the SKETCHER menu. Orient the arrow to point to the inside of the sketched section and pick **Okay** from the DIRECTION menu. Now pick **UpTo Surface** from the SPEC TO menu. Now you need to identify the surface to extrude up to. For this cut, you will be sketching on the front surface of the seat, so you need to extrude up to the angled surface of the seat back. Default the view and pick the angled surface of the seat, then **OK** from the dialog window. The feature will be created.

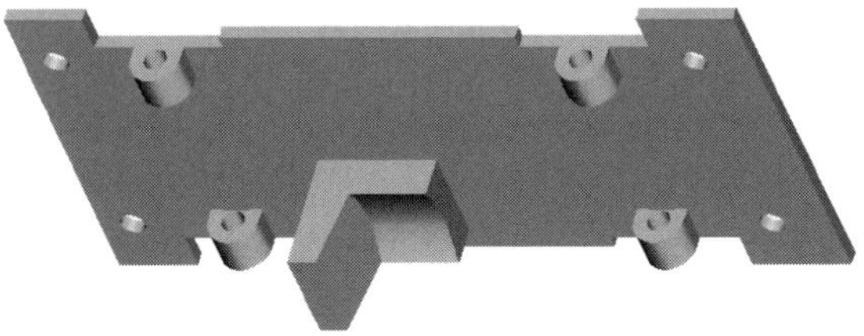

Figure 6-36
The curved cut feature on the seat bottom.

➾ **DESIGN NOTE:** *The parents of this feature are DTM3 and the seat protrusion (feature 17). DTM3 is the horizontal reference plane. The seat protrusion was used for the sketch plane, the surface to extrude to, and an edge in the section.*

✓ **TIP:** *The **UpTo Surface** option works only with closed sections.*

➾ **NOTE:** *The **UpTo Surface** option is very powerful. It has more ways to function than the **Thru Until** option. Because of a few limitations in the **Thru Until** option, you should always use the **UpTo Surface** option for features like this one.*

"Both Sides" Features with Different Options

You need to cut the same type of shape into the seat back. Because this surface is at an angle, it requires a little extra thinking. Upon examining a side view of the seat, you can see how this cut feature will be set up.

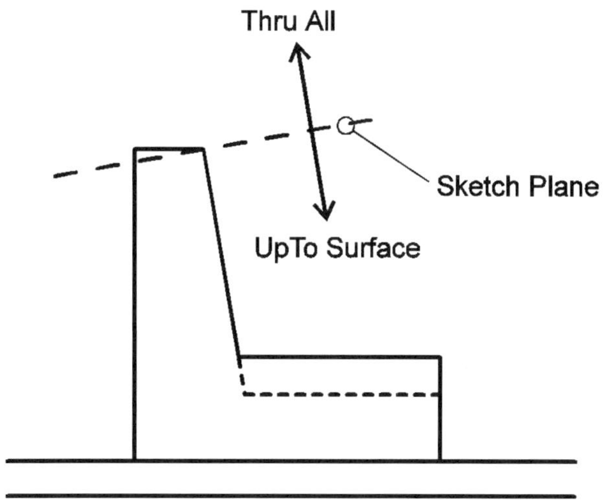

Figure 6-37
The side view of the seat showing the required feature set up.

This cut will be created to both sides of the sketch plane. One side will use the **Thru All** option, and the other side will use the **UpTo Surface** option. The sketch plane will be created "on the fly." Pick the following commands, starting at the FEAT menu:

Create ➥ Solid ➥ Cut ➥ Extrude | Solid | Done ➥ Both Sides | Done

Now pick **Make Datum** from the SETUP PLANE menu. Pick **Through** and pick the top edge of the angled surface of the seat. Pick **Normal** and pick the angled surface of the seat. This tells Pro/ENGINEER to make the datum plane through the edge of the seat back and normal (90 degrees) to the angled surface of the seat.

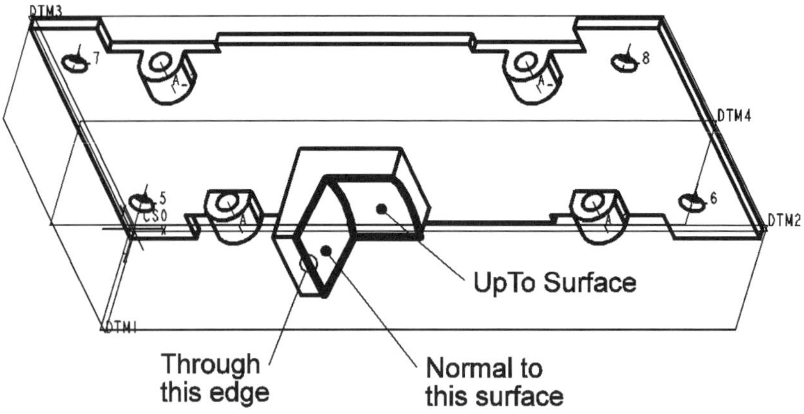

Figure 6-38
The surfaces and edges used to set up this cut feature.

Pick **Done** from the DATUM PLANE menu to create the datum plane. Pro/ENGINEER will prompt you for the direction of feature creation.

```
Arrow shows direction of feature creation. Pick FLIP
or OKAY.
```

In this case, the prompt is ambiguous because the feature is being extruded in both directions. The arrow needs to point to the side of the sketch plane that will perform the **UpTo Surface** cut. The **Thru All** side (or the first option chosen) was specified from the SPEC FROM menu. The **UpTo Surface** side (or the second option chosen) was specified from the SPEC TO menu. The arrow should point in the direction of the option chosen on the SPEC TO menu. Orient the arrow so that it points toward the seat bottom and pick **Okay** from the DIRECTION menu.

The sketch plane is oriented at an angle. You need to specify a horizontal or vertical reference plane for the sketch view. This plane must be perpendicular to the sketch plane. DTM1 and DTM3 are not valid choices for this reference, but DTM2 is. Pick **Left** and DTM2 as the vertical reference for the sketch view.

You need to use an edge of the part in this section as well. This is the same edge that the "on the fly" datum was created through. Change to the default view, pick **Geom Tools ➥ Use Edge**, and pick the top edge of the angled surface of the seat. Change back to the sketch view by picking **Sketch** from the SKETCHER menu.

Add the three-point arc to the section and define it with a radial dimension. Your section should look like the following figure.

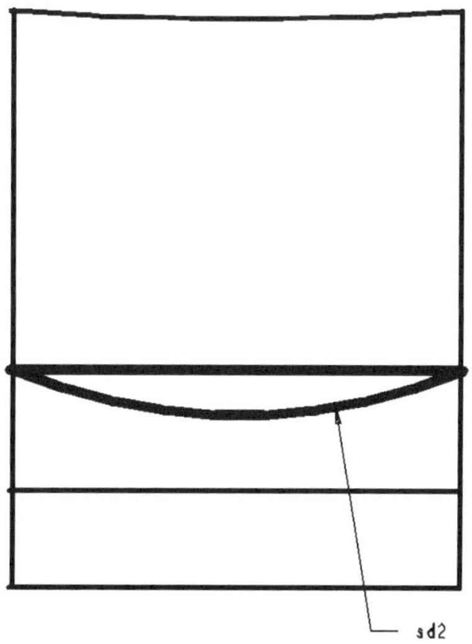

Figure 6-39
The section for this cut feature.

Regenerate the section, modify the radius value to 1, and regenerate it. Pick **Done** from the SKETCHER menu. Point the arrow toward the inside of the section to tell Pro/ENGINEER that you are removing the material on the inside of the section. Pick **Okay** from the DIRECTION menu. Pick **Thru All** from the SPEC FROM menu, then **OK** from the dialog window. This will activate the SPEC TO menu. Pick **UpTo Surface**, then **Done**. You will need to default the view to complete the next step. Do so and pick the bottom of the seat, then **Done**. The feature will be created.

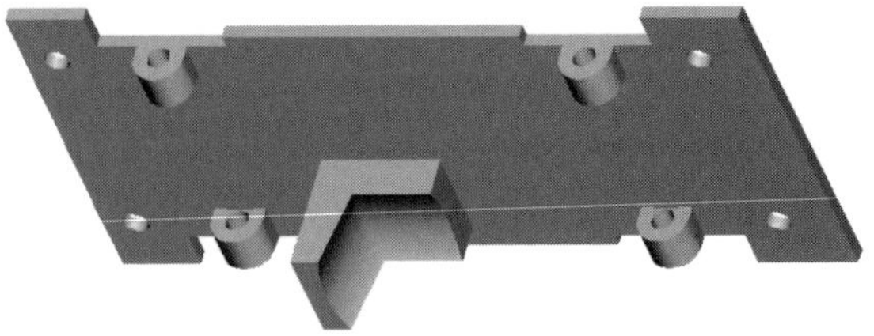

Figure 6-40
The curved cut in the seat back.

➾ **DESIGN NOTE:** *This feature has three parent features: DTM2, the seat protrusion (feature 17), and the curved seat bottom cut (feature 18). DTM2 is a parent because it was used as the vertical reference for the sketch view. The seat protrusion is a parent because its edge was used to create the sketch plane, its angled surface was used to create the sketch plane, and its edge was used in the section. The curved seat bottom cut is a parent because its surface was chosen as the "Up To" surface for this feature.*

Copying Features Using the Same References

All three seat features can be copied to the other side of the chassis. This time, you will use the **Copy** command and the **Same Refs** option. Pick the following commands from the FEAT menu:

Copy ➡ Same Refs | Select | Dependent | Done

Next, pick the three features that make the seat.

➾ **NOTE:** *You might want to try the* **Range** *option for selecting the features. In that the features are sequential, you can pick* **Range** *from the SELECT FEAT menu and type in the feature number 17 in response to the first prompt and 19 in response to the second prompt. This process will automatically select features 17, 18, and 19.*

Pick **Done** from the SELECT FEAT menu after picking these three features. All of the dimensions for the three features will appear. To create a copy of the seat on the other side of the chassis, you need to change the offset value of the first protrusion's sketch plane. Pick the dimension that represents the offset plane of the seat protrusion (nearest the coordinate system). It is .4 inches. Pick **Done** from the GP VAR DIMS menu to indicate that you do not want to change additional dimensions.

`Enter value [0.4000]:` **`1.55`** ↵

Now pick **OK** from the dialog window and the three features will now be copied.

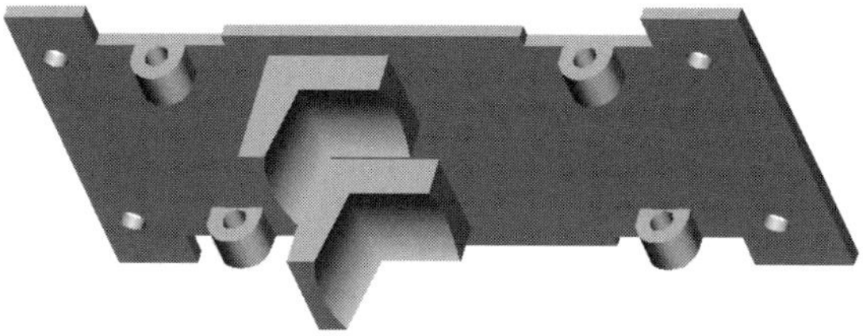

Figure 6-41
The chassis seats for two people.

NOTE: *As a result of copying all of the features simultaneously, the three new features are grouped together.*

DESIGN NOTE: *The parents of these copied features include the original features, as well as the parents of the original features.*

Making a "Thru Next" Protrusion

The toy car needs a dashboard. This will be made with an extruded protrusion. The goal is to dimension the height of the dashboard from DTM3. If you use the top surface of the chassis as the sketch plane and make a blind protrusion upward, you will not get the dimension. You can sketch the section on a plane above the chassis, and extrude it downward to the chassis.

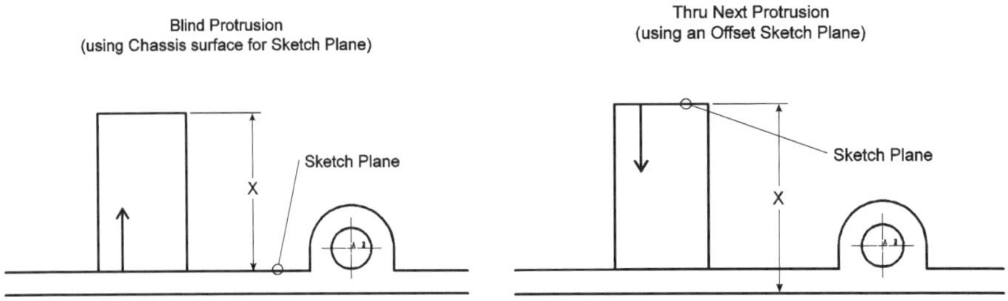

Figure 6-42
A side view showing the extrude choices and resulting dimensions.

This protrusion can be made with the **UpTo Surface** option, where you select the top surface as the surface to extrude up to. For this example, you will use a different option. The **Thru Next** extrude option allows you to extrude up to the next surface the feature would normally intersect. In fact, the feature can intersect any number of surfaces, as long as it intersects a portion of your part. To make the dashboard using the **Thru Next** option, use the information in the following section.

Feature 23, the Dashboard

Commands:

Create ➥ Solid ➥ Protrusion ➥ Extrude | Solid | Done ➥ One Side | Done

References:

Sketch plane — Make a datum plane offset from DTM3 by .8 in the positive Z direction.
Horiz/Vert reference — **Top**, DTM2.

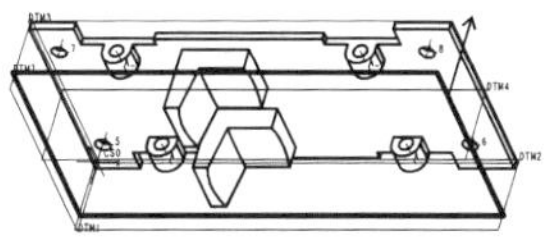

Figure 6-43

➾ ***NOTE 1:*** *Feature dimensions are shown in parentheses.*

➾ ***NOTE 2:*** *The protrusion is extruded toward you (out of the screen) and the* ***No Hidden*** *display option is being used to block out the seats and clarify the following illustration.*

1. Horizontal centerline
2. Horizontal line
3. Vertical line
4. Horizontal line
5. Three-point arc

A. Align centerline to DTM4
B. Linear dimension from the tangent of the arc to the vertical line (.4)
C. Linear dimension between the vertical line andDTM1 (3.2)
D. Linear dimension between the two horizontal lines (1.9)
E. Radial dimension of the three-point arc (1.75,

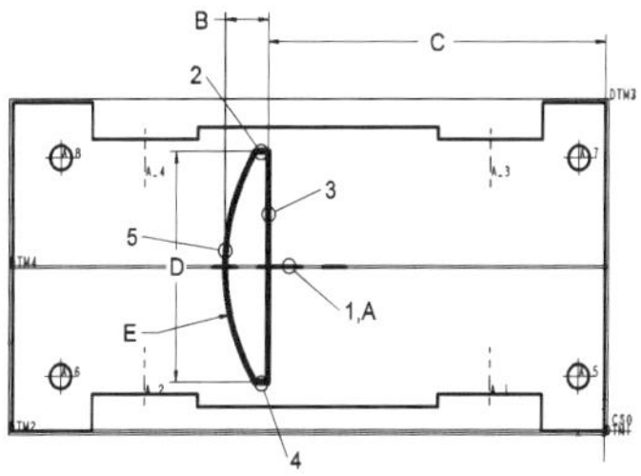

Figure 6-44

Commands:

Regenerate ➥ Modify ➥ Regenerate | Done ➥ Thru Next | Done

Then pick **OK** from the dialog window.

Parent features:

DTM3 — Sketch plane created as an offset from DTM3.
DTM2 — Horizontal reference.
DTM4 — Alignment reference.
DTM1 — Dimensioning reference.

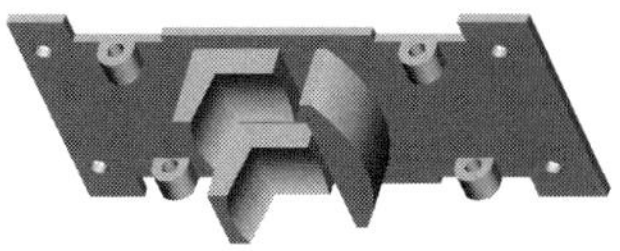

FIgure 6-45

➾ **DESIGN NOTE:** *This feature is not a child of the chassis base (feature 5). By using the* ***Thru Next*** *option, you were not required to pick the surface to extrude to, and therefore did not create a parent/child relationship to any other solid feature.*

Making a Blended Protrusion

The center console will be created with a blended protrusion. Blends are features of varied shape. The feature is created by blending between multiple sketched sections. The blend command's options control the type of blend and the behavior of the blend shape. A summary of the blend options follows. One option from each menu section must be chosen.

- **Parallel** — The blend section will contain multiple subsections. These subsections will be projected normal to the sketch plane.
- **Rotational** — Each blend section will be offset angularly about the y axis of a sketched coordinate system.
- **General** — The blend sections will be oriented with sketched coordinate systems. Orientation angles and offset values are provided for each section. Blends can be open or closed.
- **Open** — The blend feature will pass through each section, starting from the first section and ending at the last section.
- **Closed** — The blend feature will pass through all sections in order and then pass through the first section again, creating a closed shape. Parallel blends cannot be closed. Blends can be straight projections or smooth transitions.
- **Straight** — The blend surfaces will be constructed with straight lines between each section.

- **Smooth** — The blend surfaces will pass through each section and create smooth transitions between the sections. Blends can have the characteristics of extruded features.
- **Blind** — The blend feature ends at a specified distance.
- **Thru Next** — The blend feature will stop when it intersects the part.
- **Thru All** — The blend feature will be created through the entire part.
- **Thru Until** — The blend feature will be created until it intersects the specified surface.
- **From To** — The blend feature will be created between two specified surfaces.

Blend sections can be used as-is or projected.

- **Regular Sec** — The blend sections are used in planar form because they are sketched.
- **Project Sec** — The blend sections are projected onto specified surfaces before the blend feature is created. Pick the following commands to make the center console blended protrusion:

Create ➥ Solid ➥ Protrusion ➥ Blend | Solid | Done ➥ Parallel | Regular Sec | Sketch Sec | Done | Straight | Done

The sketch plane will be created with the **Make Datum** command. Make the sketch plane as an offset of DTM1. Offset the plane 1.8 inches in the positive x direction. The feature will also be created in the positive x direction (toward the dashboard). Pick **Top** and DTM3 for the horizontal reference plane. A few rules and restrictions apply to blend sections.

- Blend sections must have the same number of elements. If there are two lines and an arc in the first section, there must be three elements in each subsequent section. However, they do not have to be two lines and an arc. If you need to make a blend section with fewer elements, use a **Blend Vertex** in your section to replace one of the elements.

- Each blend section needs a start point to control the blending between sections. Start points from each section will be joined with the blend feature. Pro/ENGINEER will automatically create a start point on the endpoint of the first entity you sketch. If the start point needs to be relocated, select **Sec Tools** ➥ **Start Point** and then pick the desired location.

- Parallel blends have only one section that contains multiple subsections. Use the **Sec Tools** ➥ **Toggle** command to switch between these subsections. Sketch this section first.

➾ ***NOTE:*** *Feature dimensions are shown in parentheses.*

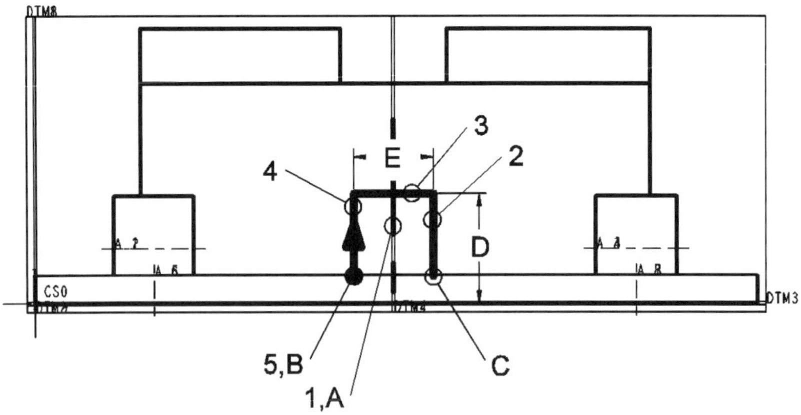

Figure 6-46

The first subsection for the center console feature.

1. Vertical centerline
2. Vertical line
3. Horizontal line
4. Vertical line
5. Start point

A. Align the centerline to DTM4

B. Align the endpoint of this vertical line to the top surface of the chassis

C. Align the endpoint of this vertical line to the top surface of the chassis

D. Linear dimension from the horizontal line to DTM3 (.4)

E. Linear line length dimension of the horizontal line (.3)

NOTE: *Do not be lazy. Change to the default view so that you can align the endpoints to the top surface of the chassis rather than to the edge shown in the sketch view. If you already did this without reading this note, congratulations!*

Regenerate the section. You should see the following message appear:

```
Must have more than one section. Select "Sec Tools",
"Toggle" for next section.
```

Now select **Toggle** from the SEC TOOLS menu. The first section will turn gray. You can now sketch this section.

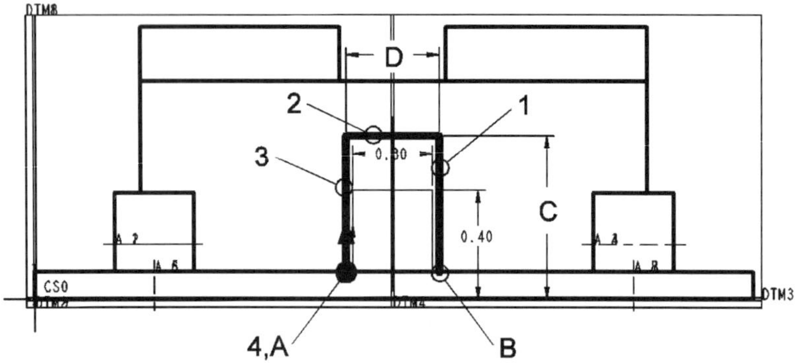

Figure 6-47
The second subsection for the center console feature.

1. Vertical line
2. Horizontal line
3. Vertical line
4. Start point

A. Align the endpoint of this vertical line to the top surface of the chassis (align in default view)

B. Align the endpoint of this vertical line to the top surface of the chassis (align in default view)

C. Linear dimension from the horizontal line to DTM3 (.6)

D. Linear line length dimension of the horizontal line (.35)

NOTE: *You do not need a centerline in the second subsection.*

Pick **Regenerate** from the SKETCHER menu. Your section should now be complete. Check to ensure that you have input the proper dimensions and that the start points are in the correct location. Pick **Done** from the SKETCHER menu.

Pro/ENGINEER will prompt you for the direction in which you would like material added. The arrow prompt in this case should point to the inside of the section. Now enter **Blind** from the DEPTH menu, then the depth of the blend between the two blend subsections.

```
Enter DEPTH for section 2 [2.8500]: 1.4 ↵
```

Then pick **OK** from the dialog window. This depth will place the second section coincident with the planar surface of the dashboard. Your chassis should look like the following figure.

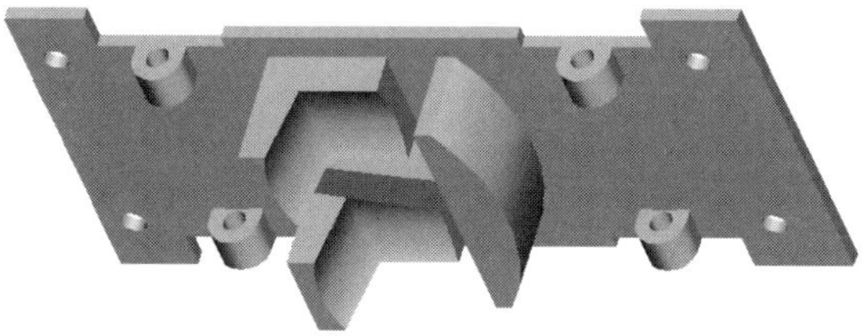

Figure 6-48
The chassis with the center console feature.

The design intent is for the console to *always* be coincident to the dashboard. If you modified the blend depth, this criterion would not be met. Situations like this are best handled by relations. You should write a relation that will drive the depth of the blend using the dashboard dimension. Pick **Relations** from the PART menu. Then pick the dashboard and console protrusions.

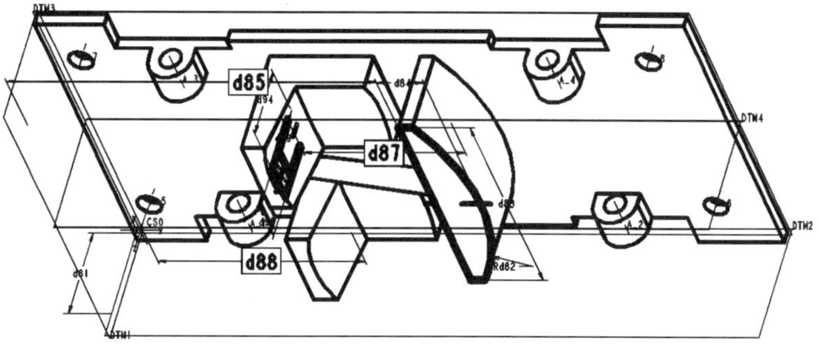

Figure 6-49
The center console and dashboard dimensions in variable form.

If you subtract the console's offset sketch plane dimension from the dashboard's DTM1 offset dimension, you will determine the proper depth for the blend. Pick **Add** from the RELATIONS menu. Pro/ENGINEER will prompt you for the relation.

```
Enter RELATION [QUIT]: d87=d85-d88 ↵
Enter RELATION [QUIT]: ↵
```

Pick **Switch Dim** and then pick **Done** from MODEL REL menu. Regenerate the part. You should see the following message:

```
Part "CHASSIS" not changed since last regen.
```

This message means that the addition of the relation did not modify the value of the blend depth. This is a good sign, indicating that the dimension for the depth was entered properly the first time. No change was required.

> **DESIGN NOTE:** *You may wonder why this relation was necessary. The chassis looked fine, and the relation did not update any dimensions. The reason the relation was added was that if the dashboard undergoes modification, you do not lose the design intent of the console.*

> **NOTE 1:** *You might notice that this relation could be avoided by making the blend start at the dashboard and project back, between the seats. There are two main differences in using this method: (1) you do not get a driving dimension between DTM1 and the nearest end of the console, and (2) the console feature becomes a child of the dashboard feature.*

> **NOTE 2:** *It is your decision as to which method you believe is correct. The method shown gives the greatest flexibility in the design because the parent/child relationship between the dashboard and the console does not exist. There is usually more than one method for creating a given feature in Pro/ENGINEER.*

➾ **DESIGN NOTE:** *The parent features of the console are DTM1, DTM3, DTM4, and feature 5 (the base protrusion). DTM1 is a parent because the sketch plane was created as an offset from it. DTM3 is a parent because it was used as the horizontal reference and as a dimensioning reference. DTM4 is a parent because the centerline in the section was aligned to it. Feature 5 is a parent because the section endpoints were aligned to the top surface of the chassis base.*

SAVE

Rounding Edges

The main features of the chassis have been completed. Now you need to round edges. Just as you did with the wheel, place the rounds toward the end of the design session. The four short edges of the chassis base need to be rounded. The front edges will have a larger radius than the back edges. Using the **Multi Const** round feature, you can place rounds of different radii in one feature. Pick the following commands, starting at the FEAT menu:

Create ➥ Solid ➥ Round ➥ Simple | Done ➥ Variable | Edge Chain | Done | Tangent Chain

Pick the following four edges:

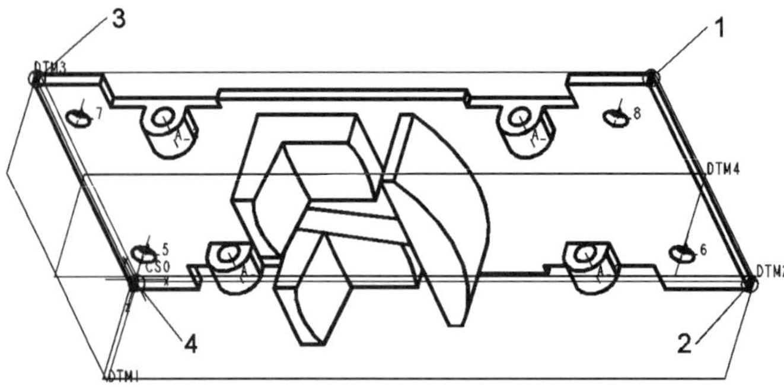

Figure 6-50
The edges of the chassis base that should be selected for rounding.

1. Radius value is .35
2. Radius value is .35
3. Radius value is .05
4. Radius value is .05

When you are finished picking these edges, select **Done** from the EDGE SEL menu. Because this is a **Multi Const** round, Pro/ENGINEER will prompt you for the radius of each edge you selected. Pro/ENGINEER will highlight the edge in cyan when it is prompting you for the dimension. Pick **Enter** from the RADIUS VAL menu to type in the dimension for the highlighted edge. After a dimension has been entered, it will appear in the RADIUS VAL menu for you to select when that radius applies to other edges. This saves you a little time when many of the edges in the **Multi Const** round have the same value. After you have given Pro/ENGINEER the value of the fourth edge, the rounds will be created.

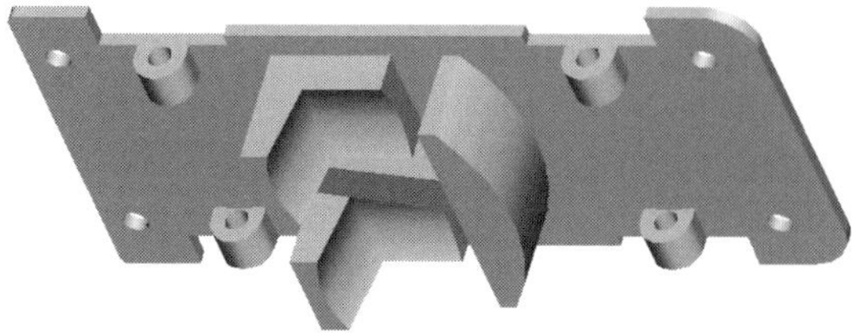

Figure 6-51
The rounded edges of the chassis base.

➾ ***DESIGN NOTE:*** *The only parent feature of this round is feature 5, the chassis base.*

The front, vertical edges of both of the seats will have the same radius. Pick the following commands:

Create ➥ Solid ➥ Round ➥ Simple | Done ➥ Constant | Edge Chain | Done | Tangent Chain

Now pick the four edges sequentially, starting with the edge on the left side of the driver's seat.

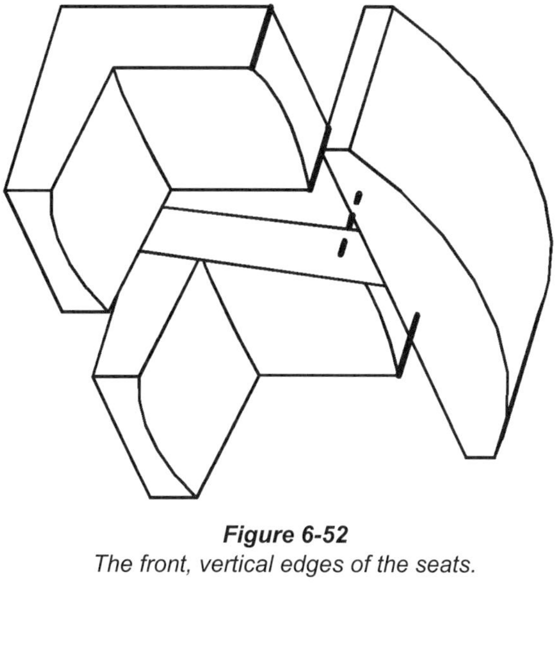

Figure 6-52
The front, vertical edges of the seats.

✓ **TIP:** *The dimension for several rounds placed with a* ***Constant*** *radius value is attached to the last edge you select.*

Look ahead to figure 11-24 to see that the right, front edge of the passenger seat should have the round attached to it. Pick the corresponding edge for that round last. Pick **Done** from the CHAIN menu. Pick **Enter** from the RADIUS TYPE menu and **New Value** from the SEL VALUE menu. Pro/ENGINEER will prompt you for this radius value.

```
Enter RADIUS [.0150]: .08 ↵
```

Then pick **OK** from the dialog window. The rounds will be created.

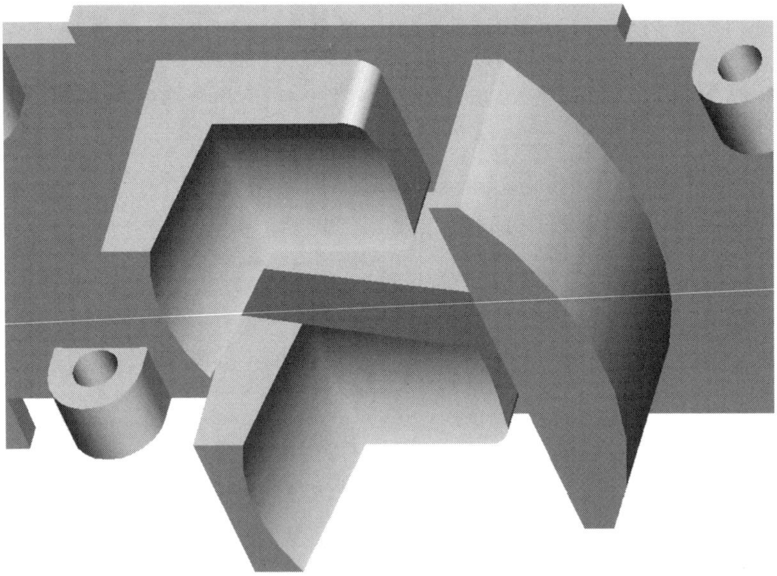

Figure 6-53
The rounds on the front, vertical edges of the seats.

➾ **DESIGN NOTE:** *The parent features of these rounds are the two seat protrusions (features 17 and 20) and the two seat bottom cuts (features 18 and 21).*

The vertical edges on the seat backs will have the same radius. Pick the following commands:

Create ➥ Solid ➥ Round ➥ Simple | Done ➥ Constant | Edge Chain | Done | Tangent Chain

Now pick the eight edges indicated in the next figure.

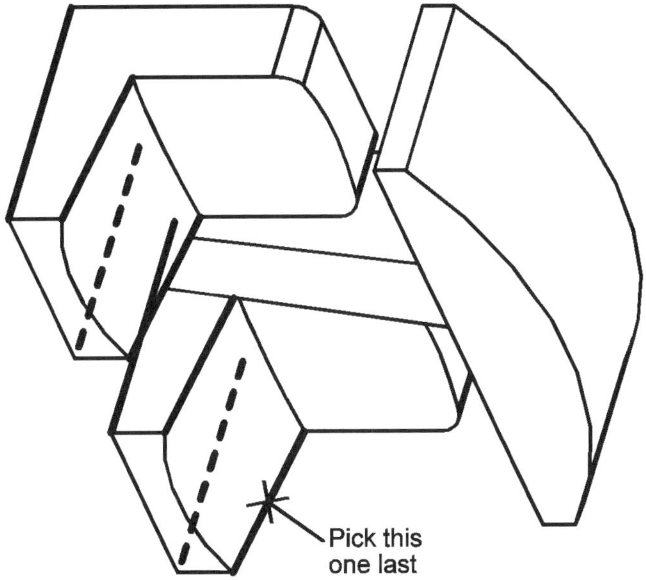

Figure 6-54
The vertical edges of the seat backs.

Pick **Done** from the CHAIN menu. Pick **Enter** from the RADIUS TYPE menu and **New Value** from the SEL VALUE menu. Pro/ENGINEER will prompt you for this radius value.

`Enter RADIUS [.0150]:` **`.05`** ↵

Then pick **OK** from the dialog window. The rounds will be created.

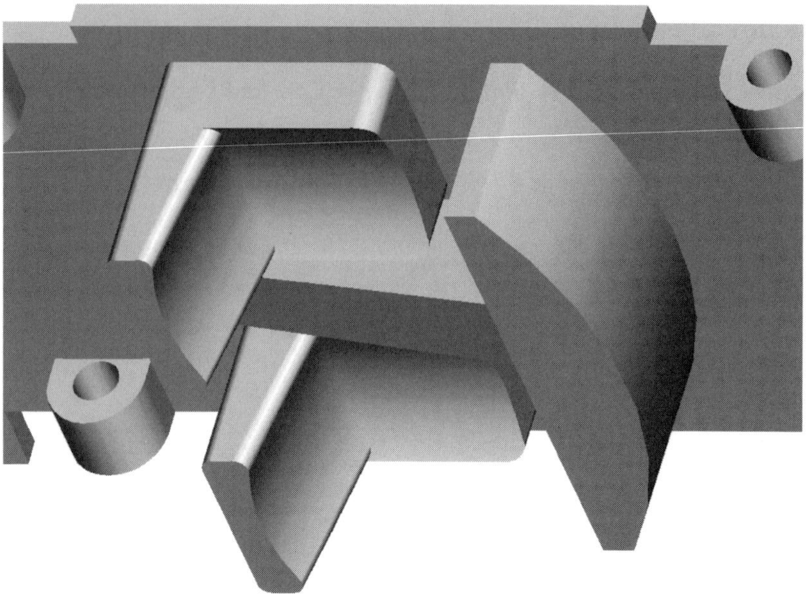

Figure 6-55
The rounds on the seat backs.

➾ **DESIGN NOTE:** *The parent features of these rounds are the two seat protrusions (features 17 and 20) and the two seat back cuts (features 19 and 22).*

Finishing a Round

As the last step to rounding the seats, you will round the continuous chain of tangent edges that surround the top of the seat backs and the seat bottoms. If you were to select these edges independently, you would need to pick 26 edges. Pro/ENGINEER offers a better way of selecting tangent edges. Pick the following commands, starting at the FEAT menu:

Create ➡ Solid ➡ Round ➡ Simple | Done ➡ Constant | Edge Chain | Done

Now pick **Surf Chain** from the CHAIN menu. Pick the surface on the seat backs and pick **Select All** from the CHAIN OPT menu. When you select one edge tangent to other edges, you will select the entire chain of tangent edges. (Choose the chain for the passenger seat bottom last.) Pick **From-To** from the CHAIN OPT menu. Now pick the points indicated in the following figure.

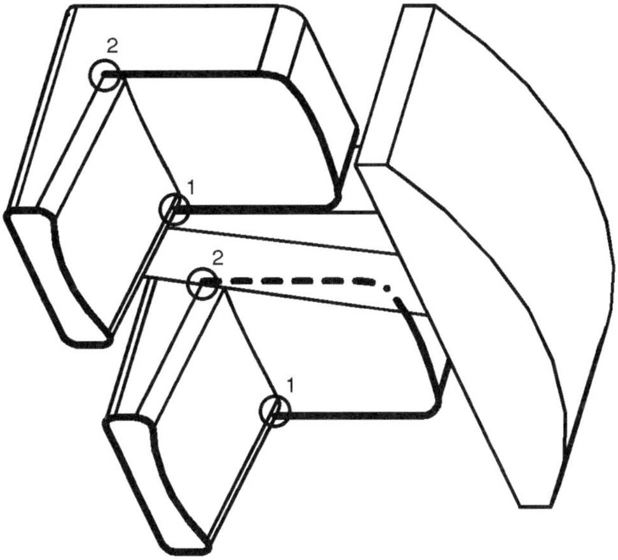

Figure 6-56
The chain edges to be rounded.

Pick **Next** from the CHOOSE menu and the edges on the seat bottom will highlight. Pick **Accept** from the CHOOSE menu. Pick **Done** from the EDGE SEL menu. Pick **Enter** from the RADIUS TYPE menu and **New Value** from the SEL VALUE menu. Pro/ENGINEER will prompt you for this radius value.

```
Enter RADIUS [.0150]: .05 ↵
```

Pick **OK** from the dialog window. The rounds will be created.

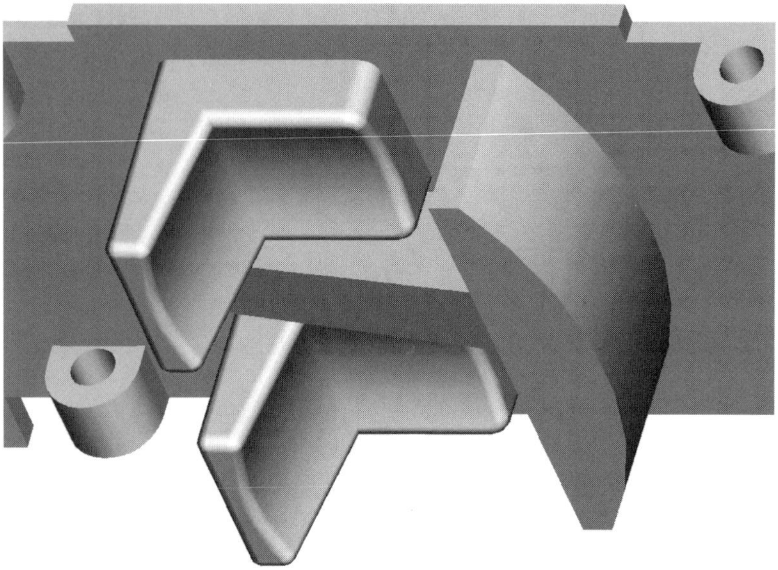

Figure 6-57
The completed seat rounds.

DESIGN NOTE: *The parent features of these rounds are the two seat protrusions (features 17 and 20), the two seat bottom cuts (features 18 and 21), and the two seat back cuts (features 19 and 22). The two prior seat rounds (features 26 and 27) might be parents, depending on the version of Pro/ENGINEER you use.*

Rounding Additional Features

The outside, vertical edges of the dashboard and the top edges of the console will have the same radius. Pick the following commands:

Create ➡ Solid ➡ Round ➡ Simple | Done ➡ Constant | Edge Chain | Done | Tangent Chain

Now pick the four edges in the next figure. (Choose the one that is shown as a hidden line last.)

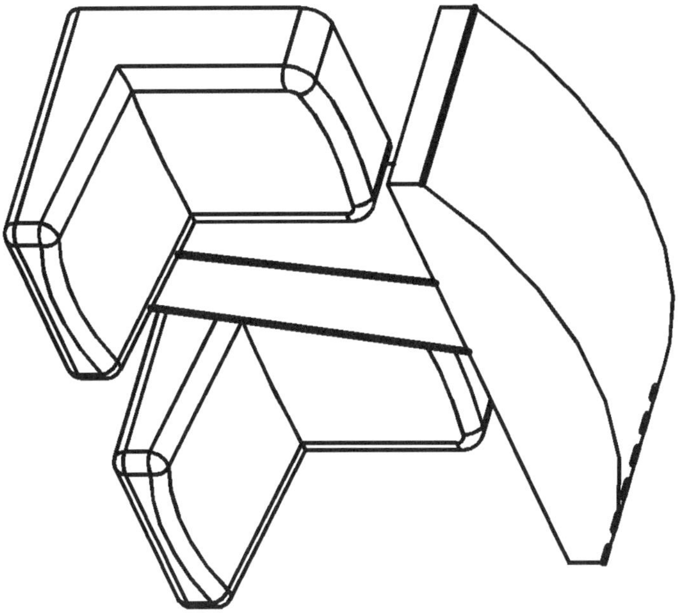

Figure 6-58
The edges of the dashboard and console to be rounded.

Pick **Done** from the CHAIN menu. Pick **Enter** from the RADIUS TYPE menu and **New Value** from the SEL VALUE menu. Pro/ENGINEER will prompt you for this radius value.

`Enter RADIUS [.0150]:` **`.05`** ↵

Pick **OK** from the dialog window. The rounds will be created.

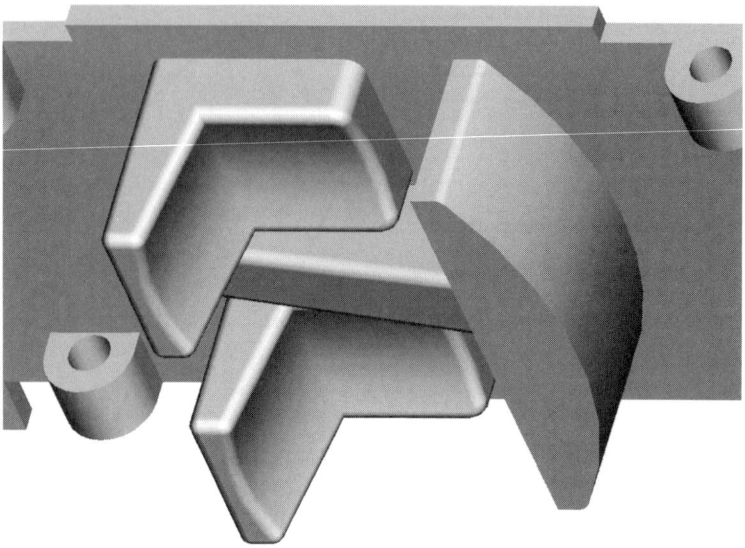

Figure 6-59
The rounds on the dashboard and console.

DESIGN NOTE: *The parent features of these rounds are the dashboard and the console (features 23 and 24).*

Using Text in Features

As a final touch, you can autograph the chassis. You will place your name in raised letters on the bottom of the chassis. This is a basic, extruded protrusion, but the sketch will contain text rather than lines and arcs. Pick the following commands to begin placing this feature:

Create ➥ Solid ➥ Protrusion ➥ Extrude | Solid | Done ➥ One Side | Done

Pick the bottom surface of the chassis as the sketch plane. Make this protrusion in the negative z direction. Pick **Top** and DTM2 for the horizontal reference plane.

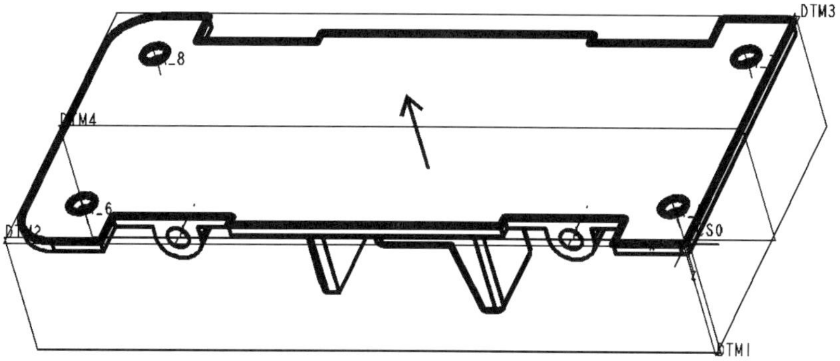

Figure 6-60
The sketch plane and the direction of feature creation.

To begin placing the text, pick **Adv Geometry** ➥ **Text** from the GEOMETRY menu. Pro/ENGINEER will prompt you for the text.

`Enter text:` **`Your Name`** ↵

Now Pro/ENGINEER wants you to draw a box. The text will be placed within the box. Place the box by picking two opposite corners, as shown in the next figure.

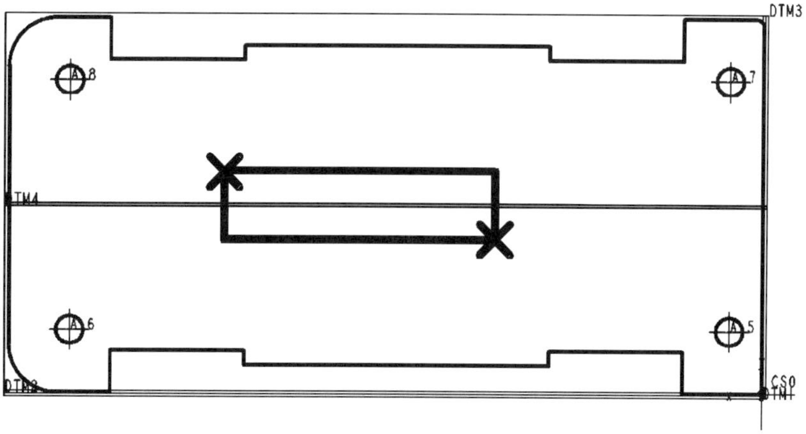

Figure 6-61
The box for the text.

Pro/ENGINEER will now prompt you again for text. Press return to stop placing text in the section.

```
Enter text: ↵
```

You need to constrain text in the Sketcher the same way you constrain other sketched entities. Pro/ENGINEER has assumed values for the text height and text width based on the box you created. To modify these values, pick **Modify** from the SKETCHER menu. Pick the text. The MOD SEC TEXT menu will appear. From this menu, you can modify any attribute of the text. Pick **Text Style**. A dialog window will appear. You will need to pick in the appropriate window for the text height value.

```
Enter new value for text height [0.6752]: .5 ↵
```

Now pick **Text Width** within the dialog window. The text width is given as ratio of the height value. In this example, we want the text width to be 40 percent of the text height.

```
Enter new value for text width factor [0.3671]: .4 ↵
```

Pick **Apply** from the dialog window. Then pick **Done** from the MOD SEC TEXT menu. Now that you have constrained the text, you need to locate it with respect to the part. Text is located with dimensions just like other sketched entities. Pick **Dimension** from the SKETCHER menu. Dimensions to text are placed at the lower left-hand corner of the text. To place a dimension, pick the text and then pick the item you want to dimension to. Place the dimension with the middle mouse button. Dimension the text to DTM1 and DTM4 in your section.

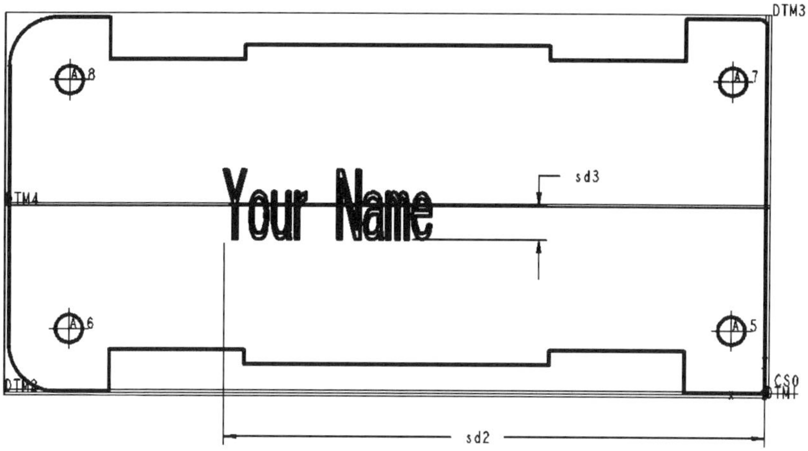

Figure 6-62
The text dimensions.

Modify the horizontal dimension to be 4. Modify the vertical dimension to be .25 (this will center the text on the part vertically). Regenerate the section. Pick **Done** from the SKETCHER menu, pick **Blind** from the SPEC TO menu, and then **Done**. Pro/ENGINEER will prompt you for the depth of the protrusion.

```
Enter depth [1.57]: .05 ↵
```

Pick **OK** from the dialog window.

> ➾ ***DESIGN NOTE:*** *The parent features of the autograph are DTM1, DTM2, DTM4, and the base protrusion (feature 5). DTM1 is a dimensional reference. DTM2 is the horizontal reference. DTM4 is a dimensional reference. The bottom surface of the base protrusion was used as the sketch plane.*

Your name will be placed in raised letters on the bottom of the chassis. Congratulations on your second part!

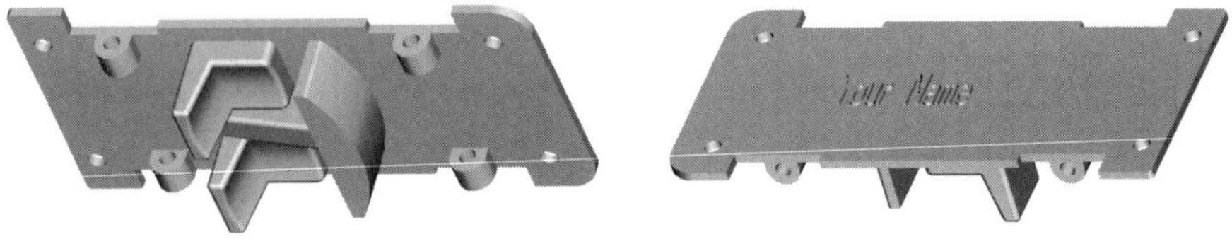

Figure 6-63
The finished chassis.

SAVE

Large Part Creation Review

Three major topics were covered in this chapter:

- Extruded feature options
- Copying features
- 3D operations in Sketcher

Understanding these topics will allow you to design 90 percent of your parts in Pro/ENGINEER.

Extruded Feature Options

After completing this part you should have a good understanding of the various extrude options. These options give the extruded **Protrusion** and extruded **Cut** commands a great deal of flexibility. You should use these commands to create most of your basic features. A summary of the extrude options used in this book follows:

- **Blind** — The section will be extruded to a specific depth value you provide.

- **Thru All** — The section will be extruded until it passes through your entire part.
- **Thru Next** — The section will be extruded until it completely intersects the part.
- **UpTo Surface** — The section will be extruded until it intersects the specified surface.

Almost all of your extruded features will use one of these options. Knowing when to apply these options is half the effort in creating a design with Pro/ENGINEER.

Copying Features

There are three very useful ways of copying features. All were used to build the chassis part.

- **Mirror** — Mirror features about a planar surface.
- **Same Refs** — Using the same references (sketch plane, dimensioning references, and so on), copy features to another location by providing a few new dimensions.
- **New Refs** — Copy features by changing references and dimensions.

Use these commands to help you create symmetric features. Your design will develop a little more quickly, and intelligence (common dimensions) between similar features of your part will be added. When you modify one copy, the others will update accordingly.

Performing Sketcher Operations in a 3D View

You probably noticed that quite a few alignments were performed in the default view during sketcher sessions. This technique is widely overlooked by users. When parts are large (with many features), reorienting views can be a time-consuming operation. Most people do not take the time to reorient, and regret it later. Do not take shortcuts that cause unwarranted parent/child relationships. Many operations in SKETCHER can be accomplished in the default view (or any other 3D view) when you take the time. Your sketcher default can be set to 2D or 3D, and you can sketch in either mode.

- **Align** — When you need to align to a surface, do not settle for an "equivalent" edge. This edge could be a member of a feature that should not be a parent (such as the tangent edge of a round).

- **Dimension** — Dimensions can be placed in the default view. When you need to dimension to a flat, part surface (not a datum plane), you should dimension it in the default view.

- **Use Edge** — This is the killer. Most people do not change to the default view before using an edge in a section. Consequently, they end up creating parent/child relationships they do not even know they have. Change to a 3D view before using edges so that you know exactly which feature's edge you are using.

Conclusion

If you follow these guidelines you will be well on your way to becoming a successful and seasoned user. Pro/ENGINEER is not a drafting tool, and the best users are not necessarily the fastest. Take your time to create the features properly. Do not cheat and cause yourself problems when parts have to be changed.

You have the option of choosing more than one method for creating a given feature in Pro/ENGINEER. You will want to choose the method that offers flexibility while it maintains your design intent. In the next chapter you will have more opportunities to put your new skills to use by following a basic part methodology.

Chapter 7
Designing a Part with a Complex Shape

A Primer for Molded Part Design in Pro/ENGINEER

Introduction

This chapter introduces you to methods used for complex part designs. Commands and concepts covered in this chapter include the following:

- ❐ Swept cuts and protrusions
- ❐ Thin features
- ❐ The shell utility
- ❐ Draft features

- Ribs
- More hole options
- Surface features
- Chamfers
- Offset features

The body for the car is a plastic molded part. This design contains features such as ribs, bosses, and draft, which are common to most molded parts.

Figure 7-1
The finished car body.

Basic Part Methodology

Now that you have created a couple of parts, you should begin to see a few trends. There is a basic part methodology you should follow. This methodology places important features early and less important features toward the end.

STEP 1 — Place your datum planes and coordinate system.

STEP 2 — Place your first protrusion.

STEP 3 — Add the major features of the part.

STEP 4 — Add the finishing touches.

By following this generic methodology, your important features will not become children of your minor detail features. This just makes good sense. The methodology is even more important when you are designing a molded part. Molded parts generally have a large number of detail features, such as draft and rounds, and usually include a shell feature, which gives the part a constant thickness. An expanded methodology for molded parts would look like the following:

STEP 1 — Place your datum planes and coordinate system.

STEP 2 — Place your first protrusion.

STEP 3 — Add the features that need to be reflected in the shell feature.

STEP 4 — Create the shell feature.

STEP 5 — Add the major features to the shelled part.

STEP 6 — Add the finishing touches.

Using this methodology in combination with good feature-building skills will make your part designs modifiable and easily maintained.

Starting a Molded Part

Select **Part** from the MODE menu. Select **Create** from the ENTER-PART menu and type in the name of the part.

```
Enter Part name [PRT0001]: body ↵
```

Just as you did with the wheel and the chassis, you should begin the body by creating the default datum planes and coordinate system. The first protrusion and the midplane datum will follow these initial datums.

Placing the Part Datums

Place your basic part datums with these commands:

Feature ➥ Create ➥ Datum ➥ Plane ➥ Default ➥ Create ➥ Datum ➥ Coord Sys ➥ Default | Done

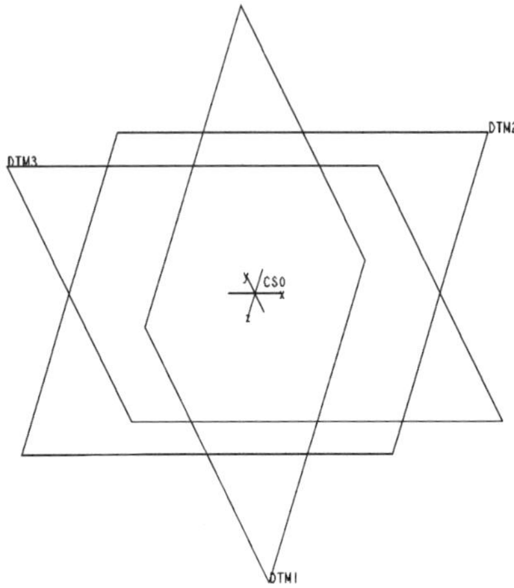

Figure 7-2
The datums for the car body.

➾ ***NOTE:*** *If you decide that you do not like the default coordinate system's orientation, you can build a different one using the* ***3 Planes*** *option. This is the reason the datum planes are created first.*

The Base Protrusion

The first protrusion will reflect the overall envelope of the car when viewed from the top. It will have three straight sides (left, right, and back) and one curved side (front). Use this information to create the first protrusion.

Feature 5, the First Protrusion

Commands:

Create ➡ Solid ➡ Protrusion ➡ Extrude | Solid | Done ➡ One Side | Done

References:

Sketch plane — DTM3.
Horiz/Vert reference — **Top**, DTM2.

➾ ***NOTE:*** *Feature dimensions are shown in parentheses.*

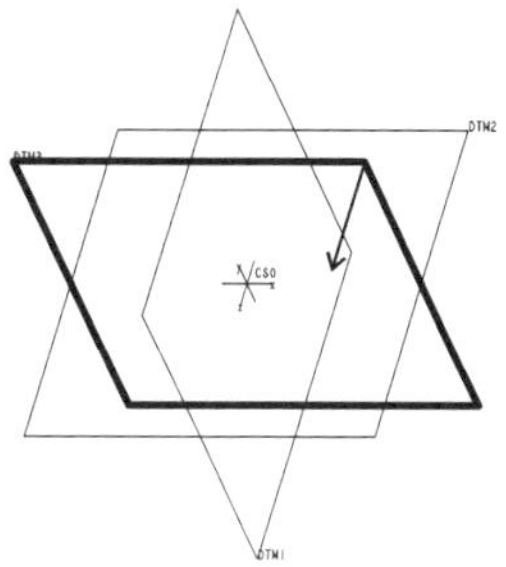

Figure 7-3

1. Horizontal centerline
2. Horizontal line
3. Vertical line
4. Horizontal line
5. Three-point arc

A. Align lower horizontal line to DTM2
B. Align vertical line to DTM1
C. Linear dimension from the tangent of the arc to DTM1 (6)
D. Linear dimension between the two horizontal lines (3)
E. Radial dimension of the three-point arc (8)

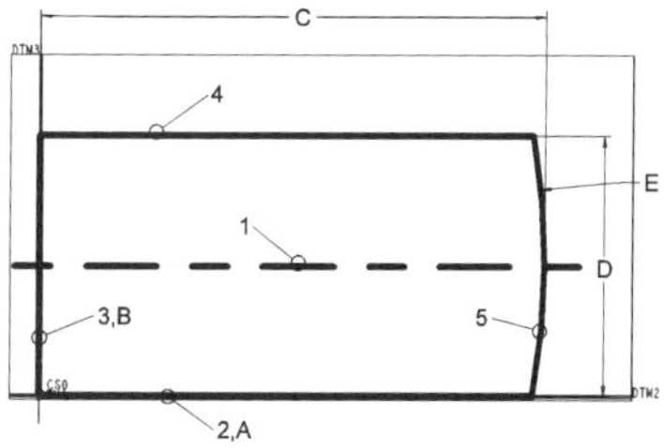

Figure 7-4

Regenerate ➡ Modify ➡ Regenerate ➡ Blind | Done

Extrusion depth is .8
Then pick **OK** from the dialog window.

Parent features:

DTM3 — Sketch plane.
DTM2 — Horizontal reference and alignment reference.
DTM1 — Alignment and dimensioning reference.

Figure 7-5

The Midplane Datum

Many features in the car body are symmetric about the center of the car. Just as you did with the chassis, you need to create a midplane datum that will represent the center of the body. Use the same procedure as before to create the datum plane.

Create ➥ Datum ➥ Plane ➥ Offset

Offset this plane from DTM2. After selecting DTM2, pick **Enter Value** from the OFFSET menu and enter the offset distance in the positive y direction.

Figure 7-6
Offset DTM2 in the positive Y direction.

```
Enter offset in the indicated direction, <ESC> to
quit [0.17]: 1 ↵
```

Select **Done** from the DATUM PLANE menu to create this datum plane. Select **Done** from the FEATURE menu to return to the PART menu. You need to drive the offset distance of this midplane datum using the width of the first protrusion. Pick **Relations** from the PART menu to write this relation between these dimensions. Pick the first protrusion and the midplane datum to show their dimensions in variable form.

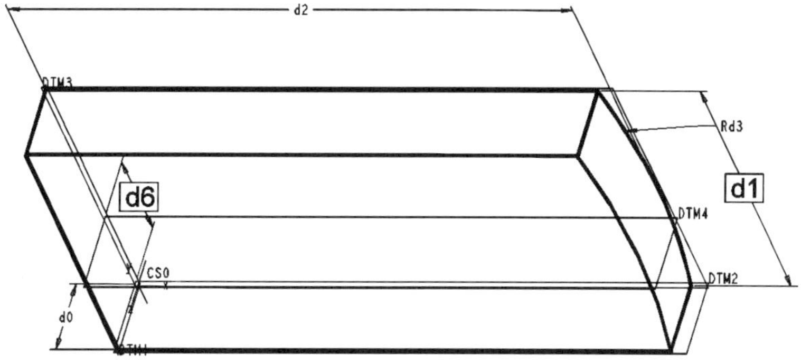

Figure 7-7
The dimensions of the first protrusion and DTM4 in variable form.

Select **Add** from the RELATIONS menu to enter the equation. You need to make the offset distance of DTM4 equal to half the height of the first protrusion. In this example, `d1` is the height of the first protrusion, and `d6` is the offset distance of DTM4. Remember, your dimension variables may be different.

```
Enter RELATION [QUIT]: d6=d1/2 ↵
```

Press <Enter> again to stop entering equations. Pick **Switch Dim** to return to number format. Select **Done** from the MODEL REL menu. From the PART menu, pick **Regenerate**. This will make DTM4 move to the center of your car body.

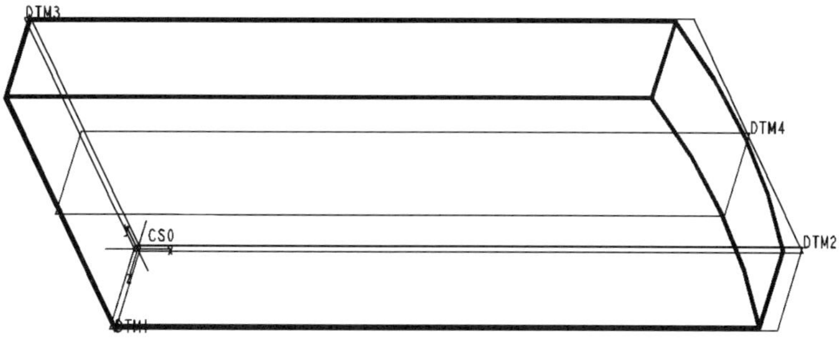

Figure 7-8
DTM4 in the center of the car body.

DESIGN NOTE: *Because of the relation, DTM4's position will always represent the center of the car body. If the width changes, DTM4 will update accordingly.*

Shaping the Molded Part

The block of material you created represents the extents of the car body. To make it look a little sportier, you need to cut away some of this material and add some curvature. You will create cuts on each side of the body and then place two cuts on the front to slope the front end and hood.

The Side Cut

The side cut will be projected in the same direction as the first protrusion. You will simply sketch a three-point arc and scoop away some material from the side of the body.

Feature 7, the Side Cut

Commands:

Create ➥ Solid ➥ Cut ➥ Extrude | Solid | Done ➥ One Side | Done

Figure 7-9

NOTE: *Feature dimensions are shown in parentheses.*

References:

Sketch plane — DTM3.
Horiz/Vert reference — **Top**, DTM2.

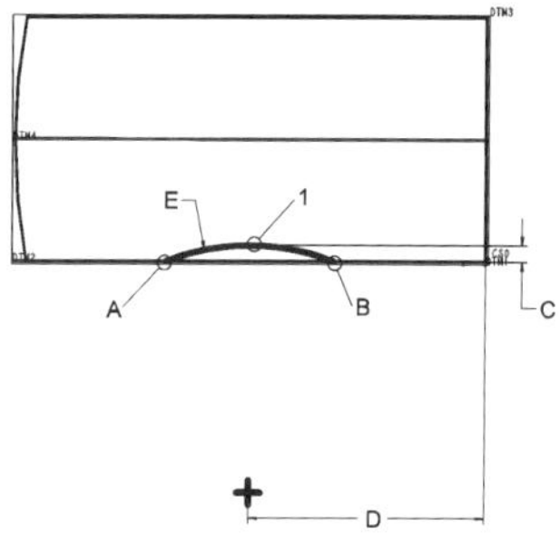

Figure 7-10

1. Three-point arc
A. Align left arc endpoint to DTM2
B. Align right arc endpoint to DTM2
C. Linear dimension from the tangent of the arc to DTM2 (.2)
D. Linear dimension from the arc center to DTM1 (3)
E. Radial dimension of the three-point arc (3)

Commands:

Regenerate ➥ Modify ➥ Regenerate | Done | Okay ➥ Thru All | Done | Done

Pick **OK** from the dialog window.

Parent features:

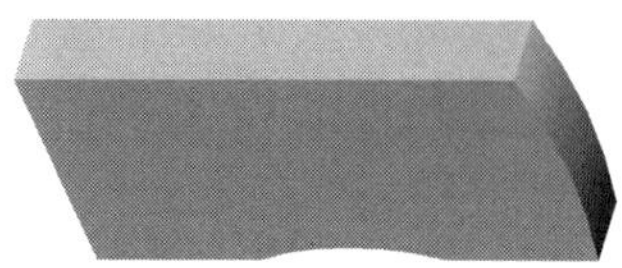

Figure 7-11

DTM3 — Sketch plane.
DTM2 — Horizontal reference, alignment reference, and dimensioning reference.
DTM1 — Dimensioning reference.

Mirroring the Side Cut

To make a matching cut on the other side, you should mirror the previous feature. Pick the following commands, starting at the FEAT menu:

Copy → Mirror | Select | Dependent | Done

Now pick the side cut (feature 7), and select **Done**.

Choose DTM4 as the plane to mirror about. The cut should be mirrored to the other side and should look like the following illustration.

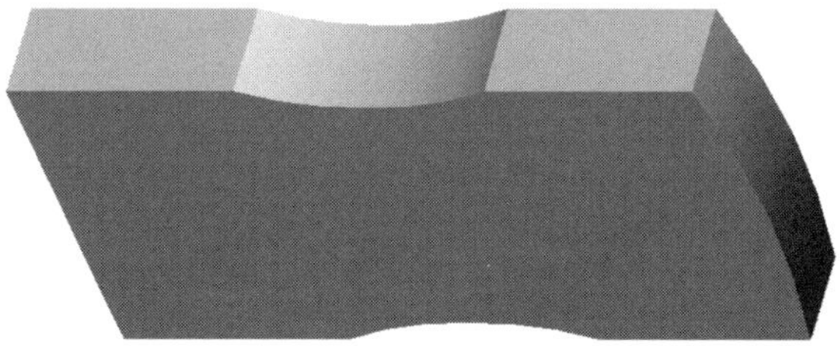

Figure 7-12
The mirrored side cut.

Swept Cuts

The next feature will be a curved cut that follows the shape of the front end. These types of shapes are easily created with swept features. Sweeps are features that have a trajectory and a section that is swept along that trajectory. The sections are swept such that they are perpendicular to the trajectory at all times. Pick the following commands, starting at the FEAT menu:

Create ➥ Solid ➥ Cut ➥ Sweep | Solid | Done

The options associated with sweep features follow:

- **Sketch Traj** — This indicates that you will sketch the trajectory curve.
- **Select Traj** — This will allow you to select part edges or datum curves to use as the trajectory.

> ✓ ***TIP:*** *If your trajectory is planar, you should choose the* ***Sketch Traj*** *option. This will give you a little more control during the feature creation process. Use the* ***Select Traj*** *option only when your trajectory is a complex 3D curve.*

For this example, you should pick **Sketch Traj ➥ Done** because the trajectory is planar. Now you need to set up the sketch plane for the trajectory. This cut will use the front, curved edge of the body that lies in DTM3 for the trajectory. That means DTM3 should be the sketch plane. Pick DTM3 as the sketch plane and view the sketch plane in the negative z direction. Select **Top** and DTM2 for the horizontal reference. Trajectories are sketched the same as every other section in Pro/ENGINEER. The only difference is that a start point must be specified somewhere on the trajectory. The start point is significant because it tells Pro/ENGINEER where you will sketch the section that will be swept.

For this trajectory, you should pick **Geom Tools ➥ Use Edge** and select the front, curved edge that lies in DTM3. Remember to change to the default view before selecting the edge so that you know which one you are selecting.

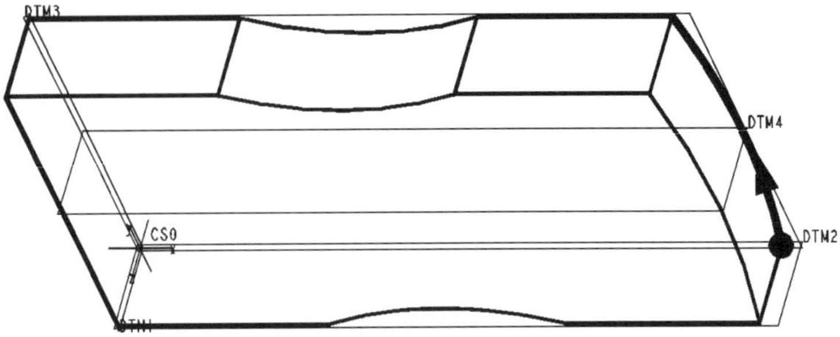

Figure 7-13
The edge to use for the trajectory.

Now place the start point at the lower endpoint of the arc using the **Sec Tools ➡ Start Point** command. Pick **Regenerate | Done**.

> ➾ **NOTE:** *You can regenerate sections when you are at any view orientation, not just the sketch view.*

When you have successfully regenerated the trajectory, the ATTRIBUTES menu appears. This menu contains the following two options:

- ❐ **Merge Ends** — The sweep feature will be extended beyond the endpoints of the trajectory until it passes completely through the part.
- ❐ **Free Ends** — The sweep feature will stop perpendicular to the trajectory and at the endpoints of the trajectory.

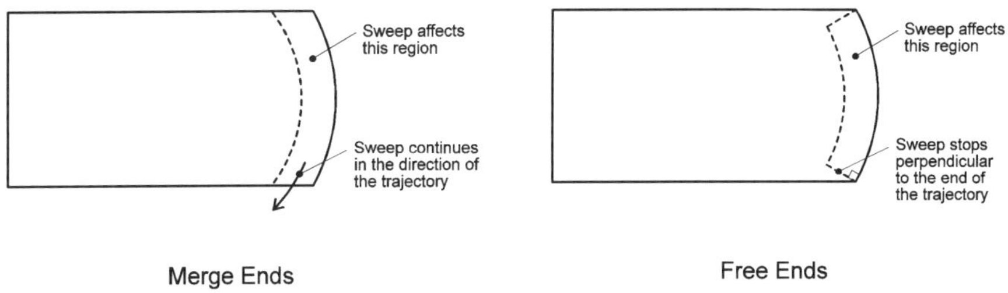

Figure 7-14
The effects of the Merge Ends and Free Ends options.

As seen in the previous figure, you should choose **Merge Ends** | **Done** for this feature. That will make the cut extend through the sides of the body beyond the endpoints of the trajectory. After you make this selection, Pro/ENGINEER will automatically orient the view for you to sketch the section to be swept.

The section view will look a little odd because you are viewing the body in an orientation where the start point is facing normal to the screen (pointing outward from the screen). This makes the body appear at an angle.

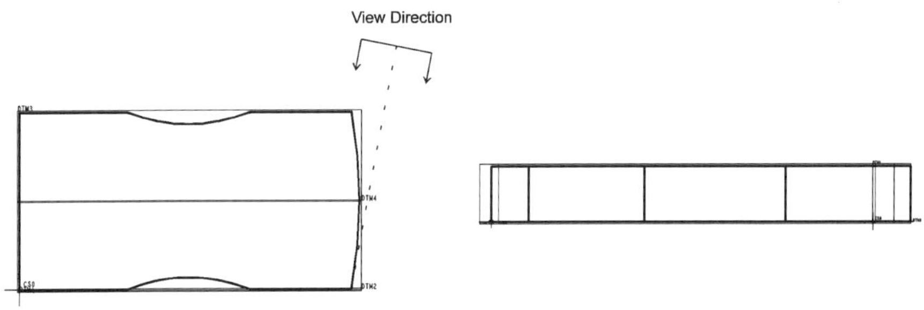

Figure 7-15
The view orientation for the section.

You should notice two light blue centerlines (called cross hairs) in the section. These centerlines represent the location of the trajectory in the sketch plane. When you sketch your section, you should use these centerlines to help you maintain a sense of direction and location. As mentioned earlier, this section will be an arc. Sketch the section as described next.

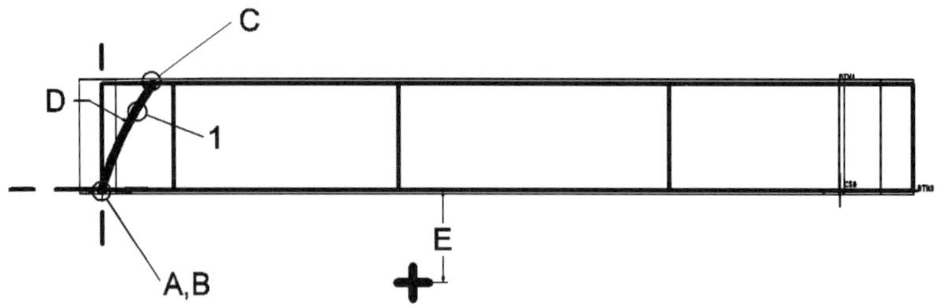

Figure 7-16
The section that will be swept along the trajectory.

1. Three-point arc

A. Align the lower endpoint to the horizontal cross hair

B. Align the lower endpoint to the vertical cross hair

C. Align the upper endpoint to the top surface of the body (align using the default view)

D. Radial dimension of the three-point arc (2.5)

E. Dimension from the arc center to DTM3 (.7)

Commands:
Regenerate ➡ Modify ➡ Regenerate | Done | Flip ➡ Okay | Done

Flip the arrow to cut the material away from the arc center, and then select **Okay**. Then select **Done** and **OK** (from the dialog window), and the swept cut will be created.

> ➾ ***NOTE:*** *Actions A and B in the previous list are automatic functions in Pro/ENGINEER, but manually aligning will reinforce your understanding of this procedure.*

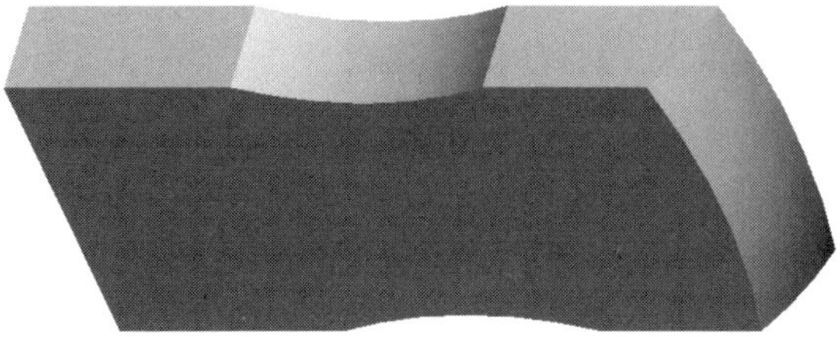

Figure 7-17
The completed swept cut.

➾ ***DESIGN NOTE 1:*** *The parent features of this swept cut are DTM2, DTM3, and feature 5 (the first protrusion). DTM2 is the horizontal reference for the trajectory sketch plane. DTM3 is the trajectory sketch plane and a dimensioning reference. The first protrusion is a parent feature because one of its edges is used for the trajectory section and because the top surface is an alignment reference.*

➾ ***DESIGN NOTE 2:*** *You may have noticed that this feature could have been produced with a rotational cut. The feature was not made this way because it would be too difficult to control the amount of rotation. It is not a 90-degree or 180-degree rotation angle. The rotation angle is equivalent to the included angle of the front cylindrical surface. By using an edge of the cylindrical surface for the trajectory, you have properly maintained your design intent without adding an extra dimension for the rotation angle of the cut.*

Tangent Cuts

The cut that forms the slope of the hood will be extruded in a side view. Use the following information to create this feature.

Feature 10, the Hood Cut

Commands:

Create ➡ Solid ➡ Cut ➡ Extrude | Solid | Done ➡ One Side | Done

References:

Sketch plane — DTM2.
Horiz/Vert reference — **Top**, DTM3.

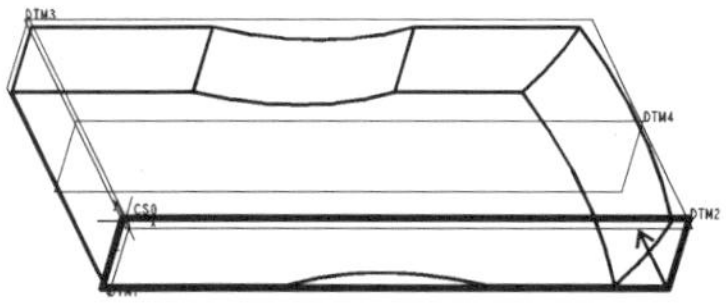

FIGURE 7-18

➪ ***NOTE:*** *Feature dimensions are shown in parentheses.*

1. Vertical centerline
2. Three-point arc (start arc at centerline)

A. Align upper arc endpoint to the top surface of the body (align using the default view)
B. Align lower arc endpoint to silhouette edge of the swept cut (align in the sketch view)
C. Linear dimension from the arc center to DTM1 (4.2)
D. Radial dimension of the three-point arc (5)

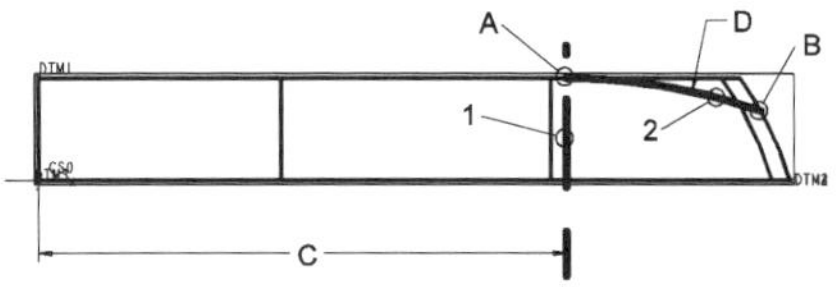

FIGURE 7-19

Commands:

Regenerate ➡ Modify ➡ Regenerate | Done | Flip ➡ Okay | Done | Thru All | Done

Then pick **OK** from the dialog window.

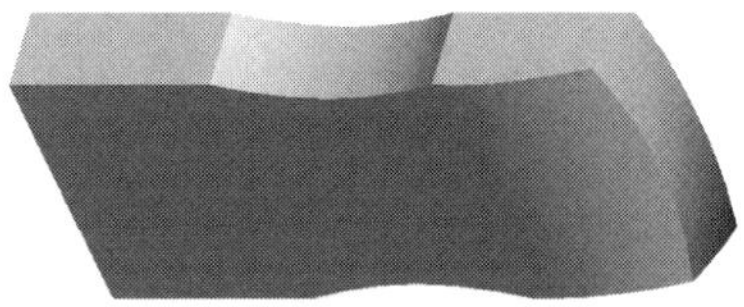

FIGURE 7-20

Parent features:

DTM2 — Sketch plane.
DTM3 — Horizontal reference.
DTM1 — Dimensioning reference.
Feature 5 — Alignment reference.
Feature 9 — Alignment reference.

➾ **NOTE:** *The edge of the swept cut you aligned is called a silhouette edge because it is not an actual part or surface edge. That edge appears only in that particular view of the part. These types of edges are valid for alignment purposes.*

✓ **TIP:** *The vertical centerline in the section will force the three-point arc to be tangent to the top surface of the body. The arc center and endpoint are assumed to lie on the centerline. The centerline is assumed to be vertical. The upper endpoint of the arc is aligned to the top surface. When these assumptions and alignments are applied, the upper endpoint of the arc is forced to be tangent to the horizontal surface of the body. This demonstrates a handy technique for forcing tangency between part geometry and sketched geometry.*

SAVE

Remaining Pre-shell Features

The part is almost ready to shell. Look at the part and try to imagine which features need to be in place before you generate the shell model. One obvious problem is that the car body looks more like a box than a curvy sports car. It definitely needs some smoothing. The other missing feature is draft.

When you design molded or cast parts, you need to consider draft. Draft is a slight angle on the sides of your part that allows the part to be removed from the mold. In Pro/ENGINEER, you can easily add draft to the sides of your parts. Adding draft as a feature is a lot easier than trying to create your part with draft from scratch. Sometimes you might consider trying to place the draft in a section, but most of the time you should just use the **Draft** feature.You need to consider draft *before* you round edges on your parts. To understand this concept, consider the following examples.

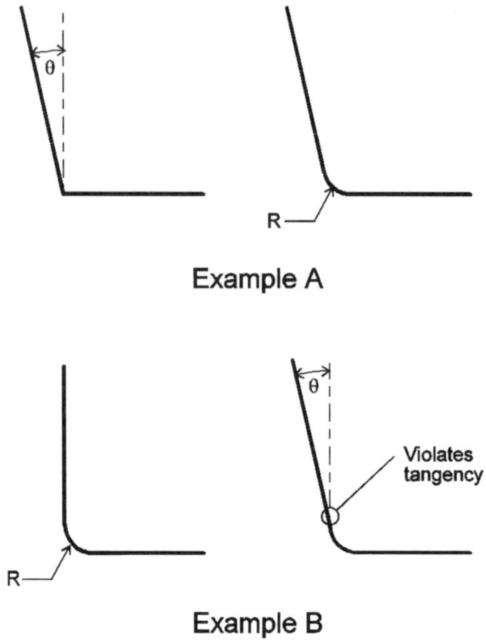

Figure 7-21
The difference between drafting first (example A) and rounding first (example B).

Example A shows that the draft is placed before the round, and everything looks good. Example B shows that the round edge prevents the draft from being placed. You cannot place draft when it violates a tangency condition of a previous feature. It is fairly easy to maintain this order; just be careful when you are designing molded parts and try to place your draft before you round the part's edges.

Placing Draft on the Sides of the Molded Part

The sides of the car body need to be drafted before you place the rounds that will smooth its shape. The draft should be created such that the body's width dimension is maintained at the bottom of the car and the side surfaces draft inward, removing material from the body. The **Draft** command is one of the many commands listed under **Tweak**. Pick the following commands, starting at the FEAT menu.

Create ➥ Solid ➥ Tweak ➥ Draft ➥ Neutral Pln | Done ➥ No Split | Constant | Done

Next you will see the SURF SELECT, SURF OPTIONS, and GET SELECT menus appear. These menus allow you to select the type of feature options related to creating drafts. The next step is to identify which surfaces of the car will be drafted. For this example, pick the four surfaces that make the sides of the car body and the one surface that is the rear of the car body.

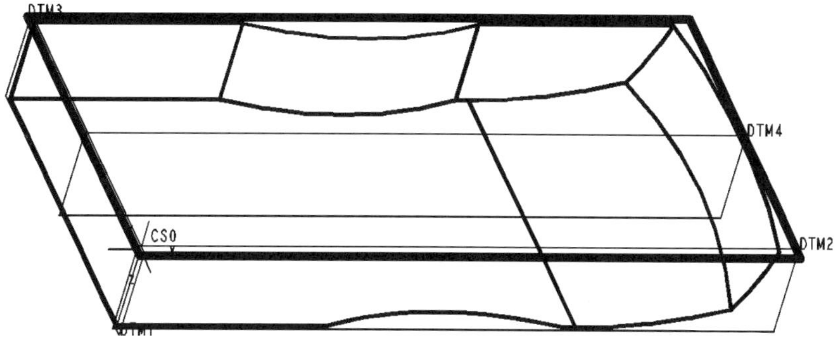

Figure 7-22
The nuetral plane for the draft.

After you have picked the surfaces you wish to add draft to, Pro/ENGINEER asks you to select a neutral (pivot) plane. This is the plane the draft surfaces will be rotated about. For the draft on the sides of the car body, you should select DTM3 (equivalent to the bottom surface of the car body).

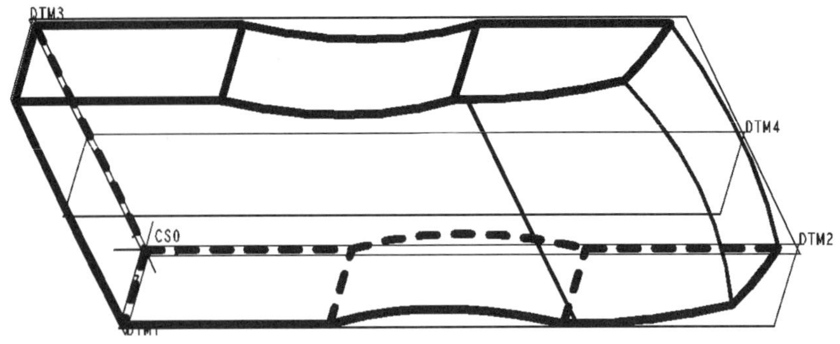

Figure 7-23
The surfaces to be drafted.

> **NOTE:** *Although the original side surface was completely split by the side cut, it still behaves as one surface.*

Select **Done Sel** from the GET SELECT menu and use **Neut Pln** from the REF DIR menu after picking the previously mentioned surfaces. Select DTM3, and Pro/ENGINEER will place a yellow arrow on the last surface you selected.

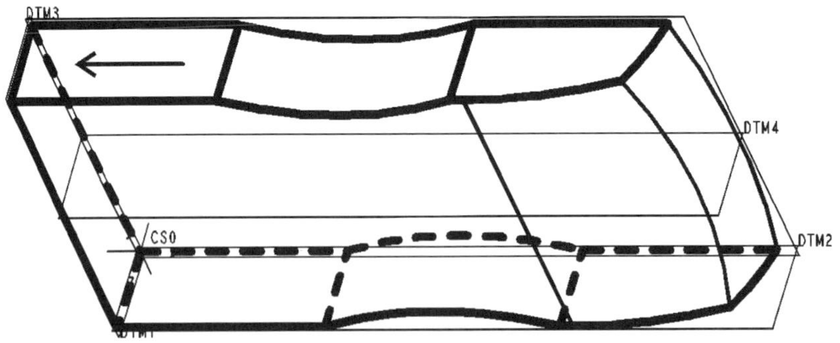

Figure 7-24
The surface rotation.

Using the right-hand rule, imagine that the arrow points in the direction of your thumb. For a positive angle, the surface will rotate in the direction of your fingers about the neutral plane. In that the arrow is pointing in the negative x direction, you need to input a negative value to rotate the surface in the proper direction (removing material). Enter the draft angle.

```
Enter draft angle (yellow arrow shows positive rota-
tion dir'n) [0.5000]: -5 ↵
```

> **NOTE:** *Draft features are limited to angles of 15 degrees or less.*

Select **OK** from the dialog window, and the draft will be created. Rotate your view so that you can see the car body from the front. Pick **View** ➡ **Orientation**. Pick **Front** and DTM1 and **Top** and DTM3. The car body should look like the following illustration.

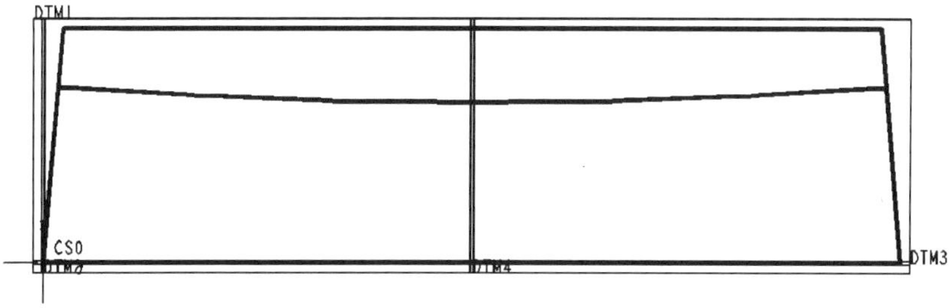

Figure 7-25
The front view of the car after drafting the side surfaces.

> **NOTE:** *If the side walls of the car body are drafted outward instead of inward, select **Modify** from the PART menu and pick one of the drafted surfaces. The angle dimension will appear. Change the dimension from 5 to –5. The draft will be created in the other direction.*

> **NOTE:** *You can also use **Apply** from the dialog window, which will allow you to see if your draft is going to be created in the desired direction. This way it can be changed before leaving the draft command mode.*

➾ **DESIGN NOTE:** *The parent features of this draft feature are DTM3, the first protrusion (feature 5), and the two side cuts (features 7 and 8). DTM3 was used as the neutral pivot plane. The surfaces of the other features were drafted by DTM3.*

Understanding the rotation arrow is a major obstacle for new users. If your draft is not created in the proper direction, modify the angle to be the negative of its current value. This will make the draft change direction.

➾ **NOTE:** *Pro/ENGINEER does not show dimensions with negative values. If you modify a dimension and provide a negative value, you are telling Pro/ENGINEER to move the feature to the other side of the dimensioning reference. Once the feature has been properly regenerated, the dimension will be shown with a positive value.*

Adding Major Rounds

Now that the draft has been created, you can place the rounds on the car body. These rounds are placed before the shell because they are large enough to be reflected on the inside walls after the shell. If the round is smaller than the shell thickness, there is no need to place it before the shell. This is the criterion you should use when determining if a round feature should be placed before or after the shell feature.

For this feature, you will place rounds on six edges of the car body. Four of these edges will receive the same radius value, and two will receive a different value. These rounds can be accomplished with one feature by using the **Variable** round option. Pick the following commands, starting at the FEAT menu.

Create ➡ Solid ➡ Round ➡ Simple | Done ➡ Variable | One by One | Done ➡ Tangent Chain

Pick the six edges in the order shown.

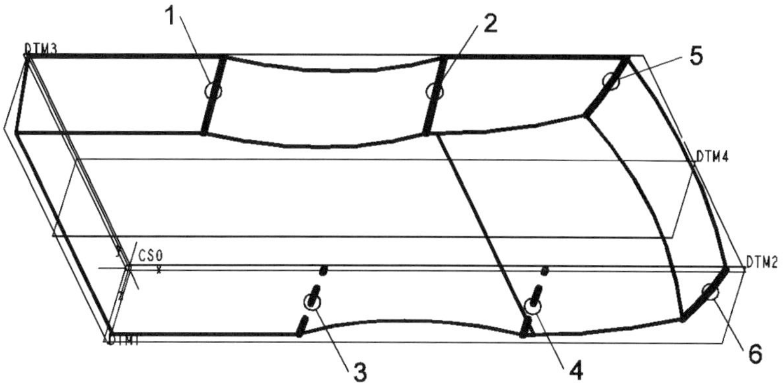

Figure 7-26
The edges to pick for the round feature.

After picking the edges, select **Done** from the CHAIN menu. Then select **Done** from the GET PNT SEL menu. Pro/ENGINEER will highlight the edges you selected in the order you picked them. Select **Enter** from the RADIUS VAL menu and provide `3` for the value.

```
Enter RADIUS [0.0160]: 3 ↵
```

Notice that the value of 3 has been added to the menu. Pick this value from the menu for the next three edges. The front corners will get a round with a .5 radius value. Pick **Enter** when the upper front edge is highlighted, and enter the value.

```
Enter RADIUS [3.0000]: .5 ↵
```

Pick the `.5` value from the menu for the last edge. The rounds will be created.

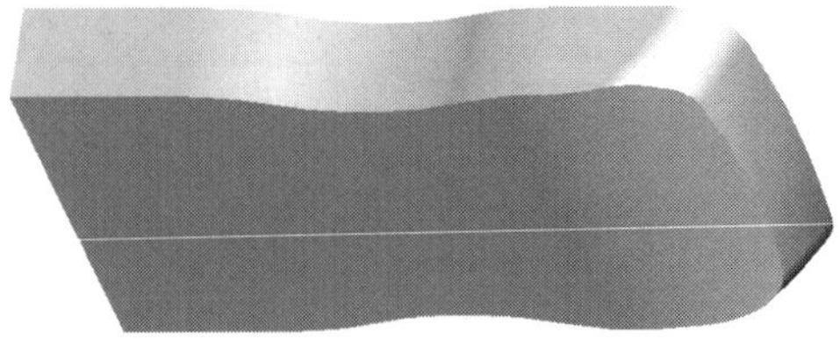

Figure 7-27
The rounds on the sides of the car body.

The next step in smoothing the body is to round the top edge. This can be achieved easily if you use the **Edge Chain** option when picking edges for rounding. Select these commands, starting at the FEAT menu.

Create ➡ Solid ➡ Round ➡ Simple | Done ➡ Constant | Edge Chain | Done

Select **Surf Chain** from the CHAIN menu. This will allow you to select an edge plus all the edges that are tangent to the one you pick. Pick the surface shown in the following illustration. Then pick **From To** from the CHAIN OPT menu. Now select the two vertices shown in the illustration. Pick **Next** from the CHOOSE menu. All of the edges except the rear one should highlight. Pick **Accept** from the CHOOSE menu.

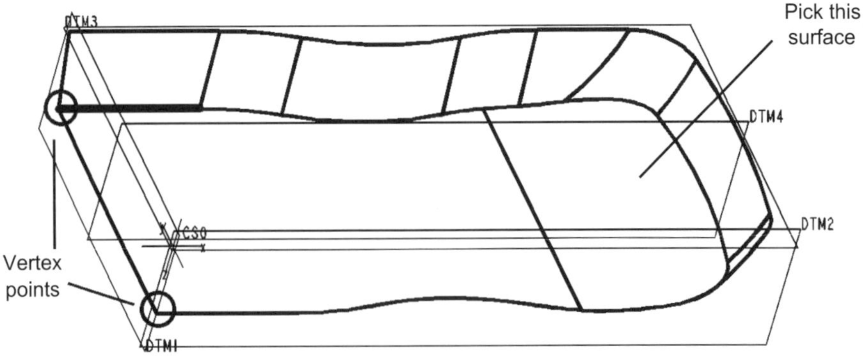

Figure 7-28
The edge to select for the rounds on the top edge of the body.

Select **Done** from the CHAIN menu after the entire chain has been highlighted. Select **Enter** from the RADIUS TYPE menu to provide the value.

```
Enter RADIUS [0.0160]: .4 ↵
```

Pick **OK** from the dialog window. The rounds will be created.

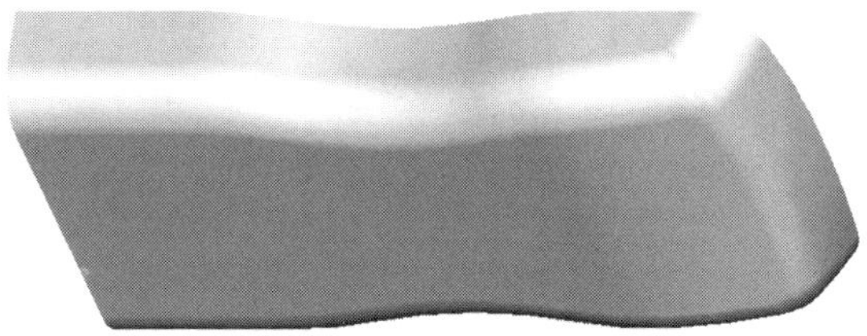

Figure 7-29
The top edges after rounding.

Finally, you need to round the back edges of the car body. These edges can be rounded in the same manner as the top edges. Pick the following commands again:

Create ➥ Solid ➥ Round ➥ Simple | Done ➥ Constant | Edge Chain | Done

Select **Surf Chain** and pick the surface and vertices shown in the following illustration.

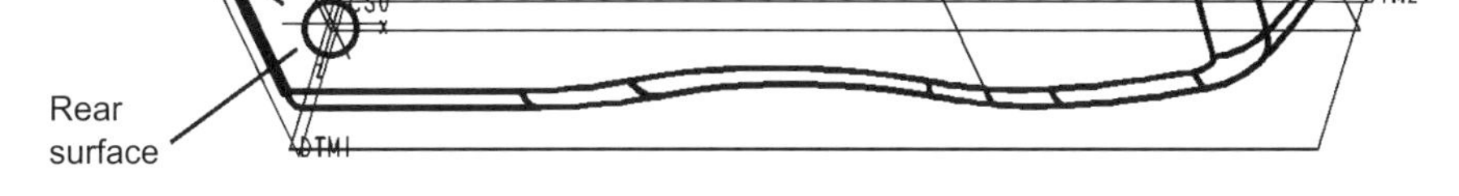

Figure 7-30
The rear edge to select for rounding.

Enter .2 for the radius value.

```
Enter RADIUS [0.0160]: .2 ↵
```

The last rounds will be created.

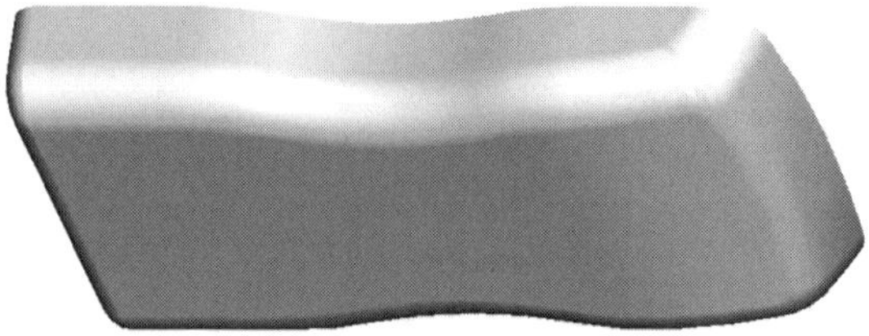

Figure 7-31
The body with all outside edges rounded.

> ➾ ***DESIGN NOTE:*** *Notice the order of the rounds. You started with the largest-value rounds and ended with the smallest. This is the best way to round your parts. It will ensure that the small radii properly wrap around the large radii.*

SAVE

Shelling the Molded Part

Your part is ready to be shelled. As mentioned previously, this is the time to specify a thickness for the part. Pro/ENGINEER removes material until the entire part has your specified thickness. To make the part into a shell, you simply need to identify the surfaces that will be removed by the shell. These surfaces are the ones your tool would remove material from if you were trying to machine the shelled part. To shell the car body, pick the following commands from the FEAT menu.

Create ➥ Solid ➥ Shell ➥ Add

Query select the bottom surface of the car. This is the surface that will be hollowed out by the shell.

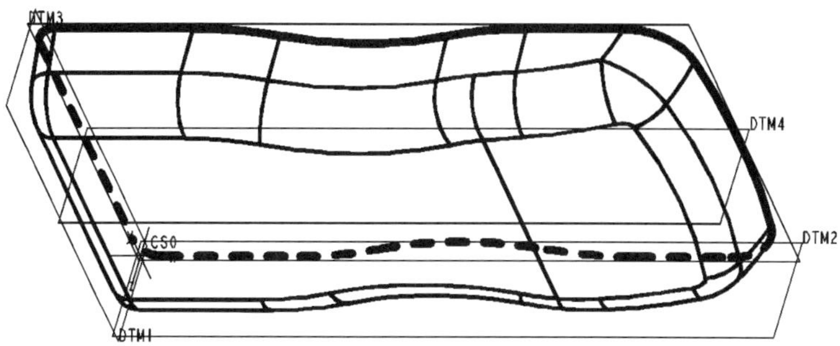

Figure 7-32
Pick the surface to remove.

You can specify more than one surface to remove, but in this case, that is the only one you need to pick. Pick **Done Refs** from the FEATURE REF menu after selecting the bottom surface. Pro/ENGINEER will prompt you for the shell thickness. Enter `.1` for the thickness value.

```
Enter thick. in range -100000.0000 to -0.0008 or
0.0008 to 3.3969 [.0849]: .1 ↵
```

Pick **OK** from the dialog window, and the shell is created. Entering a negative value for the shell thickness will add the specified thickness to the outside of your part and make your current part the empty space on the inside of the shell. Essentially, the geometry and dimensions you have created would be applied to the inside surfaces of the part. In this case, you wanted to remove only the material from the inside while retaining the part's outer dimensions.

Pro/ENGINEER will let you create shells with variable thicknesses. After inputting the basic shell thickness, you will be prompted to select surfaces that should be offset with a different value. In this example, your shell should have a constant wall thickness, so just select **Done** from the SHELL EXIT menu. The shell model will be created.

To get a better view of the shelled part, you should rotate your view. An easy way to create an opposite view is to pick **View ➡ Orientation ➡ Angles ➡ Horiz** and key in `180` for the rotation angle. The new view will appear once you select **Done/Accept** from the ANGLES menu.

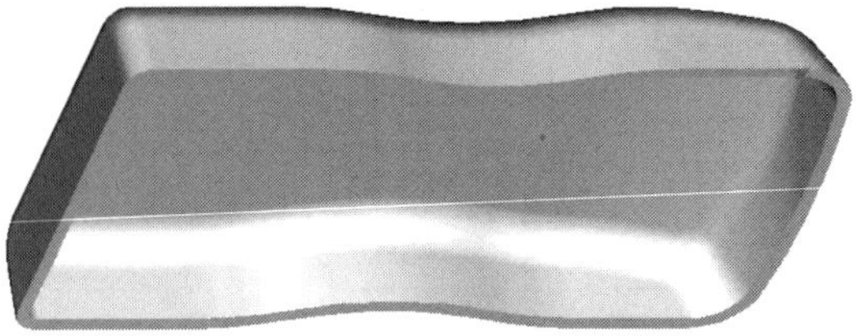

Figure 7-33
The shelled car body.

➾ **NOTE:** *Most of the remaining features will be created on the underside of the body. It would be a good idea to save this view because you will be switching to it quite often. Select* ***View*** ➥ ***Names*** ➥***Save*** *and enter the desired name.*

➾ **DESIGN NOTE:** *The only parent feature of the shell feature is the first protrusion (feature 5). Feature 5 is a parent because the bottom surface of the part belongs to this feature.*

Thinking Ahead

Before you start adding ribs or bosses to the underside of the shelled part, you should consider cutting out the wheel wells. If you put the wheel wells in now, you can cut straight through the body without intersecting any underside features.

The First Cut

Use the following information to cut out the rear wheel well.

Feature 16, the Rear Wheel Well

Commands:

Create ➥ Solid ➥ Cut ➥
Extrude | Solid | Done ➥
One Side | Done

References:

Sketch plane — DTM2.
Horiz/Vert reference — **Top**, DTM3.

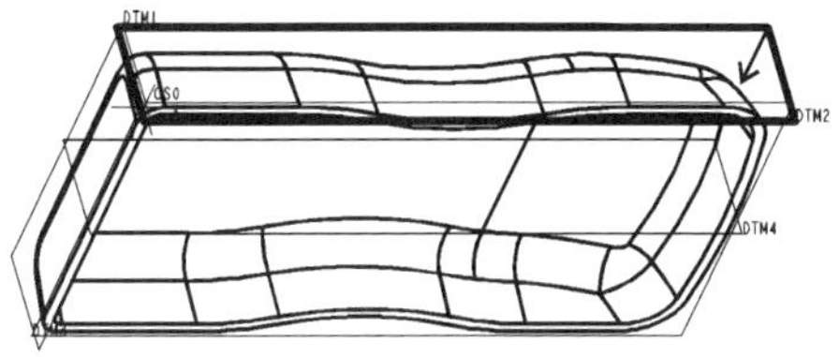

Figure 7-34

➥ ***NOTE:*** *Feature dimensions are shown in parentheses.*

1. Vertical centerline
2. Angled line
3. Tangent arc
4. Tangent line

A. Align lower endpoint of angled line to DTM3
B. Align lower endpoint of angled line to DTM3
C. Linear dimension from the centerline to DTM1 (1.25)
D. Linear dimension from the arc center to DTM3 (.2)
E. Radial dimension for the tangent arc (.45)
F. Angular dimension between the centerline and the leftmost angled line (10)
G. Angular dimension between the centerline and the rightmost angled line (10)

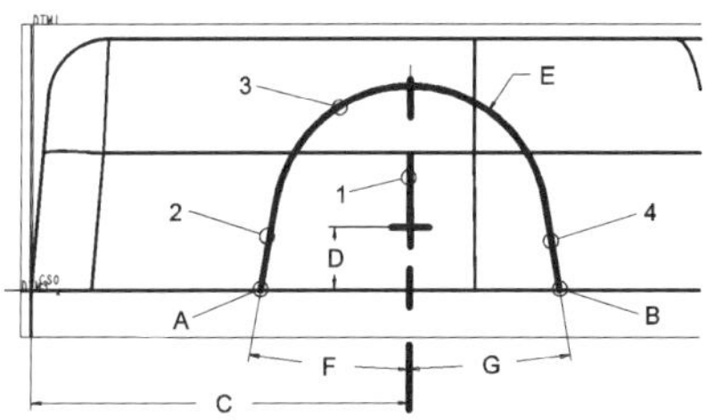

Figure 7-35

Commands:

Regenerate ➥ Modify ➥ Regenerate | Done ➥ Okay ➥ Thru All | Done

Pick **OK** from the dialog window.

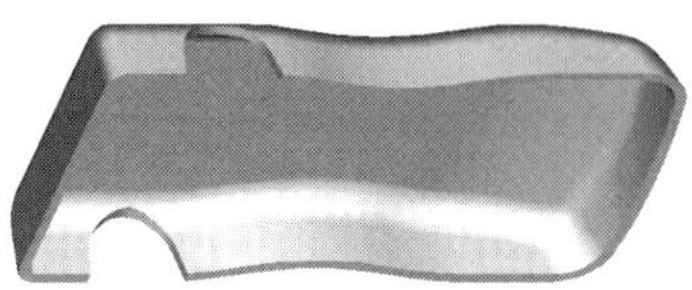

Figure 7-36

Parent features:

DTM2 — Sketch plane.
DTM3 — Horizontal reference, alignment reference, and dimensioning reference.
DTM1 — Dimensioning reference.

Copying the First Cut to Save Time

The front wheel well should be exactly like the rear wheel well, but placed in a different location. The best way to do this is to copy the first wheel well using the same references. A pattern could be used, but because our goal is to make only one copy of the original wheel well, a pattern is not necessary. Select the following commands, starting at the FEAT menu.

Copy ➥ Same Refs | Select | Dependent | Done

Pick the rear wheel well feature and pick **Done** from the SELECT FEAT menu. Now select the dimensions that should be different on the new copy. In this case, you only need to change the 1.25 dimension.

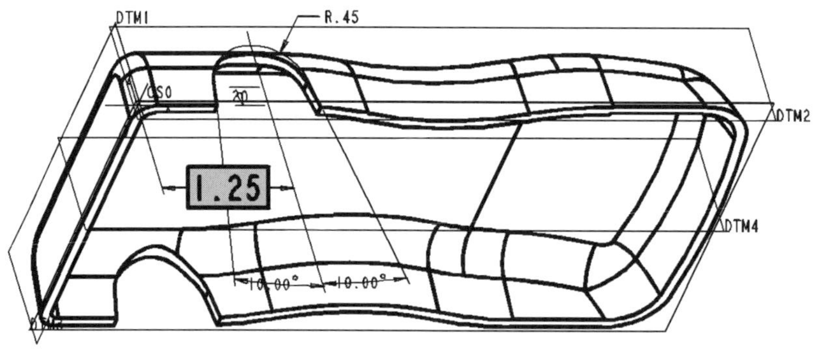

Figure 7-37
The dimension to modify for the front wheel well.

Pick the `1.25` dimension, pick **Done** from the GP VAR DIMS menu, and enter its new value.

`Enter value [1.2500]:` **4.5** ↵

Select **OK** from the dialog window to indicate that you do not want to change any other dimensions for the copy. After this, the new wheel well will be created.

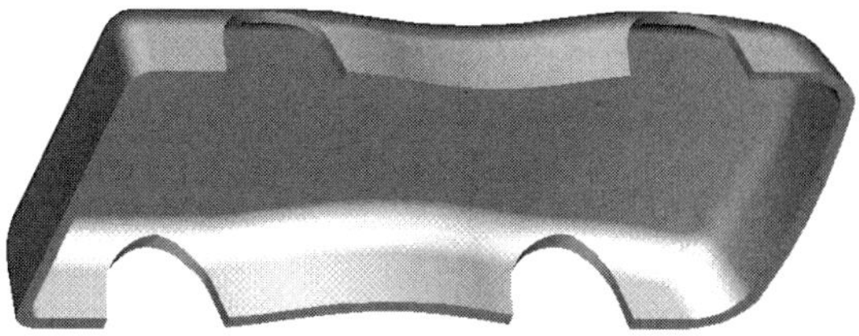

Figure 7-38
The car body with all of the wheel wells.

DESIGN NOTE: *The copied wheel well has the same parent features as the original wheel well. In addition, the dimensions for the wheel well height, radius, and side angles are all linked between the original and the copy. Altering one of these dimensions will change both wheel wells.*

SAVE

Adding Bosses

The car body is really starting to shape up. It has most of the outside features, and the wheel wells are cut through. You should start to concentrate on placing the major underside features, such as bosses and ribs. These features are important because they tie the body to the chassis.

Adding the Material for the Boss

The first boss will be placed in the right rear of the car body (near the coordinate system). Use the following procedure to make the boss.

Feature 18, the First Boss

Commands:

Create ➥ Solid ➥ Protrusion ➥ Extrude | Solid | Done ➥ One Side | Done

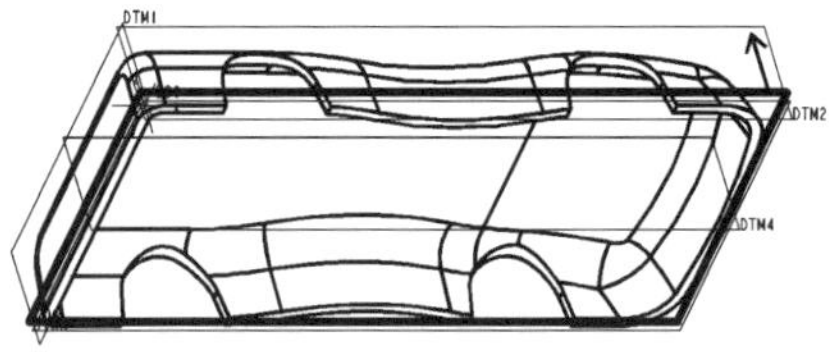

Figure 7-39

References:

Sketch plane – Make a datum plane offset from DTM3 by .1 in the positive z direction.
Horiz/Vert reference – **Top**, DTM2.

➥ ***NOTE*** *Feature dimensions are shown in parentheses.*

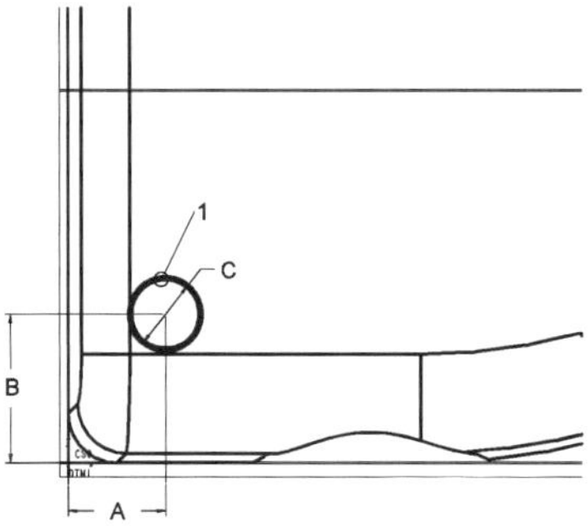

Figure 7-40

1. Circle
A. Linear dimension from the circle center to DTM1 (.4)
B. Linear dimension from the circle center to DTM2 (.6)
C. Diameter dimension for the circle (.3)

Commands:

Regenerate ➥ Modify ➥ Regenerate ➥ Thru Next | Done

Pick **OK** from the dialog window.

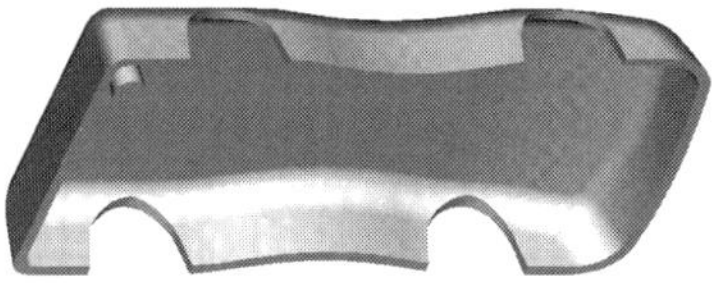

Figure 7-41

Parent features:

DTM3 — Offset sketch plane reference.
DTM2 — Horizontal reference and dimensioning reference.
DTM1 — Dimensioning reference.

➾ ***DESIGN NOTE:*** *This method of creating bosses is ideal. By using the* ***Thru Next*** *option, your boss feature is only a child of your datum plane features. This gives you the flexibility to move or copy the boss to any other location on your part as long as the boss intersects the part.*

Varying Patterns

To place the other three bosses, you will use the **Pattern** command to pattern the first boss. This pattern will be a little different from the pattern you created when you placed the holes in the chassis. Pick **Pattern** from the FEAT menu and pick the boss. Now pick **Varying | Done** from the PAT OPTIONS menu. This pattern will be varying because the front bosses intersect different surfaces from the rear bosses. The rear boss also intersects more than one surface. That alone will make **Identical** an invalid choice.

Pro/ENGINEER will prompt you for the dimension that specifies first pattern direction. Pick the `.4` dimension. Now enter the distance between this boss and the next boss in the pattern in that direction.

`Enter dimension increment:` **`4.9`** ↵

Select **Done** from the EXIT menu to stop specifying dimensions that change in the first direction. Pro/ENGINEER now asks for the total number of instances in the first direction.

```
Enter TOTAL number of instances in this direction
(including original): 2 ↵
```

The next step is to give Pro/ENGINEER the same information for the second pattern direction. Pick the `.6` dimension and enter the pattern spacing for the second direction.

`Enter dimension increment:` **`1.8`** ↵

Pick **Done** from the EXIT menu and key in the number of instances in the second direction.

```
Enter TOTAL number of instances in this direction
(including original): 2 ↵
```

The varying pattern will be created.

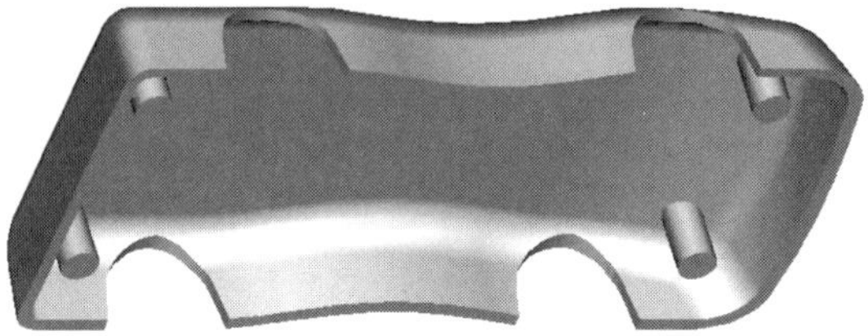

Figure 7-42
The four bosses in the underside of the car body.

Placing "Blind" Holes

A basic plastics design rule says that the hole in a boss needs to extend into the wall thickness (the shell thickness) such that the remaining material is 70 percent of the nominal wall thickness.

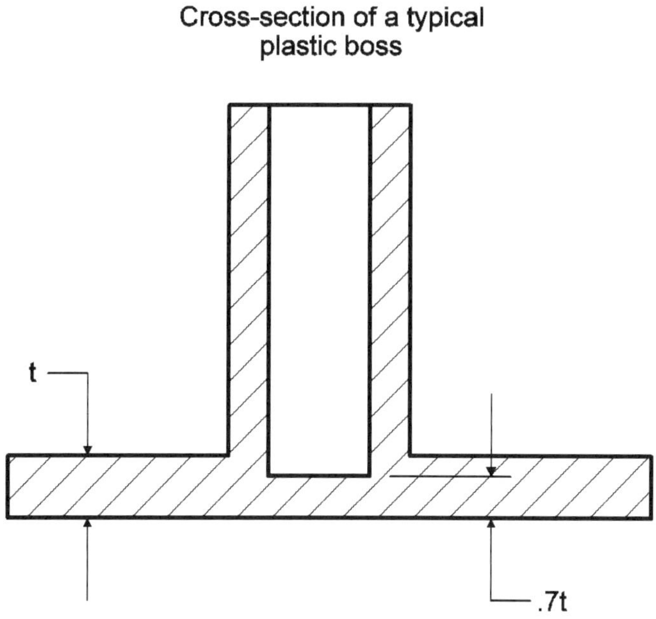

Figure 7-43
The proper method of creating a boss in a plastic boat.

On the rear bosses, this is fairly easy because the hole has a flat bottom. The hole will be placed with a blind depth (parallel to the placement plane). On the front bosses, it will take a little extra work to achieve this goal because of the curved hood shape.

Place the hole in the original boss first. Pick the following commands, starting at the FEAT menu:

Create ➡ Solid ➡ Hole ➡ Straight ➡ Done ➡ Coaxial | Done ➡ One Side | Done ➡ Blind | Done

Now pick the axis of the first boss. For the placement plane, use the planar surface on the end of the first boss.

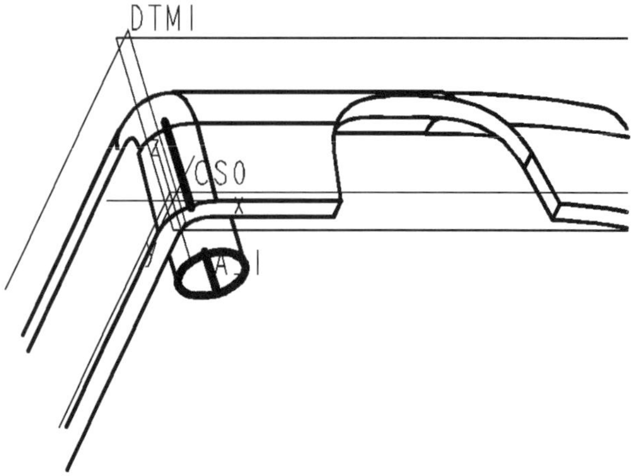

Figure 7-44
The axis and placement plane for the rear boss hole.

Finally, you need to input the depth of the hole, but you do not know what it should be. Provide a depth value you think is close.

`Enter depth [1.70]:` **`.5`** ↵

Pro/ENGINEER will prompt you for the diameter.

`Enter DIAMETER  [0.07]:` **`.15`** ↵

Pick **OK** from the dialog window, and the hole is created. The proper way to obtain the hole depth is to write a relation and calculate it. For this relation, you need to find the dimension variables for the depth of the first protrusion, the shell thickness, and the offset of the boss sketch plane. The following figure shows how the variables should stack up.

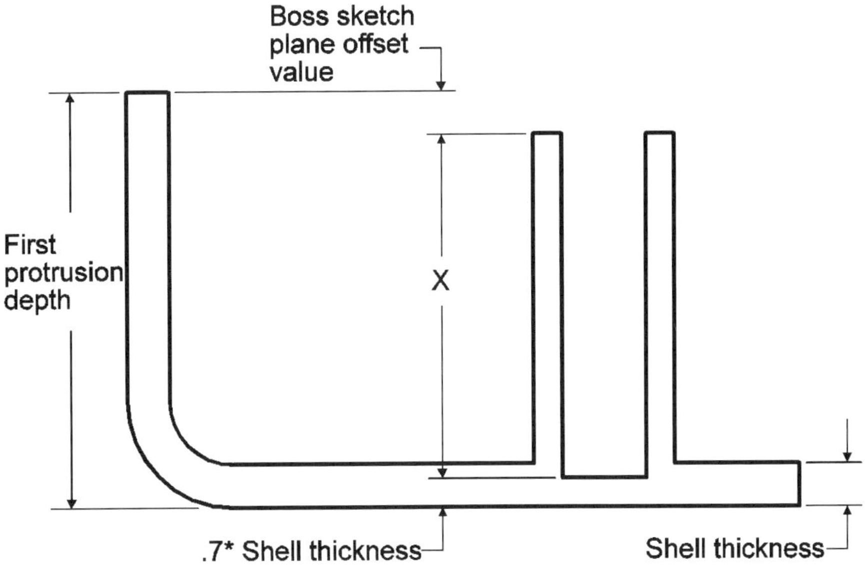

Figure 7-45
The derivation of the equation for the hole depth.

Pick **Relations** from the PART menu. Pick on the first protrusion (feature 5), the shell feature (feature 15), and the first boss (feature 18). Also pick the boss hole so that you know the variable name for the depth.

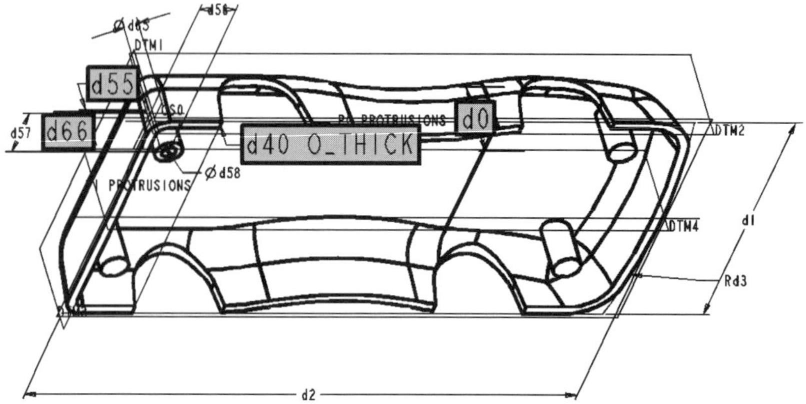

Figure 7-46
The dimension variable for the three features.

Pick **Add** from the RELATIONS menu, and type in the relation.

```
Enter RELATION [QUIT]: d66=d0-d55-(.70*d40) ↵
Enter RELATION [QUIT]: ↵
```

Pick **Show Rel** to show the relations in the part. The value for `d66` (the depth of the hole) should be .63 (or 6.300000e-01 in scientific notation). If your relation does not calculate this number, check the variables you used. Remember, your variable names may be different, depending on how you built your part. Press `q` to dismiss the relations window.

When your value is `.63`, pick **Switch Dim** to change back to number values for dimensions. Pick **Done** from the MODEL REL menu. Your depth will not be updated until you regenerate the part, so pick **Regenerate** from the PART menu. Your hole depth should be adjusted to `.63`, which follows the plastics design rule and leaves a material thickness of `.07`.

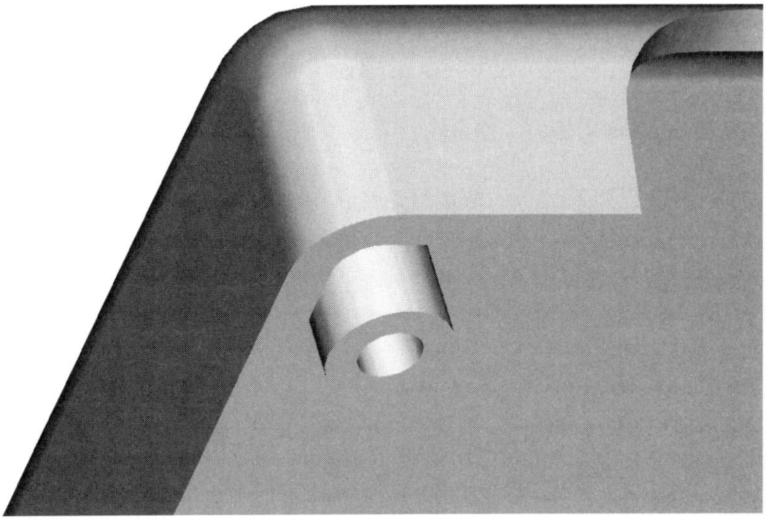

Figure 7-47
The rear boss hole.

Placing an Offset Surface Feature

The hole in the front boss needs to have a curved bottom. This bottom surface is offset from the outside hood surface by 70 percent of the wall thickness. The best way to make this hole is to create the bottom surface first and then cut the hole up to that surface.

Pro/ENGINEER can create surface geometry as well as solid geometry. Surfaces are simply features that represent a "skin" or are zero thickness. You could imagine that a surface is one of the outside surfaces of your part not connected to any particular solid. You can create surfaces in a number of ways.

These surfaces can be joined together and intersected with your solid part. They can also be used as simple construction elements that are not joined to your solid part at all.

The surface you make will be an offset of the outside surface of the hood. You will offset that surface inward by 70 percent of the wall thickness. Pick the following commands, starting at the FEAT menu:

Create ➡ Surface ➡ Offset | Done

> ➾ **NOTE:** *This method of feature creation requires the Pro/SURFACE module.*

Now pick the surface to be offset, or the outside (or top) surface of the hood.

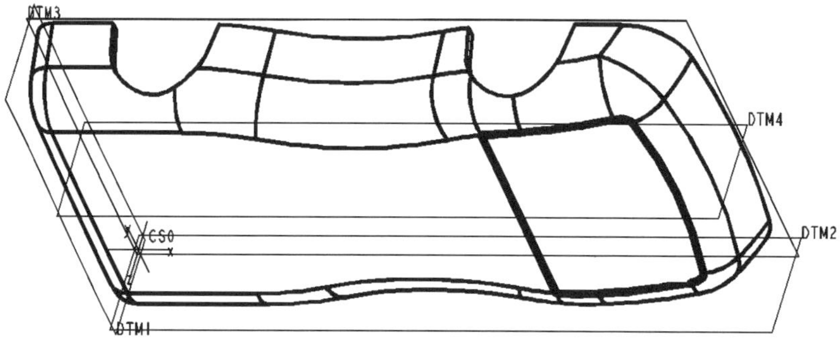

Figure 7-48
The surface to be offset.

Pro/ENGINEER will display an arrow showing the positive offset direction and prompt you for the offset distance in that direction. Input a negative number if the arrow does not point in the proper direction. You should use the thickness variable from the previous relation to get the proper offset value.

```
Enter offset distance, <ESC> to quit [0.3397]:
-.7*d40 ↵
```

> ➾ **NOTE:** *Entering an equation in response to this prompt does not add a relation to the part. Instead, it calculates a value that is*

entered. If you want this offset distance to always be 70 percent of the wall thickness, you must add a relation.

You will see the yellow boundaries of the surface. Pro/ENGINEER always shows open boundaries (free edges) of surfaces in yellow. Pick Done/Return from the SURFACE menu to return to the FEAT menu. Orient the car body so that you can see a side view. Pick **View ➥ Orientation ➥ Top** and pick DTM3. Pick **Front** and pick DTM2.

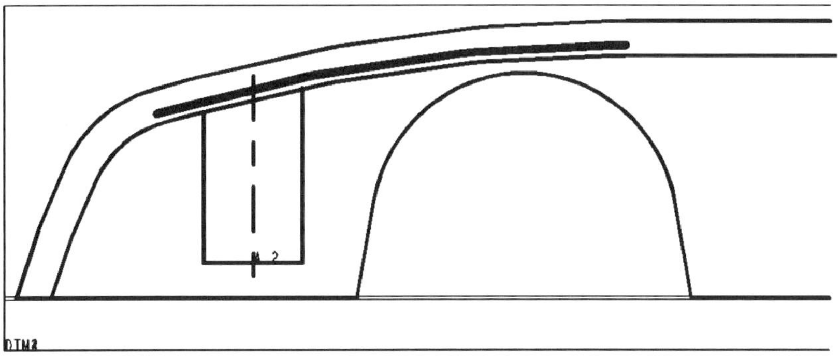

Figure 7-49
Theside view of the car body showing the offset surface.

➾ ***DESIGN NOTE:*** *The surface feature is a child of feature 10, the hood cut.*

Placing "UpTo Surface" Holes

As you can see, this offset surface is exactly what you need for the bottom of the holes in the front bosses. Now all you need to do is use that surface with the **UpTo Surface** hole feature option. Pick the following commands from the FEAT menu:

Create ➥ Solid ➥ Hole ➥ Coaxial | Done ➥ Straight ➥ Done ➥ Coaxial | Done | One Side | Done ➥ UpTo Surface | Done

Pick the axis of the front boss nearest DTM2. Pick the planar surface on the end of the same boss for the placement surface of the hole. Query select the offset surface for the surface to extrude up to. Enter the diameter of the hole.

```
Enter DIAMETER [0.07]: .15 ↵
```

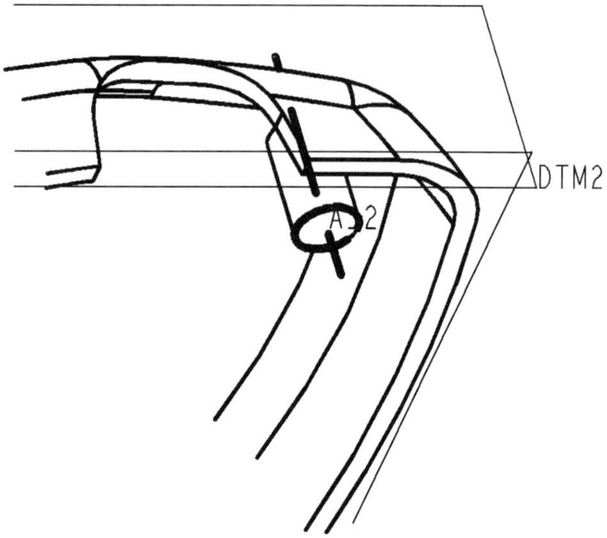

Figure 7-50
The axis and surface to use for the front boss hole.

Pick **OK** from the dialog window. The hole will be created using the offset surface as the bottom.

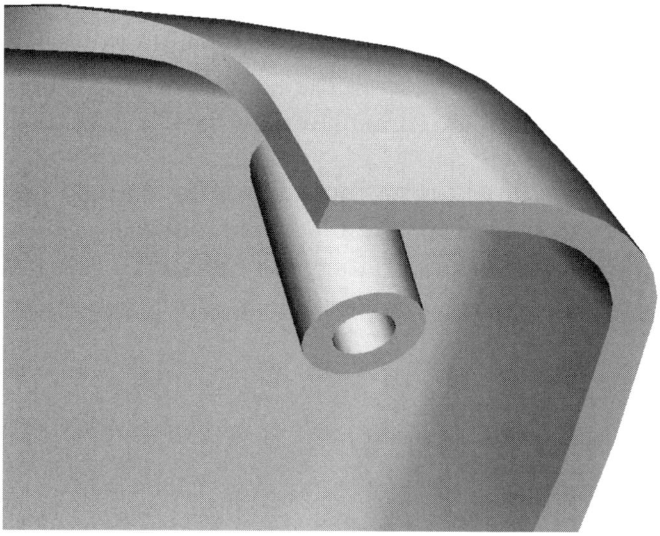

Figure 7-51
The hole in the front boss.

DESIGN NOTE: *The holes were made using the* ***Coaxial*** *option to make them stay on the boss protrusions. If the boss protrusions move, the holes will move with them. If the bosses change height, the hole depths will adjust appropriately to maintain the 70-percent wall thickness offset.*

Copying Features That Need New References

You want the same behavior from the boss holes on the other side of the body. To maintain the design intent, you can copy the left side holes to the bosses on the right side of the body. Because you placed these holes by referencing the boss axes and the boss surfaces, all you need to do is copy them and provide different references. All dimensions should remain the same on the copied holes. Pick the following commands, starting at the FEAT menu:

Copy ➥ New Refs | Select | Dependent | Done

Pick features 22 and 24, the two holes. Pick **Done** from the SELECT FEAT menu, and pick **Done** from the GP VAR DIMS menu because you do not want any of the dimensions on the copied holes to differ.

Pro/ENGINEER will begin prompting you for the new references. When you are prompted for the placement surface of the rear boss hole, pick the planar surface on the end of the other rear boss. When you are prompted for the axis, pick the axis of the other rear boss.

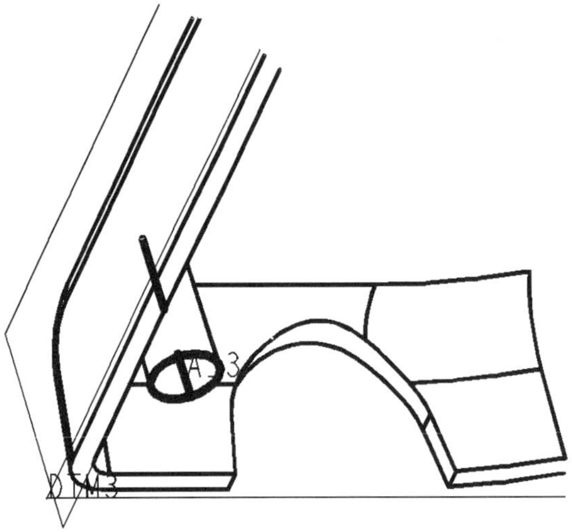

Figure 7-52
The new references for the copied rear boss hole.

Carry out the same procedure for the front boss hole copy. Pick the planar surface on the end of the other front boss and the axis of the other front boss.

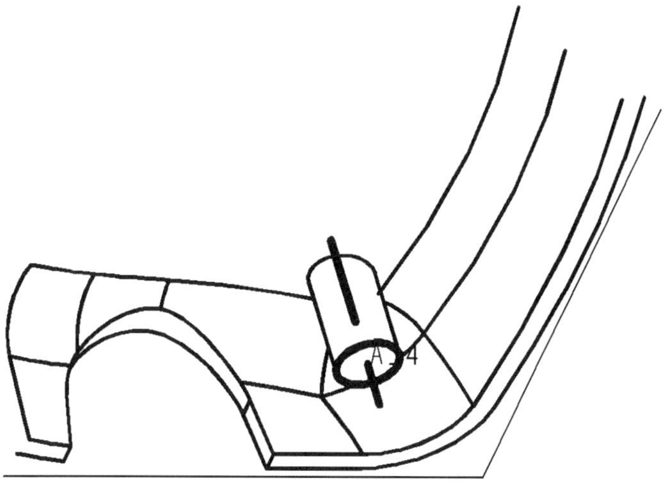

Figure 7-53
The new references for the copied front boss hole.

The final reference is the offset surface you used for the front boss hole. This reference should stay the same for the copied feature. Pick **Same** from the WHICH REF menu when you are prompted for this reference. Select **Done** from the GRP PLACE menu and the boss holes will be copied to the other side of the car body.

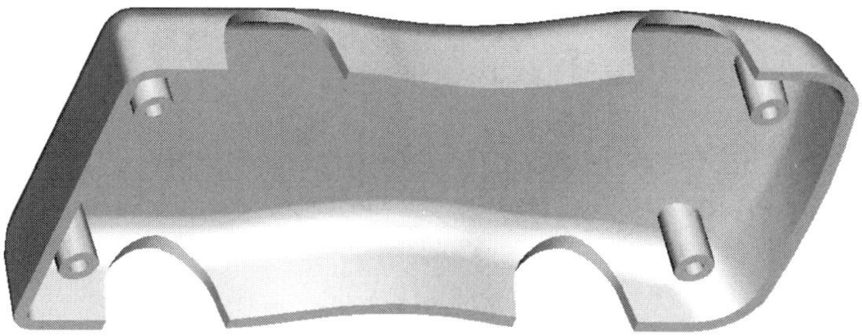

Figure 7-54
The bosses with all corresponding hole features.

DESIGN NOTE: *The new hole features have the same design intent as the original hole features, except that they will move with their own boss protrusions because you gave them different references.*

You probably will not use that surface feature anymore. If you not want to see it, you can place the surface feature on a layer and blank the layer.

Layers

Layers in Pro/ENGINEER work a little differently than layers in other CAD systems. Layers are generally used to group objects of a similar type. Common uses of layers include the following:

- **Feature Suppression by Layer** — You can place associated features on a layer, suppress that layer, and as a result, suppress many features simultaneously. You can use this to increase performance in large parts or remove detail features for an FEA (finite element analysis) model.
- **Blanking Construction Geometry** — If you use datum points, curves, or surfaces, you can place these features on layers and blank them from the display.
- **Blanking Parts in Assembly Mode** — You can place different parts on layers in Assembly mode and control the display of individual objects in the assembly.

The main difference between Pro/ENGINEER's layers and other CAD systems' layers is that objects in Pro/ENGINEER can reside on more than one layer at a time. All of the layer commands are available by selecting **Layer** from the PART menu or by selecting **View Layer Disp** from the MAIN menu if you only want to display or blank certain layers.

Creating Layers

The first thing you must do is create your layers and give them names. Select **Setup Layer ➥ Create** and enter the layer names at the prompt. The SETUP LAYER menu also provides selections to delete or rename layer names.

Placing Objects on Layers

To place an object on one or several layers, use the **Set Items** command from the LAYERS menu. This command allows you to set or unset objects from layers using the **Add Items** or the **RemoveItems** commands, respectively. To place objects on a layer, you first select the destination layer (or layers), pick the **Add Items** command, pick the type of object, and then pick the objects. Use the same procedure for removing objects from layers, but pick the **RemoveItems** command instead.

Controlling Layer Display

Pro/ENGINEER has commands for blanking and displaying layers. There is also a command for setting a layer's display to match the environment setting for hidden lines. These commands are located on the LAYER DISP menu that appears when you select **Set Display** from the LAYERS menu.

- ❒ **Display** — This command will make the selected layers display *exclusively*. This option affects only the display of assembly layers in Assembly mode.
- ❒ **Blank** — This is the command used to blank selected layers in all modes of Pro/ENGINEER.

- ❐ **Hidden** — In Assembly mode, parts assigned to hidden layers will be displayed in the Hidden Edge mode currently set in the environment.

Basically, the **Blank** command is used for 95 percent of your layer control. Just blank the layers you do not want to see. If you are in Assembly mode and have placed all of the components on individual layers, you can display a single component by using the **Display** command.

> ➾ **NOTE:** *If an object is set to be displayed and blanked at the same time, the display command will have priority.*

To see how this works, pick **Layer** from the PART menu. First you should create a layer to place the feature on. Pick **Setup Layer ➡ Create** from the LAYERS menu. Type in the name for the new layer.

```
Enter layer name [QUIT]: off_surf ↵
Enter layer name [QUIT]: ↵
```

Pick **Set Items** from the LAYERS menu. Pro/ENGINEER wants you to select the layer (or layers) to place the feature on. Pick the name **OFF_SURF** from the layer name list (you may need to scroll down by picking the inverted triangle at the bottom of the list). After you pick this layer, you will see a check mark next to it. Pick **Done Sel** from the bottom of the LAYER SEL menu after you have picked the layer. Pick **Add Items ➡ Feature** to tell Pro/ENGINEER that you want to assign a feature to that layer. Now just query select the offset surface feature (feature 23), and pick **Done Sel** from the GET SELECT menu.

Finally, pick **Set Display ➡ Blank** from the LAYERS menu and pick the layer **OFF_SURF** from the list. Pick **Done Sel** from the LAYER SEL menu and pick **Done/Return** from the LAYERS menu. Repaint the screen to make the changes take effect. Pick **Layer ➡ Set Display**

➥ **Save Status** to tell Pro/ENGINEER to save the current layer display status.

> ➾ ***NOTE:*** *Pro/ENGINEER does not automatically save the display status of layers when you save the part. If you display or blank a layer, and you want that layer change to be effective the next time you retrieve the part, pick* ***Save Status****. This will save the layer status the next time you save the part.*

Adding Ribs

The car body has two sets of ribs supporting the chassis. One set is in the front of the body, and the other set is in the back end. All ribs are the same height as the bosses. Two different methods will be used to create the ribs, one of which is the **Rib** command.

Placing a Rib Using the Rib Command

The **Rib** command will allow you to create a rib by sketching it from the side view. The sketch plane for the rib will be located at the center of the rib's thickness. After you have sketched the rib, you will provide the thickness for it. To begin placing one of the rear ribs, pick the following commands from the FEAT menu:

Create ➥ Solid ➥ Rib

The rib will be parallel to DTM2 and offset from it by 1. For the sketch plane, make a datum offset from DTM2 by 1 in the positive y direction.

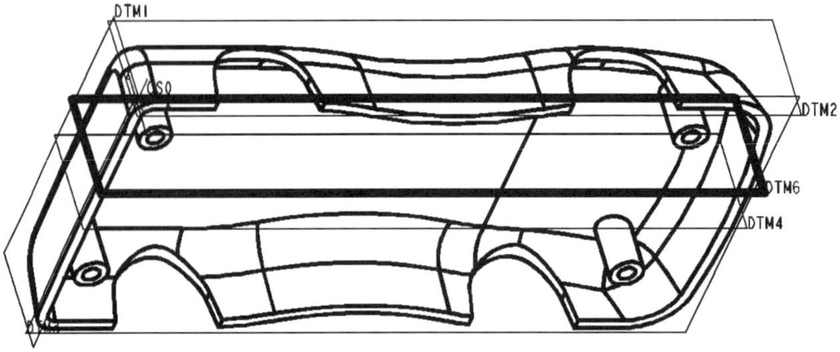

Figure 7-55
The setup for the sketch plane of the first rear rib.

Pick **Top** and DTM3 for the horizontal reference. Sketch the section in the next figure.

> ➾ ***NOTE:*** *Feature dimensions are shown in parentheses in the following list.*

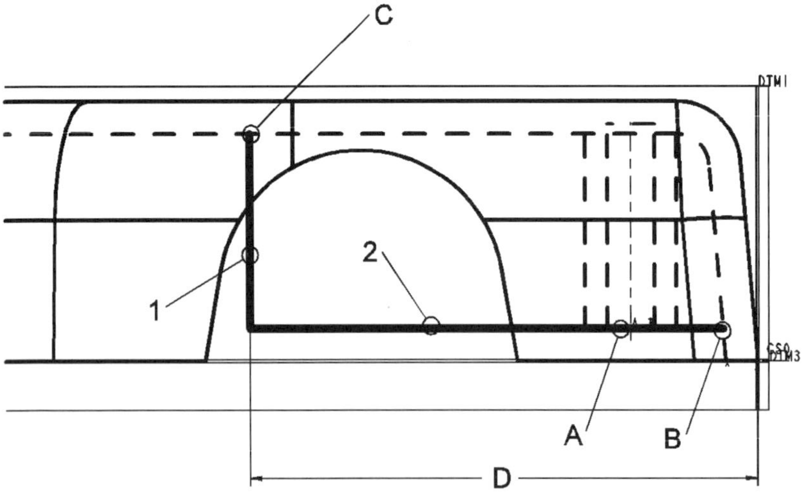

Figure 7-56
The section for the rib.

1. Vertical line
2. Horizontal line

A. Align horizontal line to the planar surface on the end of the first boss (align in a 3D view)

B. Align right endpoint of the horizontal line to the inner, shell-side surface of the rear of the body (align in a 3D view)

C. Align upper endpoint of the vertical line to the inner, shell-side surface of the trunk area of the body (align in a 3D view)

D. Linear dimension from the vertical line to DTM1 (1.6)

Commands:

Regenerate ➥ Modify ➥ Regenerate | Done

Orient the arrow toward the inside of the rib and pick **Okay** from the DIRECTION menu. Pro/ENGINEER will prompt you for the rib thickness.

`Input rib thickness [0.26]:` **`.07`** ↵

The rib feature will be created.

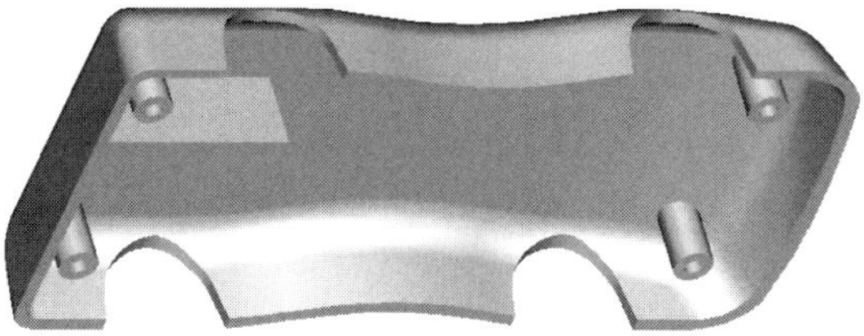

Figure 7-57
The body with the first rear rib.

➾ **DESIGN NOTE:** *The rib feature is a child of many other features. It is a child of DTM1, DTM2, and DTM3 because of the sketch plane setup and the linear dimension. It is a child of the first boss (feature 18) because the horizontal line in the section was aligned to that feature. When you aligned the endpoints of the section, you made it a child of the first protrusion (feature 5), the draft (feature 11), and the shell (feature 15). As you can see, it does not take much to make a rib feature dependent on many other features.*

Placing a Rib Using a Thin Protrusion

The rear rib has many parent features. It would be better to try to minimize these references, if possible. Remember how the bosses were created? If you apply the same **Thru Next** methodology to the rib, you can reduce the number of references and thus minimize the number of parent features.

To make the section easier to draw, you will use another new option. This option is the **Thin** option and is available for protrusions as well as cuts. The **Thin** option allows you to sketch a section and apply a thickness to that section when it is used in the solid. To place the front rib, use the following information.

Feature 28, the Front Rib

Commands:

Create ➥ Solid ➥ Protrusion ➥ Extrude | Thin | Done ➥ One Side | Done

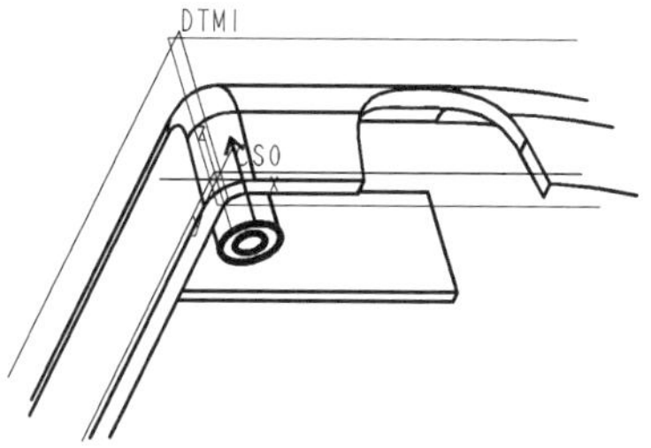

Figure 7-58

References:

Sketch plane — Use the planar surface on the end of the first boss.
Horiz/Vert reference — **Top**, DTM2.

➪ ***NOTE:*** *Feature dimensions are shown in parentheses.*

1. Horizontal line
A. Linear dimension from the left endpoint to DTM1 (4.1)
B. Linear line length dimension of the horizontal line (3)
C. Linear dimension from the horizontal line to DTM2 (1)

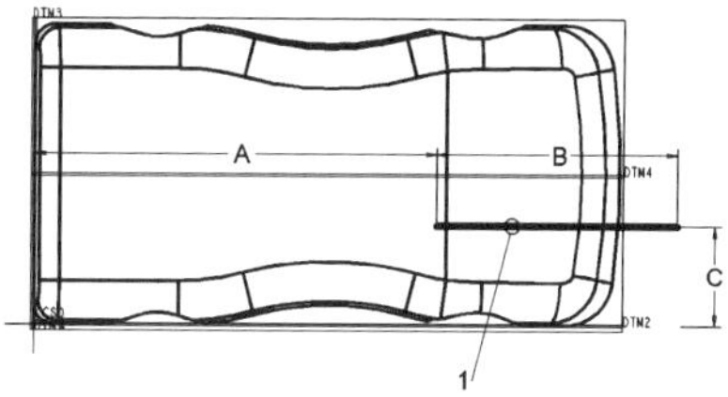

Figure 7-59

(Ignore the regeneration warning about endpoint alignments.)

Feature should be created on *both* sides of the section. Rib thickness is .07.

Commands:

Regenerate ➥ Modify ➥ Regenerate | Done | Thru Next | Done

Pick **OK** from the dialog window. The rib feature will be created.

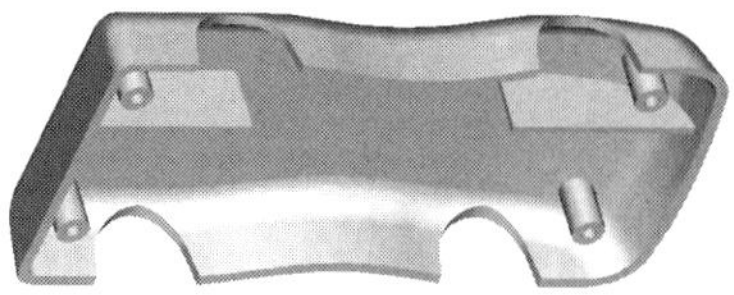

Figure 7-60

Parent features:

Feature 18 — Sketch plane.
DTM2 — Horizontal reference and dimensioning reference.
DTM1 — Dimensioning reference.

➾ **DESIGN NOTE:** *This rib creation method is extremely important to any user who makes complex parts. Notice that the rib is not a child of the shell feature (or any element of the shell). Also notice that the rib is not a child of the front swept surface of the car. You could completely change the shape of the front end of the car and it would not affect the placement of this rib. Use this method when you create ribs to avoid unnecessary parent/child relationships.*

➾ **NOTE:** *Although the section extends beyond the part, the* ***Thru Next*** *option still works. The only time this trick fails is when your section does not intersect the part. In this example, the sketch plane is located such that the section intersects the front of the car body. If the sketch plane were located on the other side of DTM3, this feature would not work.*

Mirroring the Ribs

To put copies of the ribs on the other side of the car body, you should mirror them. Pick the following commands from the FEAT menu:

Copy ➥ Mirror | Select | Dependent | Done

Pick both the rear and front ribs. Pick **Done** from the SELECT FEAT menu, and then pick DTM4 as the plane to mirror about. The features will be mirrored.

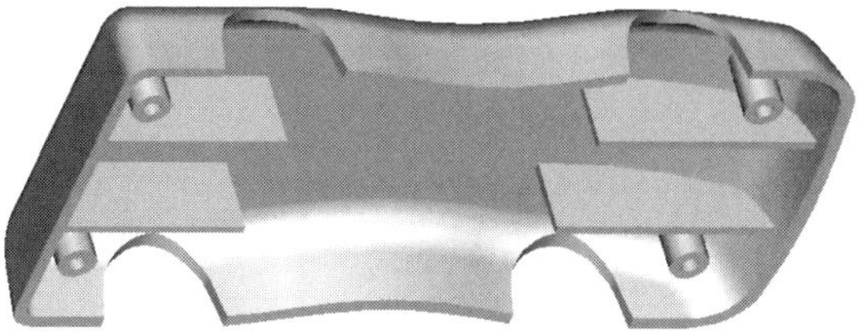

Figure 7-61
The body with all ribs in place.

SAVE

Creating Additional Extruded Cuts

The last major features to be added are for the passenger compartment. You need to cut out an area for the passengers and seats, and add a windshield and some material to fill the gap between the bucket seats.

Extruding the Cut

To remove the material for the passenger compartment, you should make a simple **Thru All** cut. Use the following information to complete the feature.

Feature 31, the Passenger Compartment Opening

Commands:

Create ➥ Solid ➥ Cut ➥ Extrude | Solid | Done ➥ One Side | Done

➾ ***NOTE:*** *Feature dimensions are shown in parentheses.*

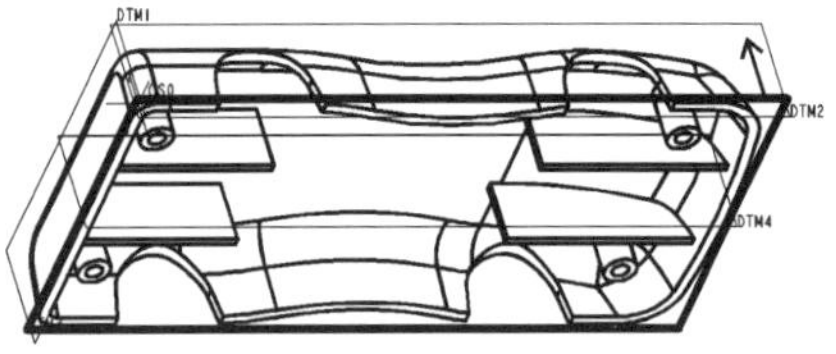

Figure 7-62

References:

Sketch plane — DTM3.
Horiz/Vert reference — **Top**, DTM2.

1. Horizontal centerline
2. Vertical line
3. Horizontal line
4. Three-point arc
5. Horizontal line

A. Align centerline to DTM4
B. Linear dimension from the vertical line to DTM1 (2)
C. Linear dimension from the tangent of the arc to DTM1 (3.8)
D. Linear line length dimension of the vertical line (2)
E. Radial dimension for the three-point arc (1.8)

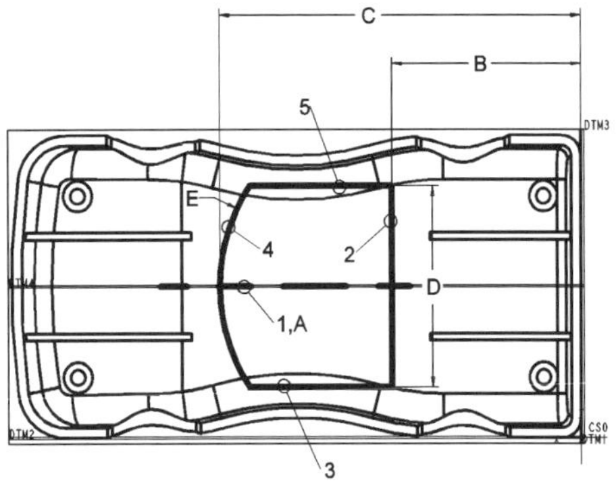

Figure 7-63

Commands:

Regenerate ➥ Modify ➥ Regenerate | Done | Okay ➥ Thru All | Done

Pick **OK** from the dialog window.

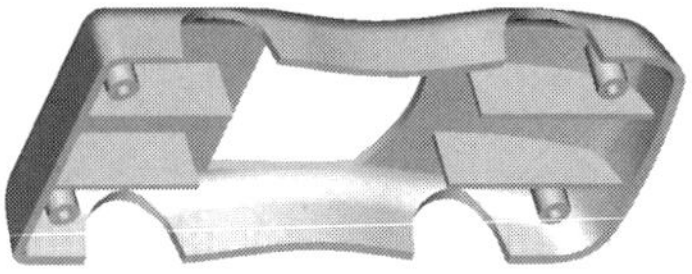

Figure 7-64

Parent features:

DTM3 — Sketch plane reference.
DTM2 — Horizontal reference.
DTM4 — Alignment reference.
DTM1 — Dimensioning reference.

NOTE: *If you cannot get the section to regenerate as shown, try adding a dimension from DTM4 to the upper horizontal line. If the section regenerates, modify all dimensions to make this one equal to 1. Regenerate the section again. Now that the section is properly centered, delete this extra dimension and regenerate the section once more.*

DESIGN NOTE: *This feature is a child of the main three datums and the midplane datum only.*

Adding a Swept Protrusion

The windshield should be built such that it follows the curved edge of the passenger compartment opening. In that the edge is curved and planar, it would be a good idea to tackle this feature with a swept protrusion. It will be created similar to the swept cut on the front end, but this time you will be adding material. Pick the following commands, starting at the FEAT menu:

Create ➡ Solid ➡ Protrusion ➡ Sweep | Solid | Done ➡ Sketch Traj ➡ Done

For the sketch plane, pick the top planar surface of the car body. Accept the default viewing direction by picking **Okay** from the DIRECTION menu. Pick **Top** and DTM2 for the horizontal reference.

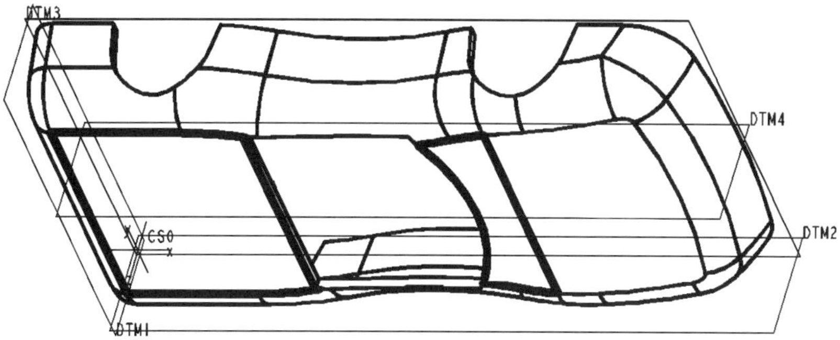

Figure 7-65
The sketch plane for the trajectory.

Change your view orientation to the default view, pick the **Geom Tools ➡ USE Edge** command, and pick the edge shown in the following figure.

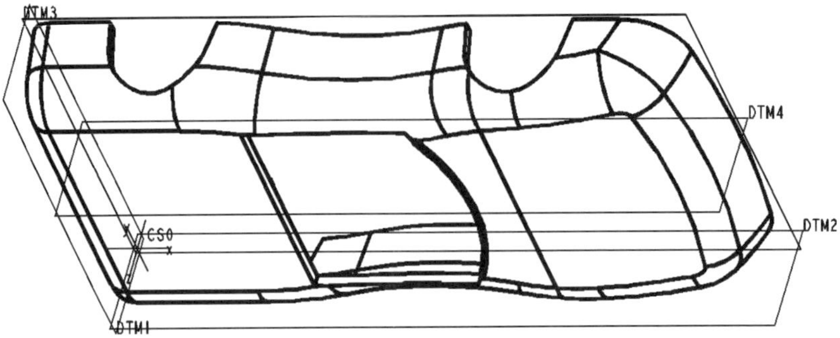

Figure 7-66
The edge to use for the trajectory.

Pick **Sketch View ➡ Mouse Sketch** from the SKETCHER menu to return to the sketch view and sketch this section.

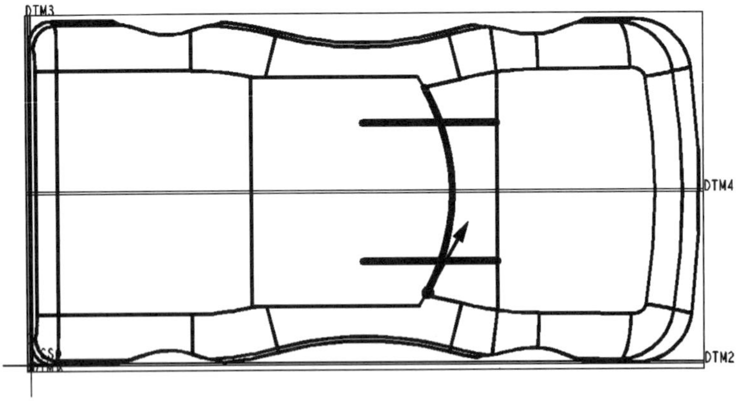

Figure 7-67
The section after adding two horizontal lines.

Trim the arc back using the two horizontal lines as your bounds. Delete the horizontal lines after you use them for the trimming operation. Place the two linear dimensions shown in the next figure.

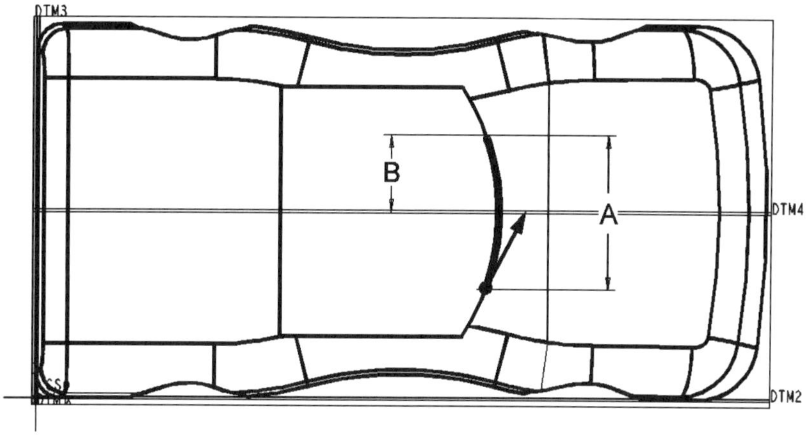

Figure 7-68
The final trajectory.

> **NOTE:** *The shorter dimension is needed because the Sketcher cannot recognize the symmetry (even with a centerline denoting the line of symmetry). If you want to ensure that the windshield stays centered on the body, you can write a relation to control the shorter dimension. Modify dimension A to* `1.7` *and dimension B to* `.85`*. Regenerate the section. If your start point is not at the bottom of the arc, place it there now and regenerate once more.*

Pick **Done** from the SKETCHER menu to proceed with the swept protrusion. In this example, you only want the windshield to be as long as the trajectory. You should select **Free Ends | Done** from the ATTRIBUTES menu to tell Pro/ENGINEER not to extend the section beyond the trajectory.

You should be looking at the left front fender of the car body (tangent to the trajectory at the start point). Sketch the following section.

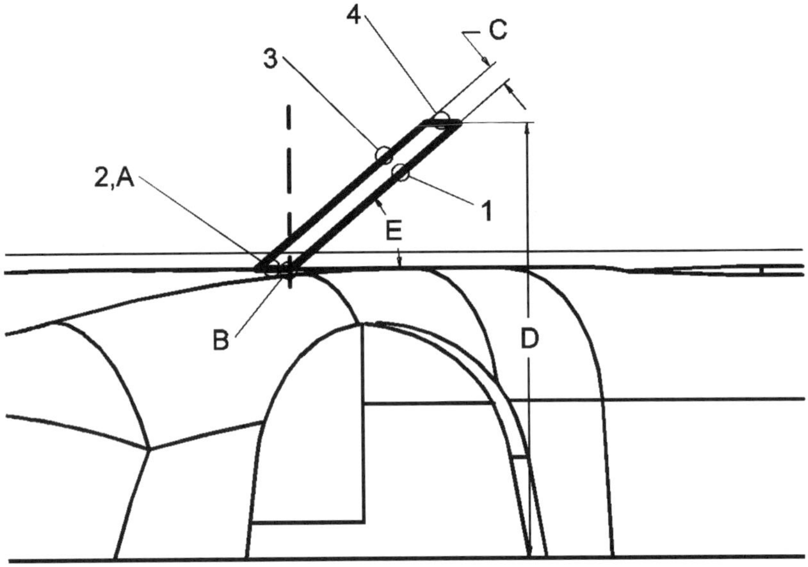

Figure 7-69
The section for the swept protrusion.

1. Angled line
2. Horizontal line
3. Angled line (parallel to the other)
4. Horizontal line

A. Align the lower horizontal line to the horizontal cross hair

B. Align the right endpoint of the lower horizontal line to the vertical cross hair

C. Linear dimension between the two angled lines (.06)

D. Linear dimension from the upper horizontal line to DTM3 (1.2)

E. Angular dimension between the rightmost angled line and the horizontal cross hair (40)

After you create this section, pick **Done** from the SKETCHER menu, and **OK** from the THIN OPT menu. Pick **OK** from the dialog window, and the swept protrusion will be created.

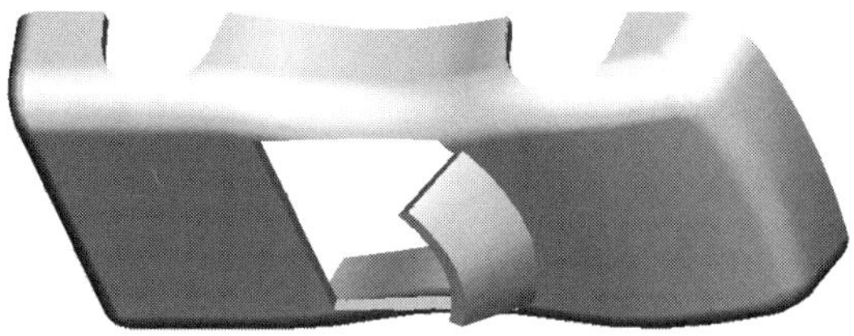

Figure 7-70
The windshield on the car body.

DESIGN NOTE: *The windshield feature is a child of DTM2, DTM3, DTM4, feature 5, and feature 31 (the passenger compartment cut). DTM2 is the horizontal reference for the trajectory. DTM3 is a dimensioning reference in the section. DTM4 is a dimensioning reference for the trajectory. Feature 5 is a parent because the trajectory was sketched on the top surface of the first protrusion. Features 5 and 31 are parents because of the part edge used in the trajectory.*

Adding a Thin Protrusion

There will be a hole between the seats if you do not put some material to fill in above the console in the chassis. This material can be created quickly with a thin, extruded protrusion. Use the following information to fill the gap between the seats.

Feature 33, the Seat Gap Filler

Commands:

Create ➥ Solid ➥ Protrusion ➥ Extrude | Thin | Done ➥ Both Sides | Done

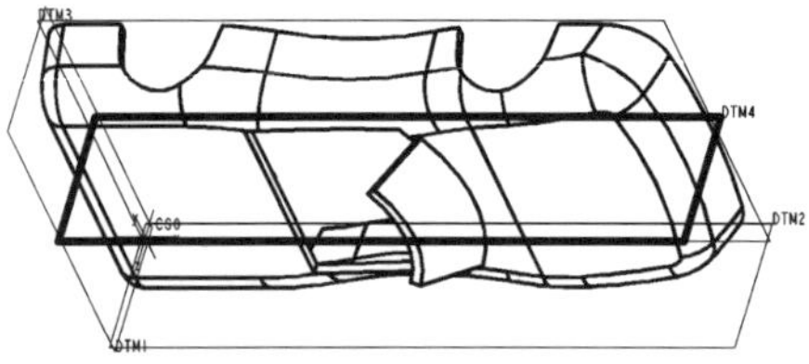

Figure 7-71

➾ ***NOTE:*** *Feature dimensions are shown in parentheses.*

References:

Sketch plane — DTM4.
Horiz/Vert reference — **Top**, DTM3.

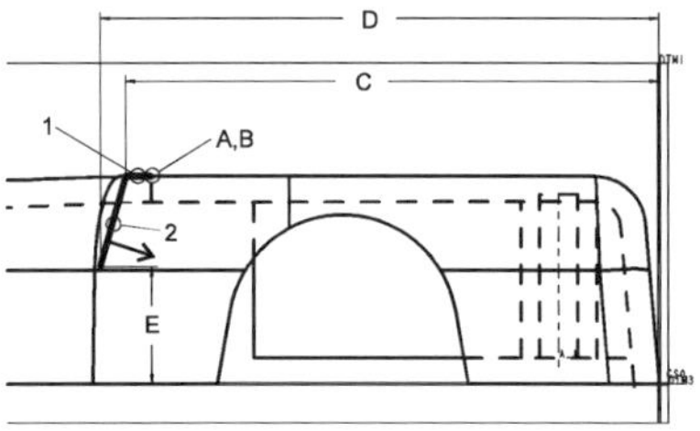

Figure 7-72

1. Horizontal line
2. Angled line

A. Align the right endpoint of the horizontal line to the back, vertical wall of the passenger compartment
B. Align the horizontal line to the top surface of the trunk
C. Linear dimension from the left endpoint of the horizontal line to DTM1 (2.1)
D. Linear dimension from the lower endpoint of the angled line to DTM1 (2.2)
E. Linear dimension from the lower end-277point of the angled line to DTM3 (.45)

Commands:

Regenerate ➥ Modify ➥ Regenerate | Done | Okay ➥ Blind | Done

Pick **OK** from the dialog window.

(Ignore the regeneration warning about endpoint alignments.)

Width of the thin feature is d40 (or .1).
Depth of the extruded feature is .3.

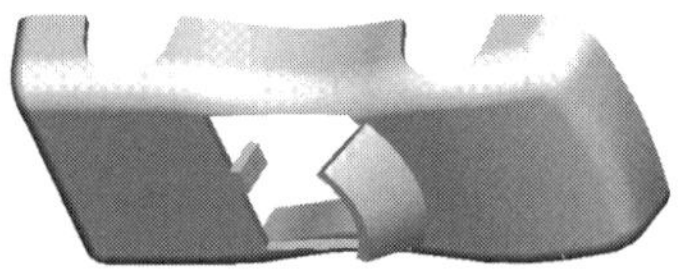

Figure 7-73

Parent features:

DTM4 — Sketch plane.
DTM3 — Horizontal reference and dimensioning reference.
DTM1 — Dimensioning reference.
Feature 5 — Alignment reference.
Feature 31 — Alignment reference.

✓ ***TIP:*** *The dimension variable* `d40` *(the shell thickness) was used as input for this feature's thickness. When you make a thin walled (or shelled) part, it is a good idea to remember the variable name for the shell thickness. This will speed things up when you make features referencing this thickness.*

➾ ***NOTE:*** *To ensure that the thickness of this feature remains equal to the shell thickness, you should create a relation.*

SAVE

Finishing Touches

Congratulations! You have come a long way if you have made it to this point. There are a few final details you need to place on this part before it is complete. A chamfer around the boss hole edges will facilitate the assembly of the actual parts. More draft features need to be placed on the ribs and bosses. A few rounds are required to smooth out the passenger compartment, and the car body definitely needs a logo.

Chamfering the Edges

We have not yet used the **Chamfer** command. You can chamfer a corner or chamfer edges of a part. Edge chamfers work much like the **Round** command because you select edges to chamfer and then enter the appropriate dimensions. There are four types of edge chamfers.

- **45 x d** — Creates a standard 45-degree chamfer on an edge between two perpendicular surfaces. The `d` is the controlling dimension and the 45-degree angle is fixed.
- **d x d** — Creates the chamfer at a distance of `d` from the edge along both adjacent surfaces.

- **d1 x d2** — Creates the chamfer at a distance of `d1` from the edge along the reference surface. The `d2` controls the distance from the edge to the chamfer on the other surface.
- **Ang x d** — Chamfers the edge based on the angle from the reference surface and the distance along the reference surface.

Chamfers are very easy to create, and the most often used chamfers are **d1 x d2** and **Ang x d** because of their versatility. To chamfer the inside edges of the boss holes, pick the following commands, starting at the FEAT menu:

Create ➥ Solid ➥ Chamfer ➥ Edge ➥ 45 x d

Enter the chamfer dimension.

```
Enter CHAMFER DIMENSION for d [0.17]: .03 ↵
```

Now pick the edges of the holes to be chamfered. You only need to pick one edge of each hole because the chamfer will automatically wrap around the entire hole.

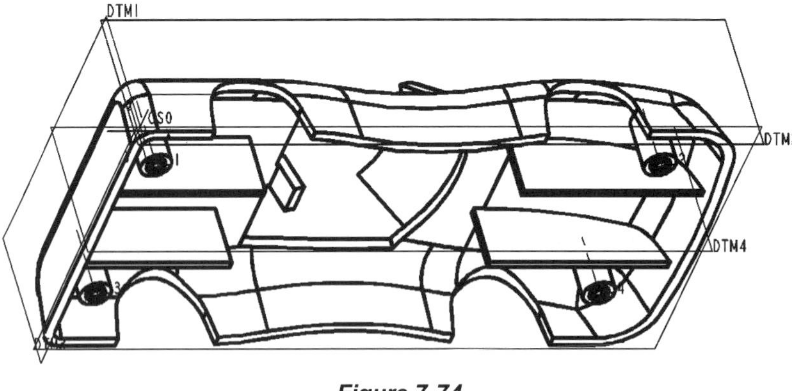

Figure 7-74
The edges to be chamfered.

After selecting the fourth edge, pick **Done Refs** from the FEATURE REF menu. Then pick **OK** from the dialog window, and the chamfers will be created.

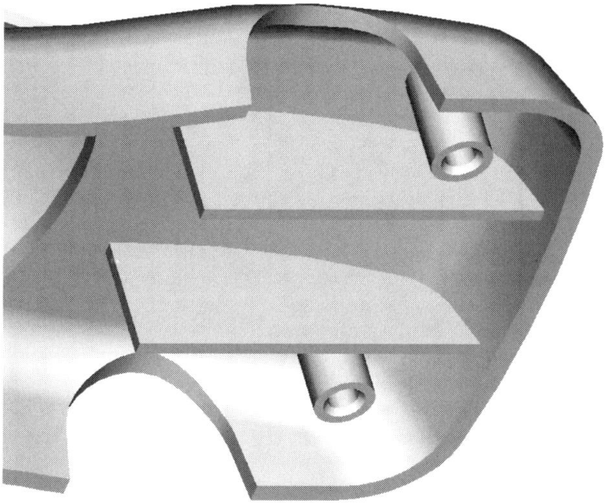

Figure 7-75
The chamfers on the boss holes.

DESIGN NOTE: *You might want to consider* ***Ang x d*** *chamfers when you place them in your designs. In this manner, you have control over the angular dimension if it needs to change.*

Drafting the Bosses

The bosses need to have draft on the inside and outside. The draft on the outside of the bosses should use the top surface of the boss as the neutral plane. Unfortunately, you cannot use this surface as the neutral plane for the draft on the inside of the holes. If you did, you would not maintain the `.15`-diameter dimension of the hole.

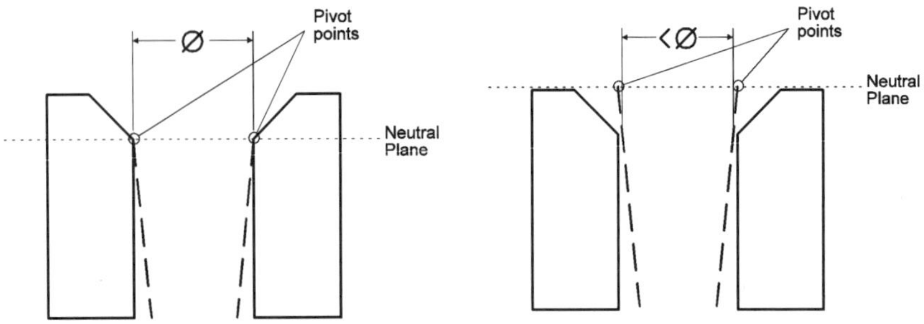

Figure 7-76
Results of using different neutral planes.

To place the draft on the outside of the bosses, pick the following commands from the FEAT menu:

Create ➥ Solid ➥ Tweak ➥ Draft ➥ Neutral Pln | Done ➥ No Split | Constant | Done ➥ Include | Indiv Surfs

Now pick one outside surface of each boss.

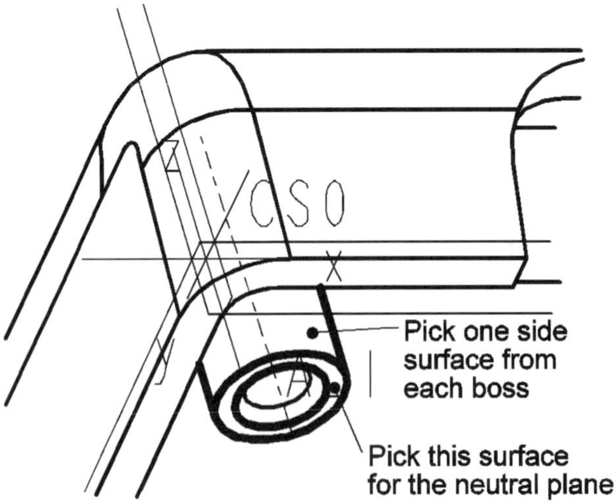

Figure 7-77
The neutral plane and the surfaces to be drafted.

Pick **Done** from the SURF SELECT menu after you have picked the last surface. Pick the top surface of the first boss as the neutral (pivot) plane. Then pick **Use Neut Pln** from the REF DIR menu. Using the arrow and the right-hand rule as your guide, enter the draft angle.

`Enter draft angle (yellow arrow shows positive rotation dir'n) [0.5000]:` **1** ↵ `Then OK from the dialog window.`

To draft the inside hole surfaces, use the same command:

Create ➥ Solid ➥ Tweak ➥ Draft ➥ Neutral Pln | Done ➥ No Split | Constant | Done | Include | Indiv Surfs

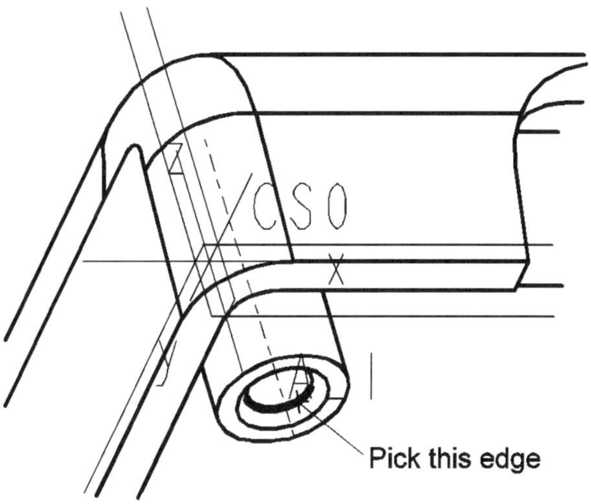

Figure 7-78

The edge to be used to locate the datum plane (neutral plane).

Now pick one hole surface from each boss.

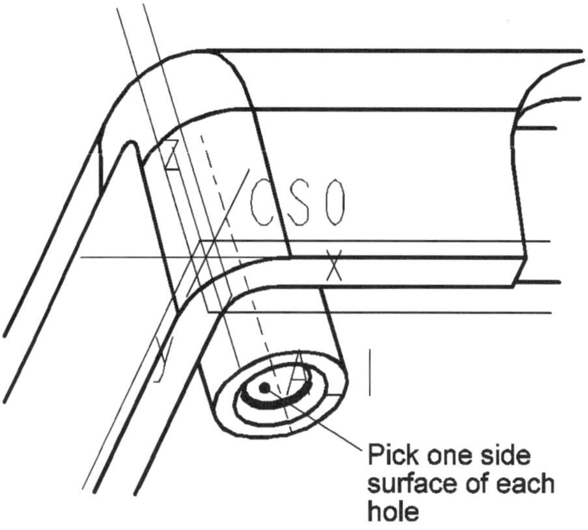

Figure 7-79
The surfaces to be drafted.

When you have picked the last surface, pick **Done Refs** from the FEATURE REFS menu. There is not an existing plane to use for the neutral plane. In this case, you need to make a datum plane "on the fly."

Pick **Make Datum** from the SETUP PLANE menu. The datum plane should be located where the hole intersects the chamfer. Pick **Through** and pick the hole edge on the first boss. Pick **Done** from the DATUM PLANE menu to place the datum plane. Then pick **NEUT PLN** from the REF DIR menu. Then enter the draft angle.

```
Enter draft angle (yellow arrow shows positive rota-
tion dir'n) [0.5000]: 1 ↵
```

The draft on the bosses is complete. Orient your view so that you are looking directly at the bottom of the car body (**Back** and DTM3, **Bottom** and DTM2). This view will tell you if the surfaces have been drafted in the correct direction.

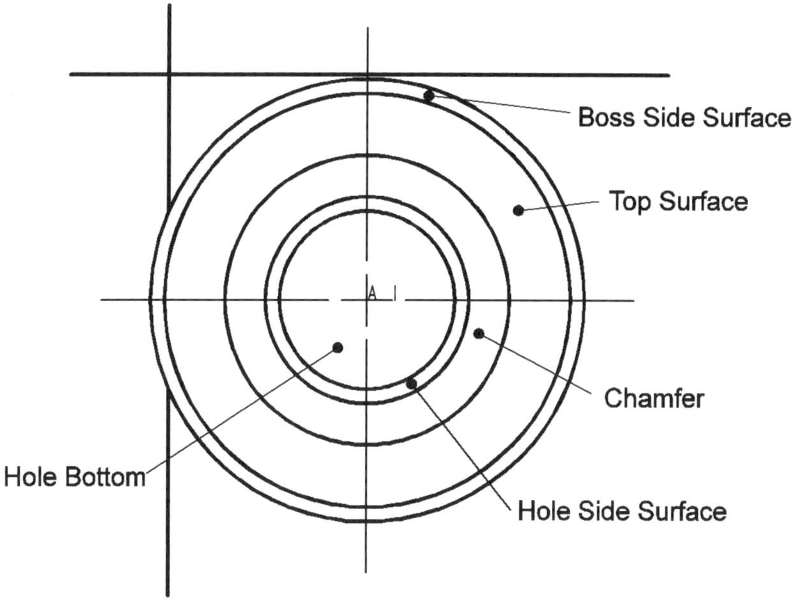

Figure 7-80
The view from the bottom of the car body.

If your bosses do not look like the previous figure, one or both of the surfaces are not drafted correctly. You will most likely need to modify the draft angle to `-1`, which would cause the direction of the draft to flip to the other side.

Drafting Ribs

Draft needs to be added to the ribs, too. Pick the same commands as before:

Create ➥ Solid ➥ Tweak ➥ Draft ➥ Neutral Pln | Done ➥ No Split | Constant | Done ➥ Include | Indiv Surfs

Pick all 12 vertical surfaces on the sides of the four ribs.

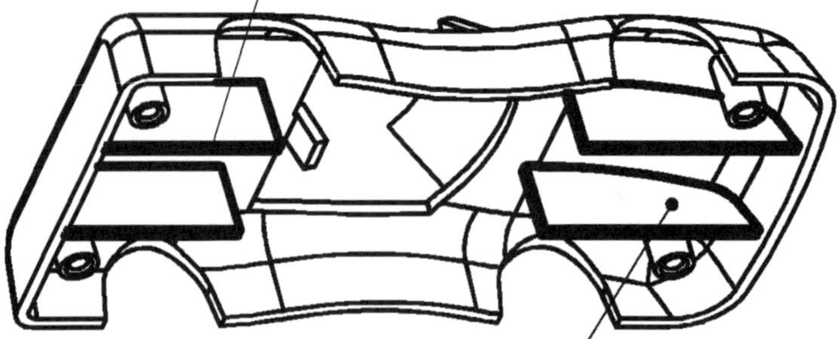

Figure 7-81
The neutral plane and the surfaces to be drafted.

Pick **Done Refs** from the SURF SELECT menu when you have finished. Pick the top surface of the first rear rib for the neutral plane. Then pick **Use Neut Pln** from the REF DIR menu. Then enter the draft angle.

```
Enter draft angle (yellow arrow shows positive rota-
tion dir'n) [0.5000]: -1 ↵
```

Then pick **OK** from the dialog window. The draft will be added to the sides of the ribs. Use the same "bottom view" method to check the draft feature. The ends of the ribs will look like trapezoids if the draft has been created properly.

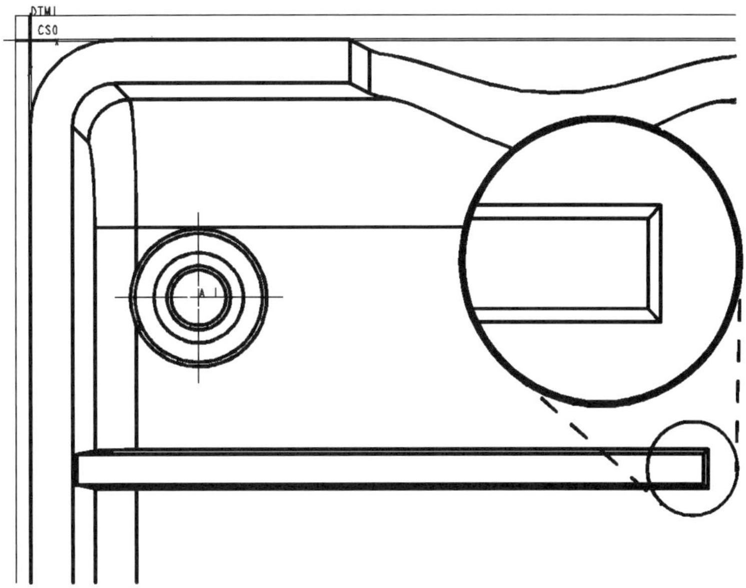

Figure 7-82
The bottom view of the car body showing the draft on a rib.

Adding the Detail Rounds

A few detail rounds need to be placed in the passenger compartment area. These rounds vary in size, but you can place them in one feature with the **Multi Const** round option. Pick the following commands, starting at the FEAT menu:

Create ➥ Solid ➥ Round ➥ Simple | Done ➥ Variable | Edge Chain | Done ➥ Tangent Chain

Pick the following seven edges:

- ❒ Four inside edges of the passenger compartment cut
- ❒ Two top edges of the windshield
- ❒ Top edge of the protrusion that fills the area between the seats

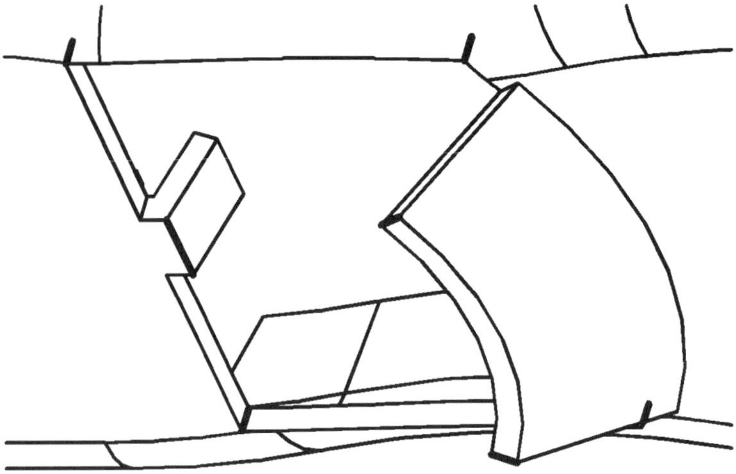

Figure 7-83
The edges to pick for this round feature.

The four inside edges of the passenger compartment cut should have a radius of `.1`. The top two edges of the windshield should be rounded with a `.2` radius. The seat gap filler should be rounded with a radius of `.1`.

Figure 7-84
The passenger compartment after rounding.

Placing an Offset Feature

Every car (toy or real) deserves a logo. To place the custom logo on the car body, you will use a tweak feature called **Offset**. Offset features are cuts or protrusions created by offsetting a surface. You can offset an entire surface or an area of a single or many surfaces. You can also control the offset direction to produce a recessed or protruded feature. For this feature, you will offset an area of the front surface of the car body and produce a recessed logo. Pick the following commands, starting at the FEAT menu:

Create ➡Solid ➡ Tweak ➡ Offset ➡Normal Off | Sket Region | SideNrmToSkt | Done

Pick the front surface of the car and **Done Sel** from the GET SELECT menu.

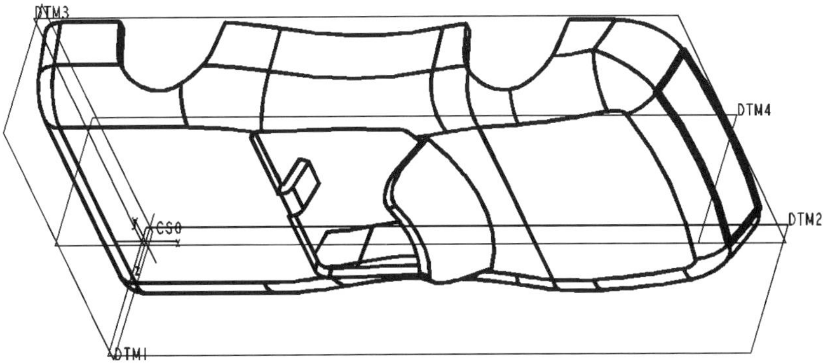

Figure 7-85
The surface that will contain the offset logo.

Use DTM1 as the sketch plane for the logo. Pick **Top** and DTM3 for the horizontal reference. Sketch this section.

➻ ***NOTE:*** *Feature dimensions are shown in parentheses.*

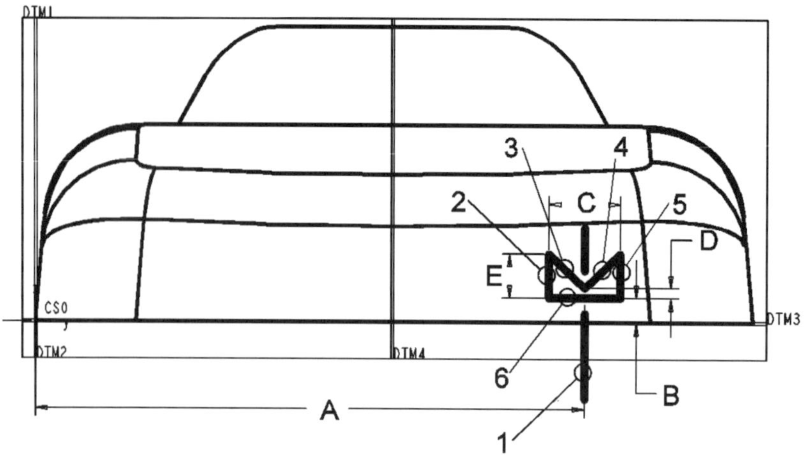

Figure 7-86
The section for the logo.

1. Vertical centerline
2. Vertical line
3. Angled line
4. Angled line
5. Vertical line
6. Horizontal line

A. Linear dimension from the centerline to DTM2 (2.3)
B. Linear dimension from the horizontal line to DTM3 (.1)
C. Linear line length dimension of the horizontal line (.3)
D. Linear dimension from the horizontal line to the lower endpoint of the angled line (.04)
E. Linear dimension from the horizontal line to the upper endpoint of the angled line (.18)

Regenerate ➡ Modify ➡ Regenerate ➡ Done | Okay

After you regenerate the section, change to the default view. Pick **Done** from the SKETCHER menu. Orient the arrow such that it points toward the inside of the logo and pick **Okay**.

The last prompt is a little confusing. It is asking for the offset dimension *and* the offset direction. If you provide a positive number, the section will be offset in the direction of the red arrow. In this example, you need to provide a negative offset value so that the offset area will be created in the negative direction (opposite of the arrow).

```
Offset in indicated direction (abs. value > 0.0007)
[0.0160]: -.03 ↵
```

The offset feature will be created on the front of the car body.

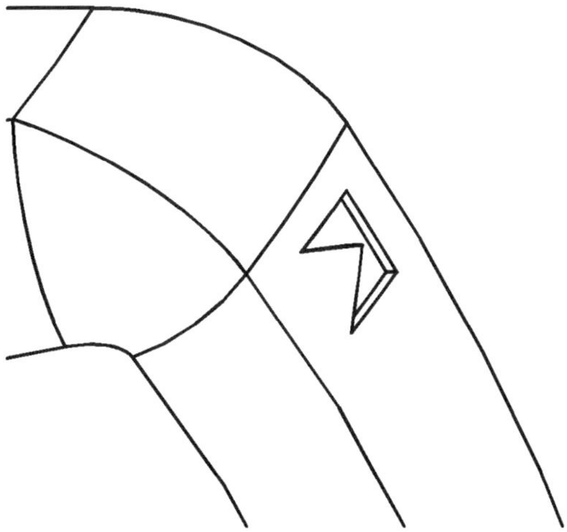

Figure 7-87
The logo on the front of the car body.

You have just completed the car body. Go celebrate!

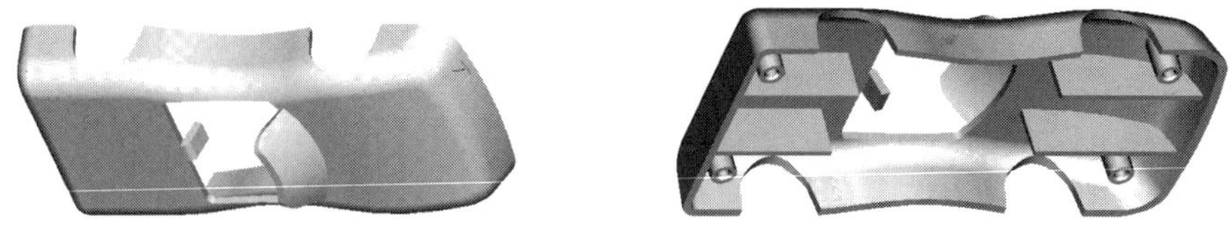

Figure 7-88
The finished car body.

SAVE

A Feature by Any Other Name...

You probably noticed that the three part examples (the wheel, chassis, and car body) did not use every feature command. There are two reasons for this.

1. Some commands are so rarely used that it is a waste of your time trying to learn them. Basic cuts and protrusions will cover 80 to 90 percent of your feature requirements. Learning these two commands and their options is a better use of your time. Consult the reference manual if you want to learn how to make a **Tweak ➥ Local Push**.

2. Many of the features are obsolete. These features were invented to solve immediate needs at a time when Pro/ENGINEER was in its infancy. As the cut and protrusion features became more adaptable, certain features were no longer required. Features that are almost never used because their functions are better served by the cut or protrusion command follow.

- **Hole** (sketched) — Revolved cut that uses a different sketching method
- **Shaft** — Revolved protrusion that uses the same setup as a sketched hole
- **Slot** — Same as a cut, but requires a closed section
- **Neck** — Revolved cut
- **Flange** — Revolved protrusion
- **Rib** (straight) — Extruded protrusion using the **Both Sides** and **Blind** options

Try not to get bogged down learning every feature. Few people have ever used the **Ear** and **ToroidalBend** features.

Conclusion

If you have followed along with the examples to create the three parts, you should be ready to tackle your own designs. Remember to pay attention to basic part design methodology as you build your parts.

STEP 1 — Place your datum planes and coordinate system.

STEP 2 — Place your first protrusion.

STEP 3 — Add the major features of the part.

STEP 4 — Add the finishing touches.

Next, pay attention to the geometry you reference when you create features. The features that contain the sketch planes, horizontal or vertical reference planes, dimensional references, and alignment references will *always* be parents of the new feature you are creating. Do not reference objects in another feature if that feature does not directly pertain to the one you are creating. Use your datum planes to set up a feature's section. Use your datum planes for dimensioning references. Align to your datum planes whenever possible. All of these techniques will help minimize unnecessary parent/child relationships and, ultimately, will make your parts easily modifiable.

Change to the default view (or another 3D view) when you are trying to align an object in the SKETCHER. If you do not, you will end up aligning to a detail feature on the other side of your part (such as a round). If you decide to delete or suppress the round, you will have to fix the references of the child feature in order to keep it. Honestly, it is easier and faster to do it right the first time. Do not take shortcuts.

Finally, try to remember the order of your features. Pay attention to the parent/child relationships as you create features. Write them down if you have to. This will help you immensely when you need to change a part. Use **Info ➥ Feature List** or **Parent Child** if you cannot remember how you built a part. The next chapter will show you ways of changing references.

Chapter 8
Changing Your Design

Understanding the Modification Process

Introduction

This chapter will teach you how to make changes to your designs. If you do not have a thorough understanding of how the design-changing commands function, you will not be able to alter your design efficiently. Many users lose work (and productivity) because they do not know how to deal with parent/child relationships in conjunction with part modifications. After reading this chapter and working through the examples, you should understand the following concepts:

- Finding the right command for your change

- Deleting features
- Suppressing and resuming features
- Using Insert mode
- Reordering features
- Modifying dimensions
- Redefining attributes, sections, schemes, and references
- Rerouting features
- Handling regeneration failures

The examples in this chapter use the wheel from Chapter 3 and the car body part created in Chapter 7. Start Part mode and retrieve the body part if it is not already the current object.

Which Command Should You Use?

There are many commands for changing a part in Pro/ENGINEER. The type of change you need to make will determine which command you use. Simple commands for deleting and rearranging features follow.

- **Delete** — To remove features from the model permanently.
- **Suppress** — To remove features from the model temporarily.
- **Resume** — To bring back suppressed features.
- **Reorder** — To rearrange the order of the features.

In addition, you can alter existing features with the following commands:

- ❒ **Modify** — To change the dimensions of a feature.
- ❒ **Redefine** — To change the feature's options, references, or section.
- ❒ **Reroute** — To change the feature's references. Proper use of these commands will let you change almost every aspect of your design.

Deleting Features

Features can be deleted from your part. Deleting a feature will completely remove it from the part.

> ✗ **WARNING:** *Deleted features cannot be undeleted. If you accidentally delete a feature, you can erase the part from memory and retrieve it again from the disk. Your part will be returned in its last saved state.*

Be sure to save the body part at this time because the changes you make in this chapter will not be shown or relied upon in the next chapter. Pick **Delete** from the FEAT menu. Pick the logo feature on the front of the car body. Pick **Done** from the SELECT FEAT menu. The logo will be deleted from the car body, and the part will regenerate automatically. To get the feature back, pick **Dbms** ➥ **Erase** ➥ **Confirm**. This will erase the current object (the car body) from memory. Begin Part mode again, and retrieve the car body. The logo feature will be restored because you saved the part before making this modification.

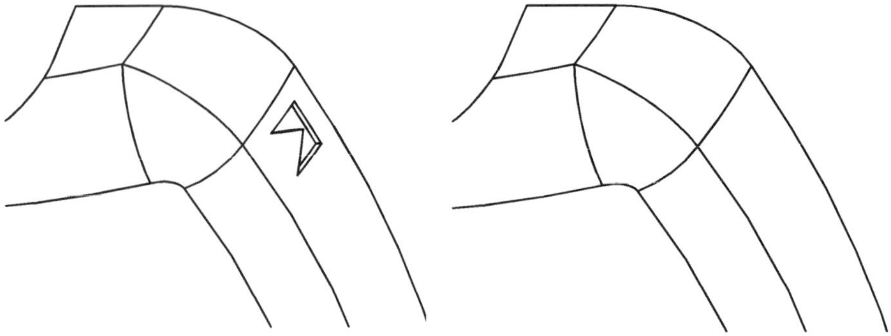

Figure 8-1
The car body before and after deleting the logo.

✓ **TIP**: *Before deleting features, you should save your part. This way, it can always be retrieved in its prior state if you accidentally delete a feature you wanted to keep.*

Managing Child Features of Deleted Features (Orphans)

Many times the features you delete will have children. When you try to delete the parent feature, Pro/ENGINEER will ask you what you want to do with the children. You have the following options:

- **Show Ref** — Show the references that make the highlighted feature a child.
- **Reroute** — Change the references of the child.
- **Mod Scheme** — Modify the dimensioning scheme of the highlighted child.
- **Delete** — Delete the highlighted child.
- **Delete All** — Delete all children of this feature.

- ❒ **Suspend** — Keep the child feature. Next time the feature regenerates, it will fail. You will need to fix its references at that time.
- ❒ **Suspend All** — Suspend all children features.
- ❒ **Info** — Get information on the highlighted feature.
- ❒ **Quit** — Quit the deletion process.

> ✓ ***TIP:*** *It is easier to deal with orphan features if you know why the feature is an orphan. If you have no clue as to why the feature is an orphan, make a note of its feature number by selecting* ***Info****, and quit the deletion process. Do not try to delete the feature until you learn which other features are children and why the others are children.*

It is up to you to pick the correct command for fixing an orphan feature. If you already know why the feature is a child, and you know that it needs to be redefined, you can select **Suspend** from the CHILD menu to keep the orphan feature. When the part regenerates, the orphan feature will fail, and you will be given the opportunity to redefine it.

Suppressing Features

Suppressed features are *temporarily* deleted. When you suppress a feature, it is removed from the part. These features can be returned to the part later when you want to bring them back. Some uses of **Suppress** are:

- ❒ **Speeding up Pro/ENGINEER** — You can suppress features to simplify a large part to speed up regeneration and display.

- **Separating stages of a part** — If you are building a part that begins as a casting, you could create all of the casting features first and then add the machining features. By using **Suppress**, you can toggle between the "as cast" and "finished" versions of the part.
- **Creating an FEA model** — Remove detail features for an FEA model.

In that **Suppress** actually removes the feature from the part, you will need to decide what to do with the children of the features you are suppressing. Pro/ENGINEER will present the CHILD menu when you try to suppress a feature that has children. This CHILD menu is the same as the first one you viewed except for two items. **Suppress** and **Suppress All** replace the choices **Delete** and **Delete All**. Pick **Suppress** from the FEAT menu. Pick the windshield. The passenger compartment rounds are children of the windshield. The rounds will highlight and the CHILD menu will appear. Pick **Suppress All** to suppress all child features of the windshield. Pick **Done** from the SELECT FEAT menu. The car body will regenerate without these features.

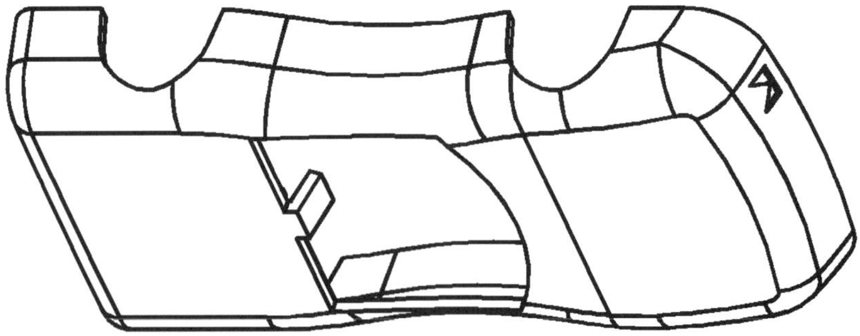

Figure 8-2
The car body with the windshield and associated features suppressed.

✓ ***TIP:*** *If there is a group of features you wish to repeatedly suppress, you should place those features on a layer and suppress the layer. Layer suppression is performed by selecting **Suppress** ➥ **Normal** ➥ **Layer** and then picking the layer name from the menu.*

Resuming Features

To bring back suppressed features, you use the **Resume** command. Selection options available when you resume features follow.

- ❐ **All** — Resumes all suppressed features.
- ❐ **Layer** — Resumes features that have been suppressed by layer.
- ❐ **Last Set** — Resumes the features suppressed during the last suppress operation.
- ❐ **Feat ID** — Resumes features when you provide their internal feature ID.

To resume the windshield and the rounds, Pick **Resume** from the FEAT menu. Pick **All** to bring back all suppressed features. The car body should not have any suppressed features now.

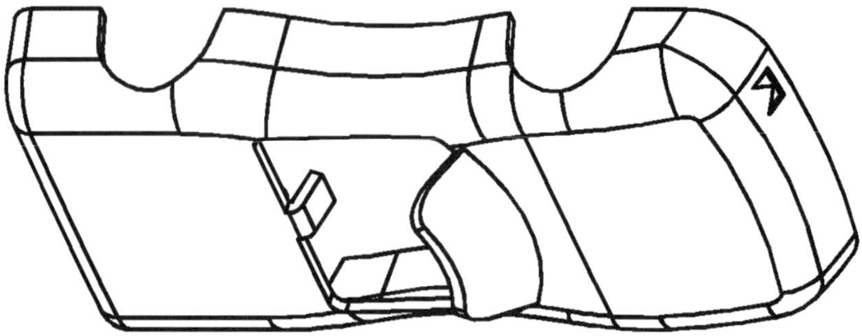

Figure 8-3
The car body with the windshield and associated features resumed.

Reordering Features

Sometimes you will need to modify your part by rearranging its features. With the **Reorder** command, you can shuffle the order of your part's features. You have the option to reorder a feature earlier or later in the part. When you are reordering a feature earlier, Pro/ENGINEER prompts you for the existing feature number to insert before. This feature number will become the feature number of the newly reordered feature. When you are reordering features later, Pro/ENGINEER prompts you for the existing feature number to insert after. This feature number will also become the new feature number of the reordered feature.

> **NOTE:** *Reordering features by feature number can be confusing because after the feature is successfully reordered, the feature numbers are reassigned. Try to keep in mind that features are neither created nor destroyed with the* ***Reorder*** *command. The total number of features will not change.*

Consider a simple part that contains four features:

Feature 1 – Protrusion

Feature 2 – Shell

Feature 3 – Cut

Feature 4 – Rib

Pretend you want to place the cut before the shell. To reorder the cut, you would pick **Reorder ➥ Earlier** and then select the cut feature. Next, you would key in 2 as the feature to insert before. The feature list should now look like the following:

Feature 1 – Protrusion

Feature 3 – Cut

Feature 2 – Shell

Feature 4 – Rib

Next, Pro/ENGINEER will reassign the feature numbers.

Feature 1 – Protrusion

Feature 2 – Cut

Feature 3 – Shell

Feature 4 – Rib

As you can see, the cut became feature 2, which corresponds to the number you entered.

➾ **NOTE:** *The same operation could have been performed by reordering the shell later (after feature 3).*

➾ **DESIGN NOTE:** *You cannot reorder a child feature before its parent. Likewise, you cannot reorder a parent feature after one of its children. Parent/child relationships determine the limits of your reordering capabilities.*

Using Insert Mode

The **Reorder** command is designed for situations in which the features you are reordering already exist. Your changes will often involve the creation of new features, which do not always belong at the end of your part (as your last features).

Pro/ENGINEER has a command called **Insert Mode** that will let you create features in the middle of your part's feature history. Using **Insert Mode** is like going back in time: it lets you create new features before existing features.

➾ **NOTE:** ***Insert Mode*** *is only the command name. It is not really a mode like Part, Drawing, or Assembly modes.*

You might think that using **Insert Mode** is like creating a feature and then reordering it earlier in the part. They are similar concepts, but **Insert Mode** is much better. When you create a feature, you reference other features. Adding a feature to the end of your part increases the risk that it will become a child of a feature you want to insert before. If this happens, you will not be able to reorder the feature as early as you wish. If you use **Insert Mode**, you decide where the feature belongs in the feature history, go to that point, and then create the feature without having to reorder it.

Practice using **Insert Mode**. For this example, you will place an air scoop on the hood of the car body. This protrusion needs to be placed before the shell because otherwise it would make the wall thickness of the part too thick in a local area. Pick **Insert Mode ➥ Activate** from the FEAT menu. Pick feature 10 (the hood cut) as the feature to insert after.

> ✓ **TIP:** *Pro/ENGINEER will only let you input the number of the feature to insert **after**. Sometimes you only know which feature you want to insert **before**. If this is the case, query select the feature you want to insert before, but* do not *accept it. Make a note of the feature number in the message window. Now pick **Sel By Menu ➥ Number**. Subtract 1 from the feature number of the feature you just queried and enter that value instead. Create the hood scoop based on the following information.*

The Hood Scoop

Commands:

Create ➥ Solid ➥ Protrusion ➥ Extrude | Solid | Done ➥ One Side | Done

➾ **NOTE:** *Feature dimensions are shown in parentheses.*

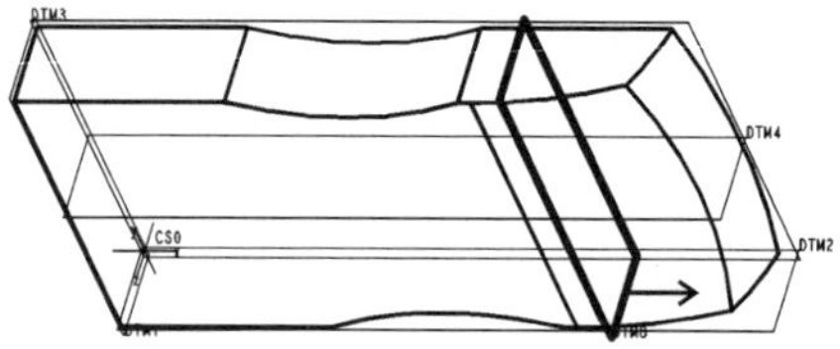

Figure 8-4

References:

Sketch plane – Make a datum plane offset from DTM1 by 4.5 in the positive x direction.
Horiz/Vert reference – **Top**, DTM3.

1. Vertical centerline
2. Three-point arc

A. Align centerline to DTM4
B. Align the endpoint of arc to the front edge of the hood
C. Align the endpoint of arc to the front edge of the hood
D. Linear dimension from the tangent of the arc to DTM3 (.9)
E. Radial dimension of the three-point arc (.5)

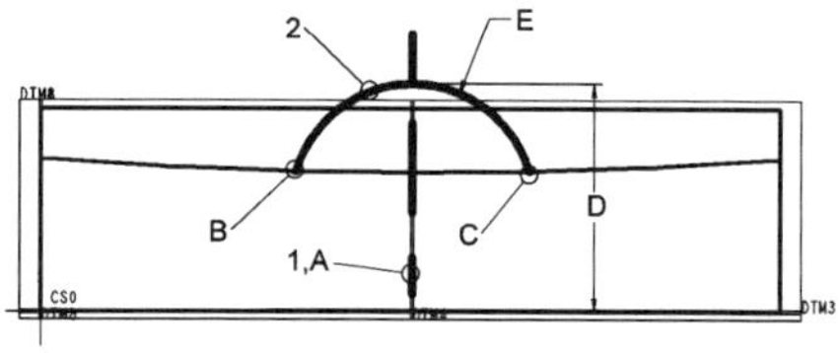

Figure 8-5

Pick **Regenerate | Modify | Regenerate | Done | Blind | Done**

Extrusion depth is .8.
Pick **OK** from the dialog window.

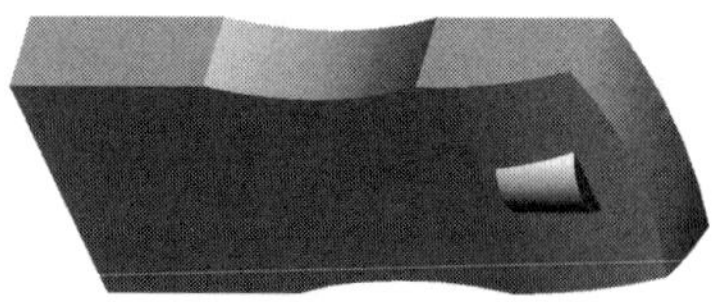

Figure 8-6

Parent features:

DTM1 – Offset sketch plane reference.
DTM3 – Horizontal reference and dimensioning reference.
DTM4 – Alignment reference.
Features 9 and 10 – Alignment reference (aligned to the edge created by these features).

After creating the hood scoop, pick **Insert Mode** ➥ **Cancel**. Pro/ENGINEER will ask you if you want to resume all of the features suppressed when you started **Insert Mode**. Press the carriage return to take the default choice of Y. All features will be restored.

✓ ***TIP 1:*** *If you do not take the default for the previous prompt and enter N instead, Pro/ENGINEER will behave as if you suppressed all of the features you do not see. At this point, you can select* ***Insert Mode*** ➥ ***Activate*** *to insert to an even earlier feature. This method saves the time of regenerating all of the suppressed features when you need to insert at an earlier point in the feature history.*

➾ ***NOTE:*** *If you are adding features with* ***Insert Mode*** *activated, the* ***Resume*** ➥ ***All*** *and* ***Resume*** ➥ ***Last Set*** *commands will cancel* ***Insert Mode****.*

➾ ***DESIGN NOTE:*** *Notice that the shell feature removed the material under the hood scoop. This is because the shell occurs after the hood scoop protrusion.*

✓ ***TIP 2:*** *You should always use* ***Insert Mode*** *when you want to place a new feature in the middle of your part's history. Do not rely on the ability to reorder it because you might not be able to reorder it to the desired point in time.*

Changing Features

Three different commands change features. The type of change you need to make will determine which command you should use. Refer to the following flowchart when you need to change a feature and are unsure of the proper command for the job.

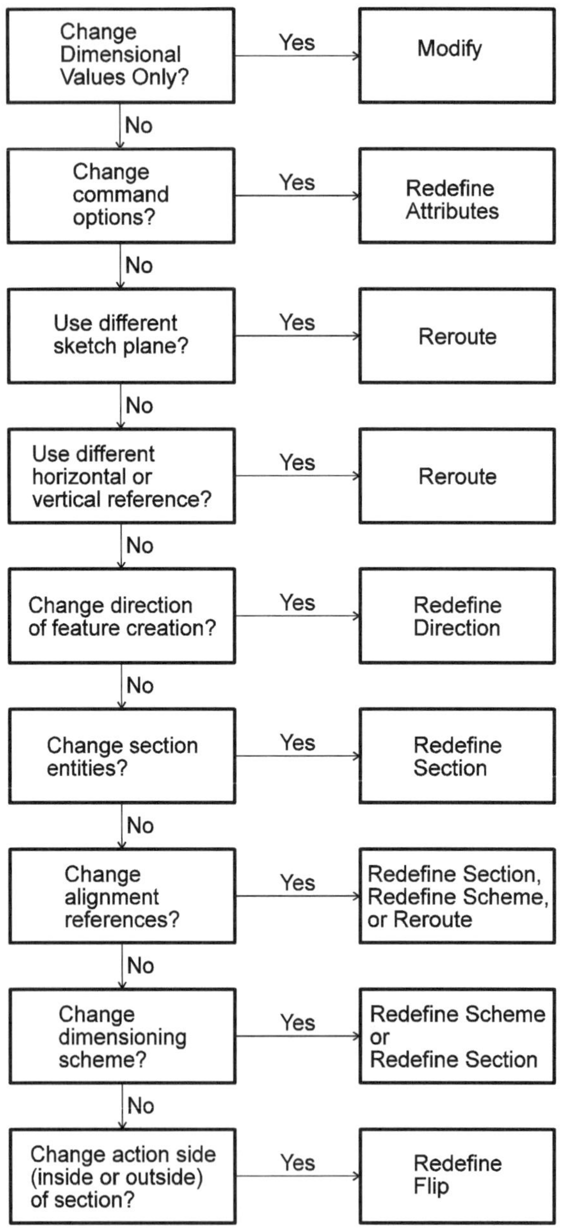

Figure 8-7
Feature modification flowchart.

Modifying Dimensions

The easiest type of change in Pro/ENGINEER is to modify a dimension. You are already familiar with how it works in the Sketcher. In Part mode, you modify features the same way. This example will use the car body part. Pick **Modify** from the PART menu, and pick the first protrusion of the car body (feature 5). Its dimensions will appear.

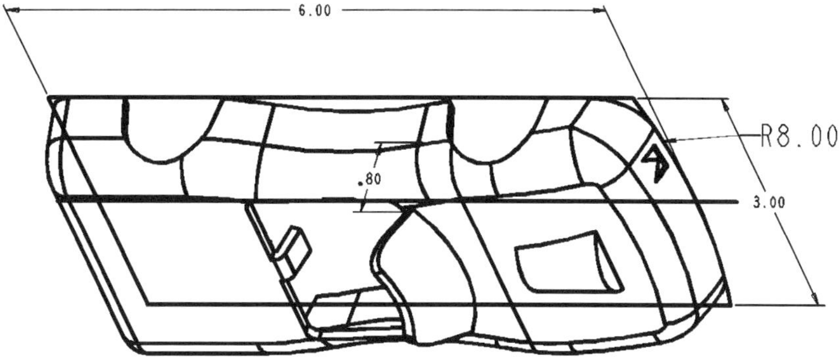

Figure 8-8
Dimensions for the car body's first protrusion.

Pick on the radial dimension with the value of 8. The dimension will highlight in red, and Pro/ENGINEER will prompt you for the new value.

```
Enter value (negative not allowed) [8.00]: 3 ↵
```

The dimension will change to 3, and will be shown in white. Unregenerated dimensions are always displayed in white. To make the part update, pick **Regenerate** from the PART menu. Now pick **Modify** and pick the same feature again so that its dimensions will display. It should now look like the following figure.

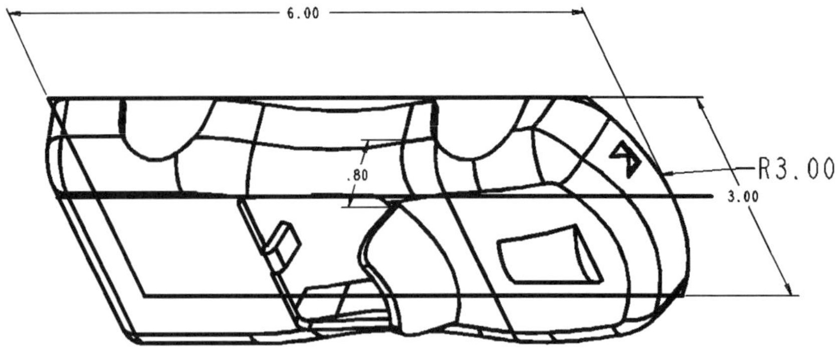

Figure 8-9
The modified car body.

You can simultaneously modify as many dimensions as you wish. You do not have to regenerate each one individually.

> ✓ **TIP:** *If you modify 20 dimensions and the part fails during regeneration, you will have a difficult time trying to determine which dimensional change is the culprit. Try to limit yourself to four or five dimensional changes. This will reduce the time it takes to identify the change that caused the failure, if there is one.*

Redefining Features

You can change almost every characteristic of a feature using the **Redefine** command. When you pick **Redefine** and pick the feature you want to change, a dialog window appears. By picking on the selection offered in the dialog window, you will see the pop-up menu dissplayed for that command requirement. For example, select a feature to redefine, pick an option from the dialog window, and define and you will see the pop-up menu appear for that option. Command options follow:

- **Attributes** – For changing command options (such as **One Side** and **Thru All**).

- **Direction** – For changing the extrude or revolve direction (from the sketch plane) of the feature.
- **Trajectory** – For changing the trajectory of a feature section, such as SWEEPS.
- **Material Side** – For changing the side to which material will be added, such as in thin feature construction.
- **Section** – For altering anything about a feature's sketched section.
- **Flip** – For changing the direction of the section's flip arrow (typically used when you accidentally point the arrow the wrong way).
- **References** – For picking different reference geometry (edges to round, surfaces to remove during a shell, and so on).
- **Boundaries** – For changing the curves of a surface created by boundaries.
- **Scheme** – For altering the dimensioning scheme of a section.
- **Curves** – For changing a curve created from a file.
- **Line Style** – For changing the line style of cosmetic features.
- **Corner round** – For changing the dimensioning scheme of a corner round.
- **Placement**– For changing the constraints of an assembly component.
- **Pattern**– For switching to a different type of pattern or changing the increment option.

Obviously, all of these options do not pertain to every feature. Pro/ ENGINEER will gray out the options that do not pertain to the feature you are trying to change. You are allowed to select as many of these options as you wish when you redefine a feature. Pro/ENGINEER will work through all of them before implementing your change.

There are limits to what you can redefine. For example, you cannot redefine a cut and make it a protrusion. Neither can you redefine a solid feature to be a thin feature or vice versa.

> ➾ **NOTE:** *If you redefine a feature that has been copied or mirrored with the* ***Independent*** *option, the copies will* not *be redefined. You can either redefine each copy or delete them and recopy the features. However, if you used the* ***Dependent*** *option, the copied features will also be redefined.*

The next few examples will use the wheel part from Chapter 3. Erase the body from memory by selecting **Dbms** ➡ **Erase** ➡ **Confirm**. Start a Part mode session and retrieve the wheel.

Changing Feature Attributes

Attributes are usually the menu items you pick at the time of feature creation before you start picking any geometry. For revolved features, the attribute is the rotation angle. For extruded features, the attributes are the **One Side** or **Both Side** choice, as well as the extrude options **Blind**, **Thru All**, and so on. The attribute of an edge round feature is the **Constant**, **Multi Const**, or **Variable** option. The attributes of a regular draft feature are **Constant** or **Variable**, **Neutral** or **Ref & Neut**, and **Unmirrored** or **Mirrored**.

Attributes are easy to change. For this example, you will change the snap relief (feature 10) to be cut toward one side of the sketch plane. Pick **Redefine** from the FEAT menu. Pick the snap relief feature. Now pick **Attributes | Depth | Done** from the REDEFINE menu. **Pick One Side** ➡ **| Done | Thru All | Done**.

Because you are changing the cut to be toward only one side of the sketch plane, you need to provide the direction of feature creation by picking **Change** from the FEATURE EDIT menu, then **Direction** from the SEL ELEMENT menu. This will allow you to orient the arrow in the positive y direction and accept it. Then pick **Done** and the part will be regenerated.

Figure 8-10
The wheel before and after the attribute change.

✓ **TIP:** *Pro/ENGINEER will regenerate the part automatically after you redefine a feature. If you need to redefine multiple features that are placed early in your part's history, you will waste time repeatedly regenerating features that might not be affected by the change. In this case, you should use* ***Insert Mode*** *to "go back in time" and make all of your modifications. When you are finished changing the part, cancel* ***Insert Mode*** *and let Pro/ENGINEER regenerate the remaining features.*

Changing Feature Sections

If you need to change a feature's section, the **Section** option will open a Sketcher session and retrieve the section. You can change anything about a feature's section when you redefine it. You can add or delete entities, create or remove alignments, and add or delete dimensions. To change the sketched entities in a section, you use the same sketching commands you are already familiar with. The only things you need to watch out for are entities that have been used as references in features that follow the one you are redefining. If you delete a sketched entity that has been used as a reference in another feature, you might delete that feature. Pro/ENGINEER will provide a warning if this is about to happen.

```
WARNING: This feature is referenced by another fea-
ture or features. Continue? [N]: Y ↵
```

Answering `Y` to this prompt will guarantee that a feature will fail later in the regeneration process. To avoid this hassle, you should try to avoid deleting these entities.

Pro/ENGINEER has a tool called **Replace** that you can use to replace referenced entities in a section. To see the effects of this command, add a .020-inch edge round feature to the external edge of the wheel hub. It should look like the following figure.

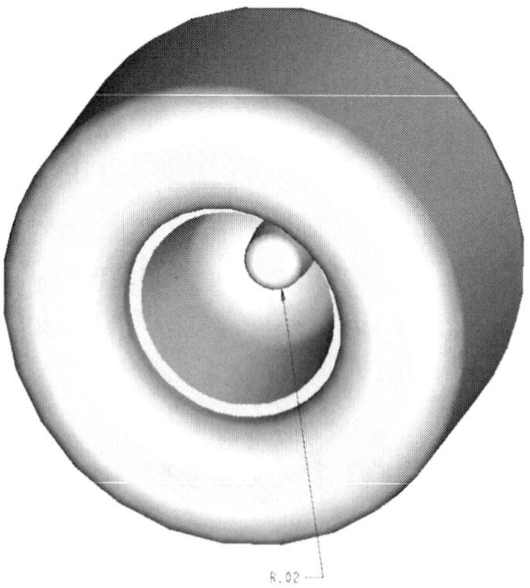

Figure 8-11
The new round on the wheel hub.

Now pick **Redefine** and pick the hub protrusion (feature 8). Pick **Section | Done** from the SEL ELEMENT menu and pick **Sketch** from the SECTION menu. After the section is displayed, Pick **Sketch** to place you in the sketch view. Sketch the two lines shown in the next figure.

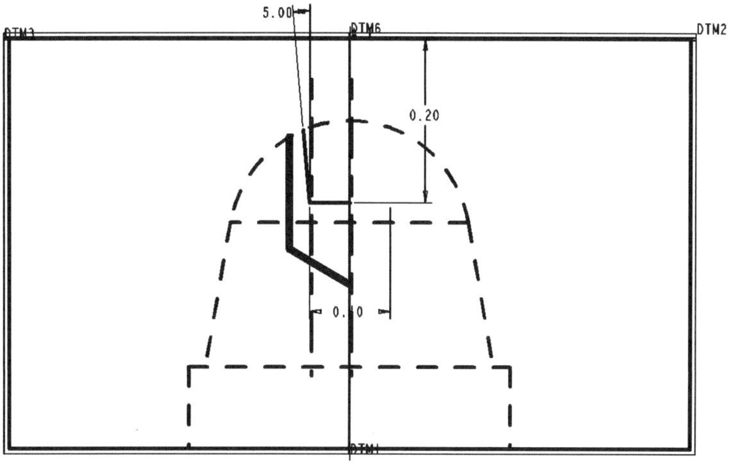

Figure 8-12
New lines in the old hub section.

You will replace both of the old sketched lines with the two new ones. Pick **Geom Tools ➥ Replace** from the SKETCHER menu. Pick the vertical line you just sketched. Now pick the angled line in the section. Pro/ENGINEER will warn you because the entity you are replacing has been aligned and dimensioned.

```
Cannot replace aligned or dimensioned entity. Delete
dimensions? [N] Y ↵
```

Answering `Y` to this prompt will delete the alignments, dimensions, and the angled line. Pick **Replace** from the GEOM TOOLS menu again. Pick the angled line you just sketched, and pick the horizontal line from the original section as the entity to be replaced. Answer `Y` to its confirmation prompt. Delete the extra centerline that passed through the left endpoint of the old horizontal line. Your section should now look like the next figure.

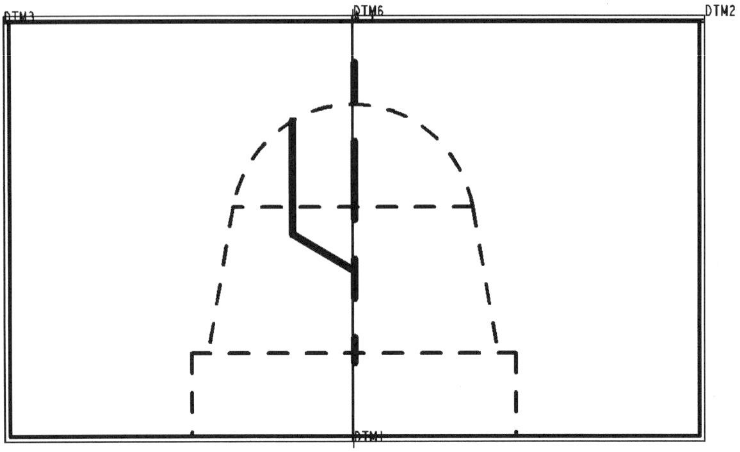

Figure 8-13
The new hub section entities.

Constrain the section.

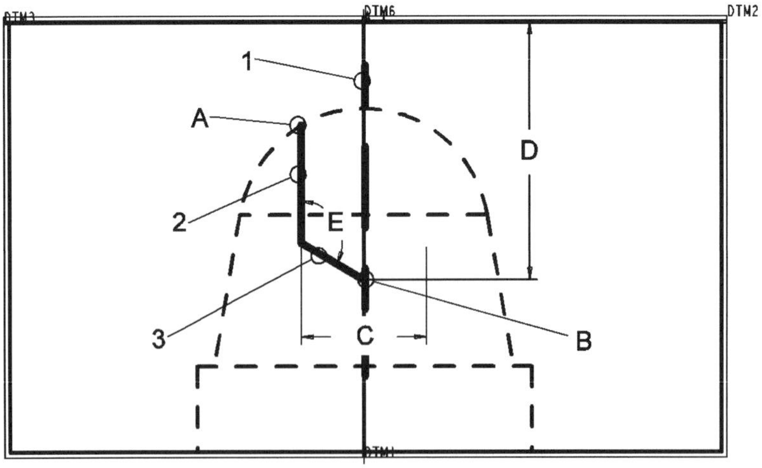

Figure 8-14
The new hub section constraints.

➺ ***NOTE:*** *Feature dimensions are shown in parentheses.*

1. Vertical centerline (a leftover from the original section and still aligned)
2. Vertical line
3. Angled line

A. Align the upper endpoint of the vertical line to the arc edge of the revolved wheel cut

B. Align the right endpoint of the angled line to the axis

C. Diameter dimension between the vertical line and the centerline (.15)

D. Linear dimension from the right endpoint of the angled line to DTM3 (.3)

E. Angular dimension between the vertical line and the angled line (120)

Regenerate the section and pick **Done** from the SKETCHER menu, then pick **Done** from the FEATURE EDIT menu. The part will be regenerated with the new shape.

Figure 8-15
The new wheel hub.

Notice that the round regenerated properly. Although the entities it was originally rounding are no longer present, it is still located in the right place. This is the power of the **Replace** command.

Changing Feature Dimensioning Schemes

There is a way to change a feature's section dimensions without affecting the sketched geometry. Although you can do this by redefining the section, you run the risk of spoiling something else. If your only goal is to change the reference edges for the dimensions in a section, and not the size or type of geometry, redefine the feature's scheme.

Pick **Redefine** from the FEAT menu, and pick the hub protrusion again (feature 8). Pick **Section | Define** from the dialog window, and then pick **Scheme | Done** from the SECTION menu when the hub section appears. Pick **Sketch** to orient the view to the section. Delete the angular dimension and add the linear dimension shown in the next figure.

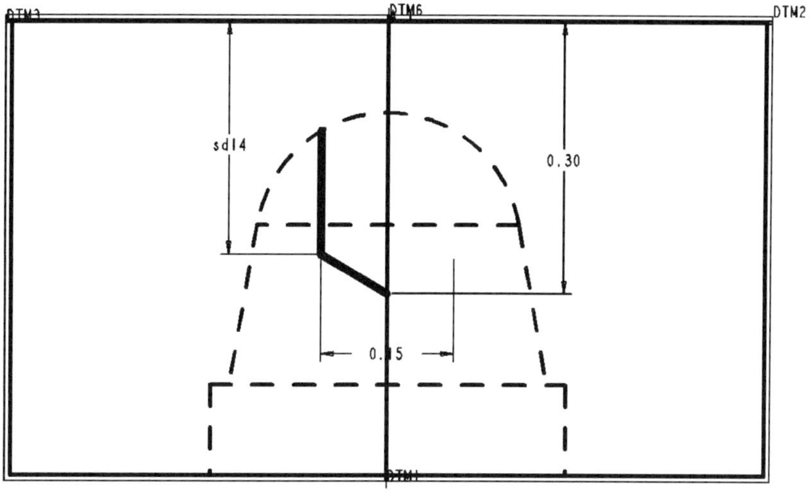

Figure 8-16
The new hub dimension.

Regenerate the section. Pick **Done** from the SKETCHER menu, then **Done** from the FEATURE EDIT menu and let the part regenerate. Now the hub is controlled by a different dimensioning scheme, but its size is exactly the same as before.

> ✓ ***TIP:*** *This technique is commonly used when you make a drawing of a part and realize that some of the features are not properly dimensioned.*

> ***NOTE:*** *You can also change alignments that were created in the section by redefining the scheme.*

Changing Feature References

The **Reference** option on the REDEFINE menu is a little confusing to new users. When you redefine a reference, you are not changing the feature's sketch plane or horizontal or vertical reference. The following are a few items you can change with the **Reference** option.

- **Extruded references** – You can pick alternate geometry if the feature was extruded with the **Thru Until**, **UpTo Pnt/Vtx**, **UpTo Curve**, and **UpTo Surface** options.

- **Shell surfaces** – You can pick alternate surfaces to remove in the shell feature, and specify surfaces at different thicknesses.

- **Rounded edges** – You can add or remove edges of an edge round feature.

- **Drafted surfaces** – You can add or remove surfaces drafted in a draft feature.

Change the references of the large rounds on the wheel. The goal is to remove the inner round.

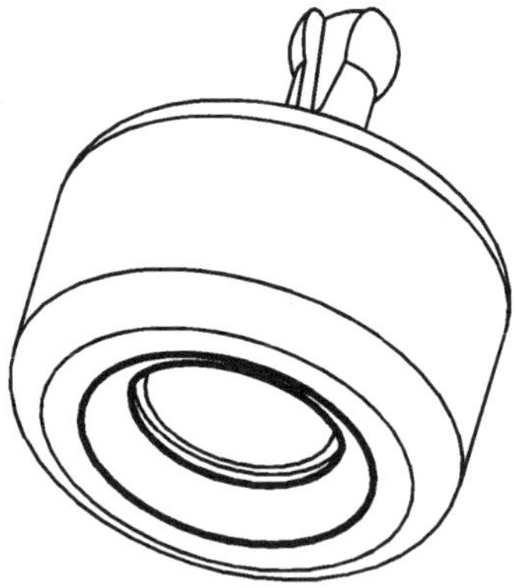

Figure 8-17
The round to remove from the round feature.

Pick **Redefine** from the FEAT menu. Pick the large rounds on the wheel (feature 12). Pick **References** | **Done** from the REDEFINE menu. The ADD RMV EDGE menu will appear with options for adding and removing edges from the round feature. Pick **Remove**. Pick the two cyan edges that create the large inside rounds on the wheel.

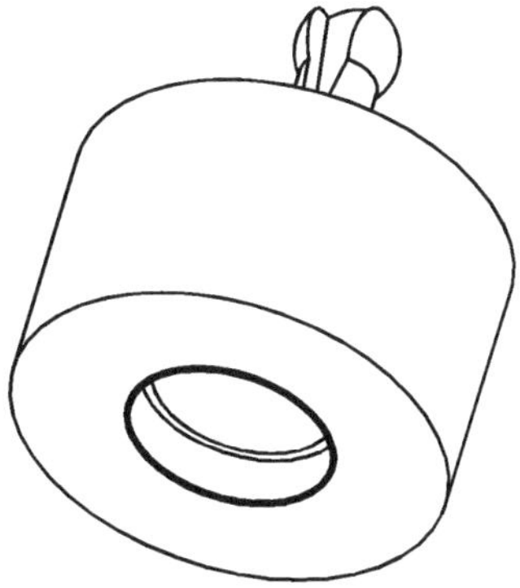

Figure 8-18
The edges to remove from the round feature.

Pick **Done** from the ADD RMV EDGE menu after selecting these edges. The wheel will regenerate and the round feature will update as expected.

> **NOTE:** *Both 180-degree edges must be removed from the set in order to remove the round.*

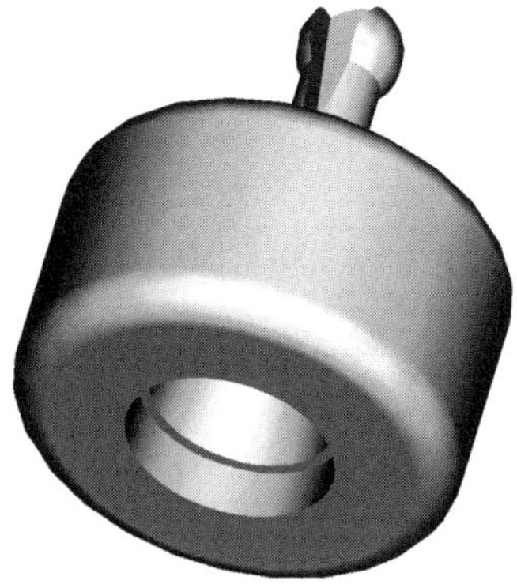

Figure 8-19
The wheel after removing the large inside rounds.

Redefining feature references is quick and easy. It will save you time if you redefine a round's references and avoid deleting the round.

Rerouting Features

Rerouting a feature is the only way to change its sketch plane or horizontal or vertical reference. The **Reroute** command takes you through the feature's references one at a time (like the **Copy ➡ New Refs** command), allowing you to keep the old reference, pick a new reference, or even create a new reference on the fly.

> ➾ ***NOTE:*** *When you pick an alternate sketch plane or horizontal or vertical reference, your selection must be parallel to the old reference.*

Reroute will also let you select different alignment or dimensioning references that were used in a feature's section.

> ✓ ***TIP:*** *If you need to change alignment or dimensioning references, you should use **Redefine ➡ Section**. You will have more options and more flexibility by redefining instead of rerouting.*

Retrieve the car body part for this example. The goal will be to move the thin protrusion between the seats (feature 33) away from the centerline of the body. Because DTM4 is the feature's sketch plane, you must reroute the feature and find a new sketch plane for it.

After retrieving the car body, pick **Reroute** from the FEAT menu. Pick the protrusion that fills the gap between the seats (feature 33). Pro/ENGINEER will prompt you to see if you want to "go back in time" to the point the feature was created.

```
Do you want to roll back the part? [N]: Y ↵
```

> ➾ ***NOTE:*** Always *enter* `Y` *to this prompt. This will prevent you from trying to reference a feature created after the one you are rerouting.*

Pro/ENGINEER will begin prompting you to select different references for the feature. The first prompt will ask you to select a different sketch plane. This is the item you want to change, but you do not have another datum plane to reroute to. In this case, you should select **Make Datum** to make one on the fly. Pick **Make Datum**. Make a datum plane offset from DTM2 by 1 inch in the positive y direction. After the datum plane is created, Pro/ENGINEER automatically accepts it as the sketch plane and continues prompting for the remaining references. Pick **Same Ref** for all remaining references. The part will regenerate with the changes.

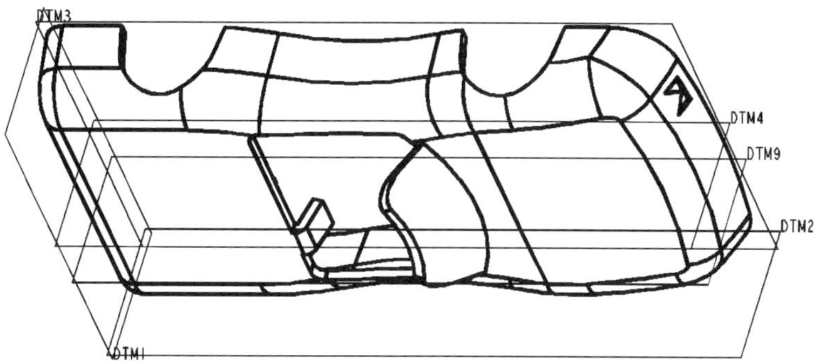

Figure 8-20
The car body after rerouting the thin protrusion.

> **NOTE:** *Datum planes created on the fly during rerouting do not behave the same way as datum planes created on the fly when you create new features. Datum planes created during rerouting are stand-alone features. They have their own feature numbers, and they are not hidden from view afterward. They are inserted prior to the feature you are rerouting. In that they have their own feature numbers, the datum plane is now feature 33 and the protrusion you just rerouted is feature 34.*

Regeneration Failures

Sometimes when you change your part, the regeneration process will fail. Failed features are handled in two ways: through the dialog window, using the Resolve option, or through the FAILED FEA-

TURE menu. Pro/ENGINEER gives you the following choices when this happens.

- **Fix Model** – Picking this option allows you to fix either the current model or a backup model. This is useful if you do not wish to take a chance on creating errors in your existing model. This option offers several other options that are standard menu picks within other Pro menus.
- **Trim Part** – Picking this option will initialize the dreaded TRIM MODEL menu.
- **Undo Changes** – Picking this option will restore the part to its previously regenerated state, if possible.
- **Investigate** – Picking this option will allow you to investigate the reason the regeneration is failing without already committing to fix the problem by selecting **Trim Part**. You still have the option to select either **Trim Part** or **Sys Recover** after selecting **Investigate**.
- **Quick Fix** – Picking this option will activate several choices for the failed feature. **Redefine**, **Recreate**, **Suppress**, and **Clip Supp** are all normal to their application under the FEAT menu.

> **NOTE:** *Once you pick* ***Trim Part****, you cannot go back to* ***Sys Recover*** *or* ***Investigate****.*

> ✓ **TIP:** *A novice user's best defense against regeneration failure is the* ***Save*** *command. You can always quit Pro/ENGINEER, restart Pro/ENGINEER, and retrieve your part from the disk in its last saved state. Just remember to save your part prior to any major modifications.*

These commands are designed to let you query your part, evaluate its current state, and fix the features causing the failure.

Using the Investigate Option

The **Investigate** option should be your first choice if you do not know why the regeneration process failed. This option and its commands (which follow) let you gather information about your part. You can also select **Current Model** or **Backup Model** for this option.

- **Diagnostics** – Examine geometry errors in the part.
- **List Changes** – Get feature information for the failed feature.
- **Backup Model** – Open another window and show the part in its last saved state.
- **Show Ref** – Shows (graphically) all references of the failed feature.
- **Roll Model** – Redisplays the model at a specified feature, or before the failed feature.

Using the System Recover Option

If you choose not to solve the problem, you can go back to the previously regenerated part (if the only thing you did was modify dimensions). When you pick **Fix Model**, you can restore any or all modified dimensions by picking **Restore**, which will activate the RESTORE menu. Options follow.

- **Dimensions** – Pick on an individual dimension to restore.
- **Parameters** – Restore a modified parameter value.
- **All Changes** – Restore all modified feature entities.
- **Relations** – Restore the relations.

You should use **Fix Model** when you caused the regeneration failure by changing many dimensional values and you do not know which dimension was the problem.

Any dimensional value that existed when the part was last saved can be displayed and modified. If a reference is missing and it is not clear what the reference had been, the references used when the part was last saved can be displayed. This information can be used to reroute the missing reference of the failed feature.

- **Show Ref** – Show the references of the failed feature.
- **Measure** – Measure the part's geometry.
- **Backup Model** – Redraw the model at its current failed state.

These commands can affect any feature, not just the failed feature. Select **Roll Model** in order to specify a failed feature. If you still have absolutely no idea why the feature failed, use the information tools such as **Diagnostics**, **List Changes**, and **Show Ref** to determine why the feature is failing.

Example Regeneration Failures

Try not to panic when the regeneration process fails, because you can usually fix the problem. The most difficult action is determining why the regeneration failed. Common causes of regeneration failures follow.

- **Unattached features** – A dimensional modification caused a feature to not intersect the part. You can usually redefine or modify the failed feature to correct this problem.
- **Lost references** – A reference of a feature no longer exists. Redefine or reroute the failed feature, if possible, to correct this failure.

- **Exceeded feature capabilities** – Features such as **Draft, Round,** and **Shell** have limitations. If you make a change that alters the geometry such that the feature will no longer work, it will fail. You will most likely have to delete the feature, and recreate it a different way.

If you currently have the wheel part in memory, erase it now. Retrieve the wheel part from the disk. The wheel will be used for all of these examples.

EXAMPLE 1:

Pick **Modify** and pick the outside tire round feature (feature 12). Change the value from .08 to .11 and pick **Regenerate**. The round feature should fail. What do you do now?

Pick **Investigate** ➡ **Current Model**. The wheel will be shown without the exterior rounds on the tire. Based on the shape of the wheel and tire, there are limitations on the size of the round. In this case, the round cannot be larger than the depth of the first wheel cut (feature 6). The depth of this extruded cut is .10. Obviously, the round cannot be .11.

In this case, you should pick **Undo Changes** to go back to the previous, fully regenerated state. This will allow you to examine your part freely and determine the next best course of action.

EXAMPLE 2:

Sometimes regenerations fail because you accidentally key in wrong values. To simulate this, pick **Modify** and pick the first protrusion (feature 4). Change the extrusion depth from .50 to .10 and regenerate the part. As always, pick **Investigate** ➡ **Current Model** to see what your part currently looks like. It should look more like a washer than a wheel.

This time, you will change the dimension back manually. Pick **Fix Model** from the RESOLVE FEAT menu. Pick **Restore** from the FIX

MODEL menu. Pick **By Feature** from the RESTORE DIM menu. Pick **Failed Feat | Done** from the SELECT FEAT menu. Pick **Regenerate** from the FIX MODEL menu. This will change the dimension back to its prior value. The wheel should be back to normal now.

EXAMPLE 3:

Try to make the wheel wider this time. Pick **Modify** and pick the first protrusion (feature 4) again. Change the value for the depth from .50 to .80 and regenerate. Pick **Investigate ➡ Current Model** to see what your part currently looks like. The wheel should look considerably wider, but that is what we wanted. What happened? Why did the revolved cut for the wheel fail?

This time you need to fix the problem. Pick **Quick Fix**.

You know you did not remove any references because all you did was modify a dimension. In that the failed feature is an "unattached" cut, you should first look at the cut's section. Pick **Redefine** from the QUICK FIX menu. Pick **Section | Define** from the dialog window, and then pick **Sketch** to get into the sketch view orientation. Next, select **Regenerate**. Now you can see what the problem is.

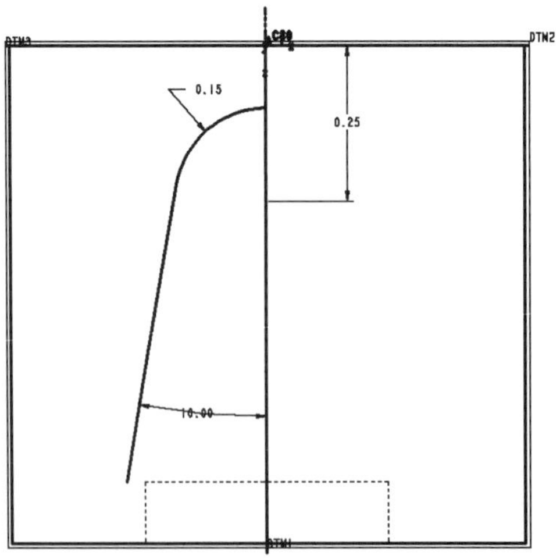

Figure 8-21

The increased width of the wheel has made this revolved cut "lose contact" with the inner surface of the wheel's extruded cut. It has become *unattached*. You could change the design of this revolved cut to "reattach" it. But what if such a change is not an option? Another way to attach the revolved cut is to make the extruded cut deeper. If the extruded cut were made .3 deeper, it would compensate for the .3 increase in wheel width. How can you change the depth of the cut now? You have already started redefining the *next* feature! Oops.

This is not a big deal. The first step is to quit the redefining process. Pick **Quit** from the SKETCHER menu. Pick **Confirm** from the CONFIRMATION menu, and you will be taken back to the FIX MODEL menu. At this point, the revolved cut is still failing. Pick **Current Model**. Pick **Modify Dims** and pick the extruded cut (feature 6). Change the depth from .1 to .4 and pick **Regenerate**. Pick **Done/ Return** and the part will regenerate properly.

> **NOTE:** *This example showed how to modify a feature that is prior to the failed feature.*

You can also redefine and reroute features prior to the failed feature. You would pick **Redefine** or **Reroute** for the feature you want to change.

> ✓ **TIP:** *If you do not know the feature number of the feature you want to change, pick the feature from the model tree unti the feature you want to change highlights. Its feature number will appear in the message window.*

Go ahead and attempt a few more failures; after all, the example parts in this book are intended to help you learn how to use Pro/ ENGINEER. That learning process includes the handling of regeneration failures. Erase all the parts from memory before continuing. The alterations you have made to the parts in this chapter will not be used in the rest of the book. You should use the parts as they were completed at the end of their respective chapters.

Conclusion

One of the biggest challenges for new users is modifying parts. Most people can start Pro/ENGINEER and create a model. The modification process often causes designs to fall apart. Design changes are inevitable. If you have a design that needs changing, there is a simple, yet effective, strategy you should follow when making the modifications: *Change the part, and do not patch it.*

If you have a boss that must be deleted, do not cut it off with another feature. Delete the boss's protrusion. If your part increases in width, edit the width dimension rather than adding another protrusion onto the end of your part. These quick solutions will cause you more problems later on. There are plenty of part modification tools within Pro/ENGINEER. If you learn how to use them effectively, you can make any change you need. A common thread among the modification commands is the effect of parent/child relationships. When you carelessly reference geometry, you affect the ability of the part to be modified. Pro/ENGINEER has many tools for changing a part, but there is no tool that can take the place of proper referencing techniques. Reread the part design chapters if you still feel uncertain about feature referencing strategies.

Chapter 9
Bringing It All Together

Assembling the Toy Car

Introduction

This chapter introduces the basic concepts behind assembling parts. We will show you how to assemble the components using several assembly constraints. Commands and concepts covered in this chapter include the following:

- ❒ Deciding which component to use as the base part
- ❒ Adding components using assembly constraints
- ❒ Making an exploded assembly

- ❐ Modifying assembly constraints
- ❐ Modifying part dimensions in Assembly mode

You will assemble the toy car. The car's appearance after assembly is shown in the next figure.

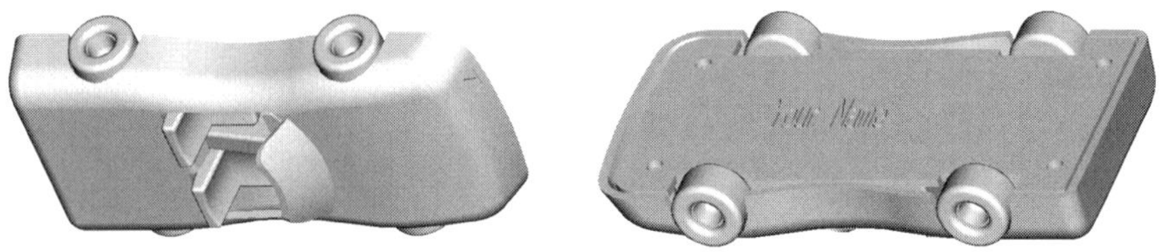

Figure 9-1
The assembled toy car.

Beginning an Assembly

When you want to create an assembly of components, you need to use Assembly mode. Select **Assembly** from the MODE menu. This will place you in Assembly mode and display the entry menu for Assembly mode. Since you are creating a new assembly, pick **Create** from that menu. A prompt for the new assembly's file name will appear in the message window.

`Enter assembly name [ASM0001]:` **`car`** ↵

This will create an assembly named "car" in memory. When you save this assembly, Pro/ENGINEER will create a file called *car.asm*.

Choosing the First Component

To start a new assembly, you need to determine which component should be the base component. The base component should be the one that most of the other components will attach to when they are assembled. It is usually the largest component. Because this is a new assembly without any existing components, Pro/ENGINEER only lets you assemble a component.

There are no constraints to specify for this first component because there are no parent components or assembly features to constrain it to. However, it will become the parent of other components when you specify how they are assembled. Pro/ENGINEER has parent/child relationships in Assembly mode just as it does in Part mode.

> ➾ **NOTE:** *Pro/ENGINEER recognizes a component as being either a part or another assembly. In the case of placing another assembly as a component, that assembly would be considered a subassembly.*

For the toy car, the first component should be the chassis because all other components attach to it. Many commands are unavailable at the start of a new assembly. Pro/ENGINEER lets you use only assembly features that add components. To begin the assembly process, pick **Component** ➡ **Assemble** from the ASSEMBLY menu. Now Pro/ENGINEER asks you to tell it which component to retrieve.

`Enter NAME [QUIT]:` **?** ↵

From the menu of components in the current directory, select **chassis.prt**.

> ✓ **TIP:** *When Pro/ENGINEER prompts you for a file name, you can usually enter <?> to display a name list menu of applicable file names.*

Now your assembly will contain the first component.

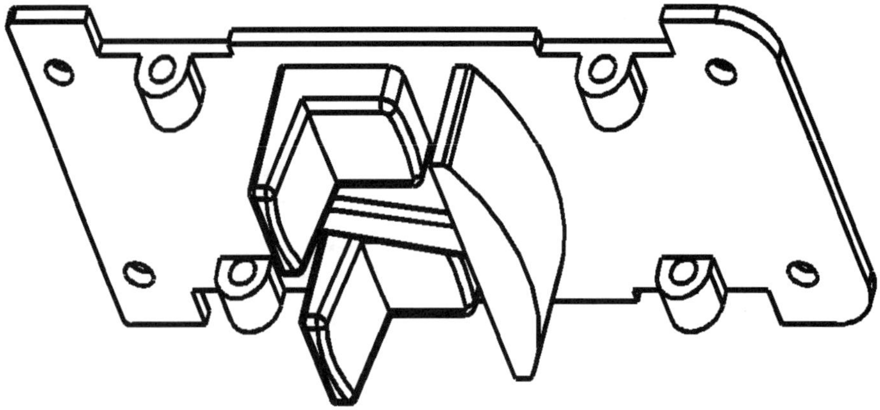

Figure 9-2
The first component.

➾ **NOTE 1:** *You will notice as you pick assemble from the COMPONENTS menu that the MODEL TREE window will appear. The model tree will provide assembly and subassembly breakdowns, and as you add components to the assembly you will see them update in the model tree. The model tree will also provide a convienent means of selecting components during various assembly functions described in this chapter. However, if you do not wish to use the model tree it may be turned off through your environment settings.*

➾ **NOTE 2:** *The default orientation of the first component will determine the default orientation of the assembly.*

✓ **TIP:** *If you want a different default orientation, you can create datum planes in the assembly prior to placing the first component. At this point, you can assemble the first component to those planes in whatever orientation you choose.*

Assembly Constraints

Constraints are used when assembling components just as constraints are used to add a feature to a part. Thus Pro/ENGINEER's parametric nature permeates the different modes. Constraints follow.

- **Mate** — Two planar surfaces can be constrained so that they are coplanar and facing each other.
- **Mate Offset** — Two planar surfaces can be constrained to be parallel, facing each other, and with a gap or interference between them.
- **Align** — Two planar surfaces can be constrained to be coplanar and facing in the same direction. Revolved surfaces can be constrained so that their axes are coaxial.
- **Align Offset** — Two planar surfaces can be constrained to be parallel, facing in the same direction, and with a gap between them.
- **Insert** — The axis of a revolved surface is made coaxial with the axis of another revolved surface.
- **Orient** — Two planar surfaces can be constrained to be parallel and facing in the same direction. There is no constraint for the gap between the surfaces.
- **Coord Sys** — Two datum coordinate systems can be made coincident to each other.
- **Tangent** — Two surfaces can be constrained to mate at their point of tangency.
- **Pnt on Srf** — Two surfaces can be constrained to mate so that a datum point on one surface is in contact with the other surface.
- **Edge on Srf** — An edge can be constrained to contact a surface.

How Assembly Constraints Work

The constraints used in placing components are evaluated every time you regenerate an assembly. When parts change size, you can regenerate the assembly, and the assembly will update accordingly. By using these logical constraints, you create parametric assemblies of your parametric parts.

Parent/Child Implications in Assembly Mode

The part geometry you select when assembling components creates relationships similar to the parent/child relationships in Part mode. Every time you reference a surface of a component during a new component's assembly, you create a parent/child relationship between the components. The parent/child relationships you create in Assembly mode have the same benefits as those created in Part mode. When a part is modified, the child components move to maintain the assembly constraints. Removing a component that is a parent has the same effect as removing a parent feature. You need to know its children components and what you want to do with them.

The Second Component

Now you need to decide which component to assemble next. You can assemble the body next, but it will hide your view for attaching the wheels to the chassis. This is not a big deal, but it just makes it harder to see the wheels. If you place the wheels next, then it will be easier for you to pick the surfaces required to place the wheels without affecting the process of placing the body on the chassis.

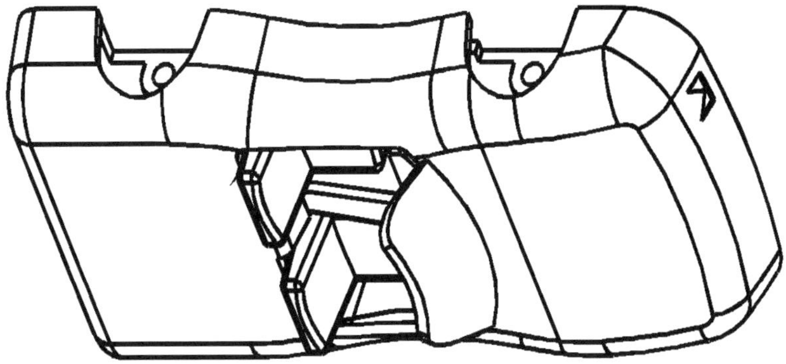

Figure 9-3
The view after making the body the second component.

Placing a Component into an Assembly

To place a wheel on the chassis, pick Assemble from the COMPONENT menu.

```
Enter Name [CHASSIS.PRT]: ? ↵
```

From the next menu that appears, select **wheel.prt**. Your design intent is to have each wheel attached to an axle lock so that the position of the axle lock *drives* the position of the wheel in the assembly. Each wheel will be individually assembled on each chassis axle lock.

You should see the graphics window for the assembly containing the chassis, and two new windows. One of the new windows, labeled COMPONENT WINDOW, contains the wheel. The other one is labeled Component Placement and it shows your assembly constraints as you place them.

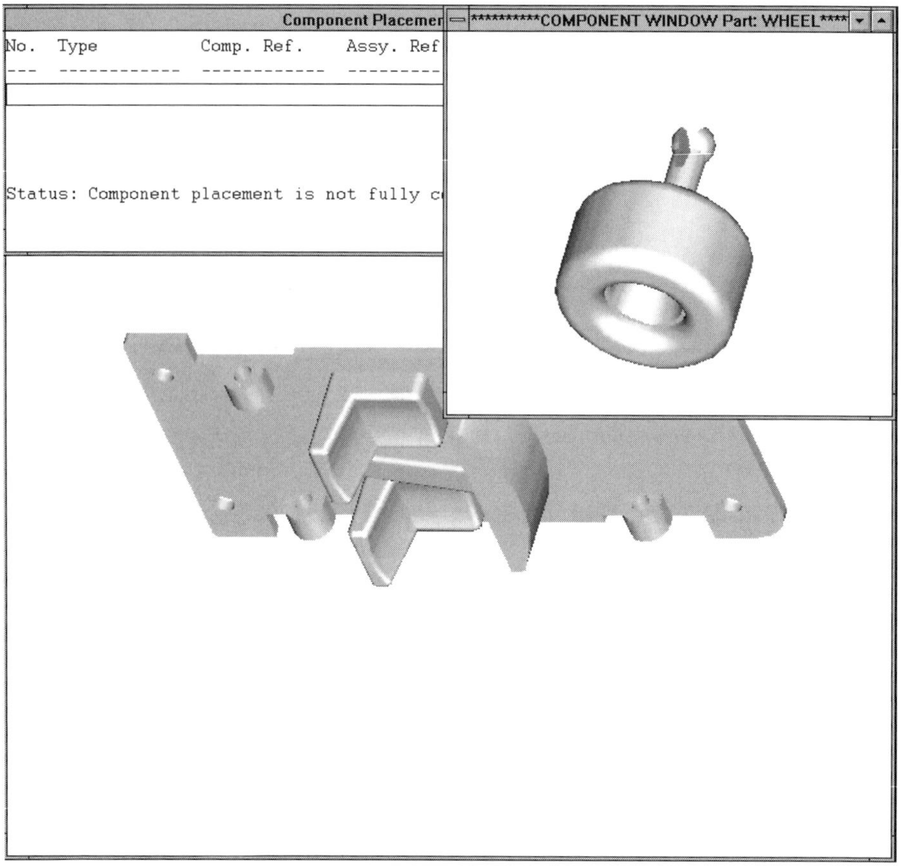

Figure 9-4
The component window with the assembly window.

NOTE: *You will notice that the COMPONENT WINDOW contains asterisks in its title bar. The asterisks indicate that it is the active window. If you select a view command, it will be performed in this window. However, if you pick **Alter View** from the MAIN VIEW menu, you can then pick in the other window. It will display the asterisks in its title bar and become the active window. This allows you to orient the assembly and component any way you wish while placing the component in the assembly.*

The procedure for attaching the wheel to the chassis in Pro/ENGINEER is the same procedure you would use if you were holding the parts in your hands. You would insert the snap end of the axle into the axle lock in the chassis until the side of the wheel touches the side of the chassis. Then you would move the wheel slightly away from the chassis so that there is a small gap to keep them from rubbing against each other.

Using the Insert Constraint

Use the following figure to select the correct surfaces.

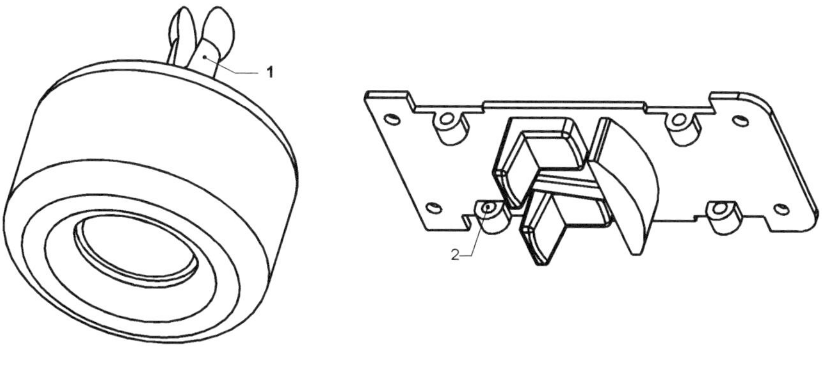

Figure 9-5
Surfaces to select to insert the wheel into the chassis.

From the PLACE menu, pick **Insert**. Next, pick the axle surface on the wheel labeled 1. Move your cursor into the main graphics window and pick the surface on the chassis labeled 2. This is the inner surface of the axle lock that will hold the axle in place. When you finish picking the geometry, you will see the constraint appear in the Component Placement window.

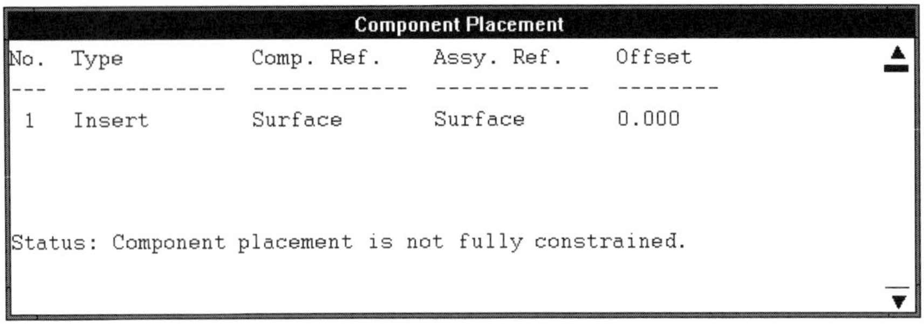

Figure 9-6
The Component Placement window after placing the first constraint.

✓ **TIP:** *On occasion the Component Placement window is pushed to the back as you pick the geometry for the constraints. You should move the Component Placement window up a bit in order to expose its border along the top edge of the Main Graphics window. You can then easily pick the Component Placement window's border to pop it to the front when you want to read it.*

Using the Mate Offset Constraint

Use the following figure to select the correct surfaces.

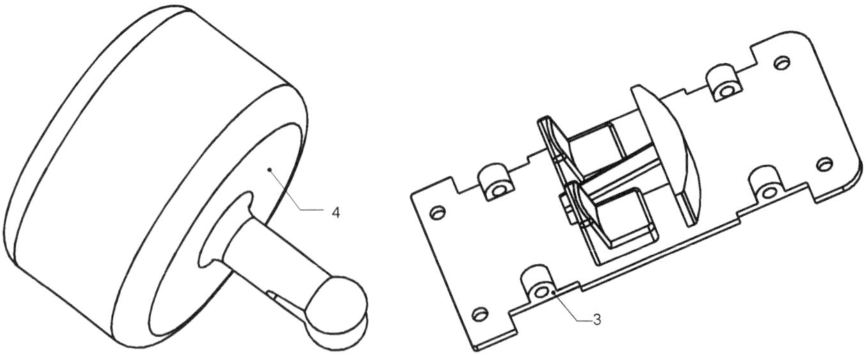

Figure 9-7
Surfaces to select to mate the wheel to the chassis with an offset gap.

To mate the side of the wheel to the side of the chassis and then maintain a small gap between them, pick **Mate Offset** from the PLACE menu. In that the COMPONENT WINDOW has been covered by the assembly window, it will be easier to pick the first surface for mating from the chassis. Pick the surface on the chassis labeled 3. You can either use **Query Sel** to select the hidden surface or you can orient the view of the part as shown in the picture so that the surface is not hidden. This surface is the side of the chassis the wheel will mate against. Now, pick on the border of the COMPONENT WINDOW to make it pop on top of the assembly window. You can now pick the surface on the wheel labeled 4. That surface is the back side of the wheel, which will mate against the side of the chassis. Pro/ENGINEER will prompt you for the offset gap. If you enter a positive value, the wheel will be offset in the direction of the arrow by the amount you specify. If you enter a negative value, the wheel will be offset in the other direction, causing an interference by the amount you specify.

`Enter offset in the indicated direction [0.0000]:` **`.08`** ↵

You have specified the gap between the side of the wheel and the chassis. Look at the Component Placement window again.

Component Placement

No.	Type	Comp. Ref.	Assy. Ref.	Offset
1	Insert	Surface	Surface	0.000
2	Mate	Surface	Surface	0.080

`Status: Component can be placed with present constraints.`

Figure 9-8
The Component Placement window after placing the second constraint.

Nothing else is required to assemble the wheel to the chassis (according to the status line in the Component Placement window). Pick **Show Placement** to see a preview of the wheel's placement in the assembly. What if you wanted the snap relief oriented horizontally instead of vertically? You could specify another constraint by picking **Add Constrnt**. You could also modify and delete previous constraints by using the **RedoConstrnt** and **Del Constrnt** commands. Use the **Next** and **Previous** commands to highlight the constraint you want to change, and then pick the **RedoConstrnt** or **Del Constrnt** command. Once you are satisfied with the wheel's placement, pick **Done** to complete the wheel's assembly.

> **NOTE:** *The wheel may seem under-constrained because you did not specify a constraint to control the wheel's rotational position. Pro/ENGINEER will assume a rotation for the wheel. If you do not like the assumed rotation, you can specify an* ***Orient*** *constraint and use DTM1 or DTM2 from the wheel to control the wheel's rotation.*

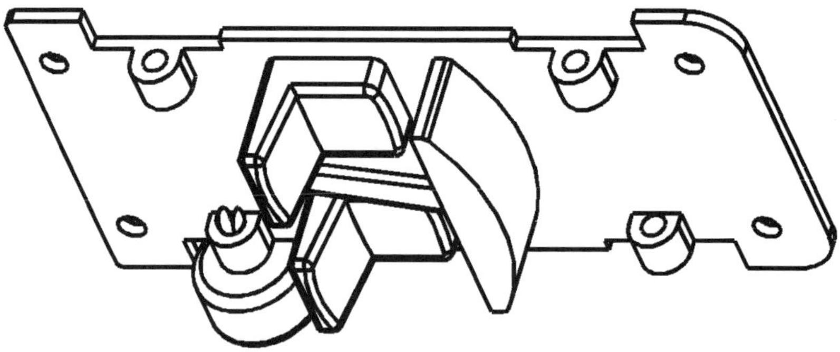

Figure 9-9
A wheel attached to the chassis.

A part can be placed into an assembly many times. To attach the other wheels, use the same process and assemble the same wheel into the three remaining axle locks.

> **DESIGN NOTE:** *When you change the wheel part, all occurrences of the wheel in the assembly will also change.*

Your assembly should look like the following figure after placing all of the wheels.

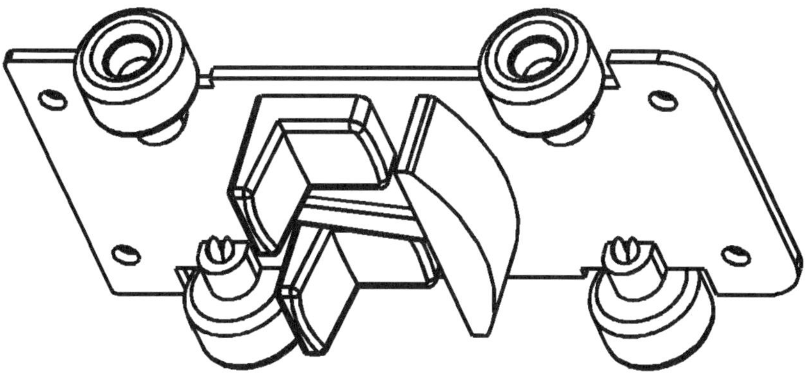

Figure 9-10
The chassis with attached wheels.

Assembly Constraint Behavior

Every time you selected a pair of entities from the chassis and the wheel, you created a design constraint between them. The type of constraint was specified by your selection from the PLACE menu.

The constraint created when you selected **Insert** ensures that the centerlines of the selected cylindrical surfaces remain coaxial. This action establishes the chassis as a parent component of the wheel.

The **Mate Offset** constraint causes the selected surfaces to be oriented so that they face toward each other with the specified gap between them. That again establishes the chassis as a parent component of the wheel.

> **DESIGN NOTE:** *If the axle lock moves up, down, forward, or back, the wheel will move with it because of the* ***Insert*** *constraint. If the axle lock moves inward or outward, the wheel will always maintain the .08" distance from the axle lock.*

Using Layers in Assembly Mode

Placing your components on individual layers in the assembly is a useful practice because it allows you to temporarily blank any parts you do not want to see. This practice is beneficial when you are working in assemblies with many components.

> ✓ **TIP:** *Blanking components you do not need to see will speed up the time it takes for Pro/ENGINEER to repaint the screen.*

Now would be a good time to make a layer called "wheels," and place all the wheels on that layer. Pick **Done** from the COMPONENT menu. Pick **Layer** ➥ **Sel Level** ➥ **Assembly** | **Top Level** ➥ **Setup Layer** ➥ **Create**. Pro/ENGINEER will prompt you for the layer name.

```
Enter layer name [QUIT]: wheels ↵
Layer WHEELS was added to CAR. Enter again [QUIT]: ↵
```

Pick **Set Items** ➥ **WHEELS** | **Done Sel** from the LAYERS menu. Pick **Add Items** ➥ **Component** | **Individual** and select all wheels, one by one. Pick **Done Sel** from the GET SELECT menu. The wheels are now associated with the layer named WHEELS.

Go ahead and blank the WHEELS layer by picking **Set Display** ➥ **Blank** ➥ **WHEELS** | **Done Sel**. Pick **Done/Return** from the LEVEL SEL menu and repaint the screen to make the wheels disappear.

SAVE

Using Simplified Representations

Pro/ENGINEER allows you to include or exclude components in the assembly. These included and excluded components can be replaced with simplified versions of subassembly parts. Different configurations of an assembly can be created and stored using these simplified representations. The major benefit of simplified representations is memory usage. If you store an assembly that has been simplified by excluding components, the excluded components are not retrieved into memory when you retrieve the assembly. This can be a useful tool when you are working in a small area of a large assembly or when your workstation has limited resources.

Adding Another Component

The only remaining task is to attach the body to the assembly. You want to mate the end surface of the bosses of the body to the top surface of the chassis. You also need to line up the holes in the body to those in the chassis. Only two sets of holes need to be aligned to fully determine how the body will be attached to the chassis. Pick **Component ➥ Assemble** starting at the ASSEMBLY menu.

```
Enter Name [WHEEL.PRT]: ? ↵
```

Now pick **body.prt** from the menu that appears. You are ready to attach the body to the chassis.

Using the Mate Constraint

Use the following figure to pick the correct surfaces from the chassis and the body.

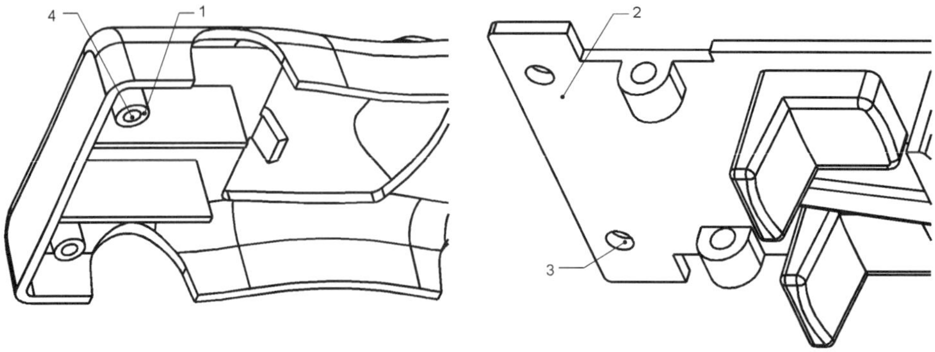

Figure 9-11
Surfaces to select for mating the body to the chassis and to align a hole in the body with a hole in the chassis.

First you will mate the body to the chassis. Pick **Mate** from the PLACE menu. The surfaces labeled 1 on the body and 2 on the chassis should be selected. The surface selected first does not matter. You can reorient and zoom as appropriate during this selection process to help you pick the proper surfaces. Do not forget that the asterisks in the title bar denote which window the view commands will act within. Use **Alter View** and then pick in the other window to change the view.

Using the Align Constraint

The second and third steps are to align the holes in two of the bosses of the body with the appropriate holes in the chassis. Pick **Align** and select the surfaces labeled 3 on the chassis and 4 on the body. Notice that Pro/ENGINEER refers to this type of alignment as an insert constraint in the Component Placement window.

Component Placement

No.	Type	Comp. Ref.	Assy. Ref.	Offset
1	Mate	Surface	Surface	0.000
2	Insert	Surface	Surface	0.000

Status: Component can be placed with present constraints.

Figure 9-12
The Component Placement window after placing the alignment constraint.

Notice also that the status line says you can stop placing constraints now. Should you? Probably not. Pro/ENGINEER can assume an orientation for the body in the same fashion it assumed an orientation for each wheel. Because the body should only go on the chassis one way, you should specify a third constraint.

Pick **Align** again. This time you will pick the hole axes instead of the hole surfaces.

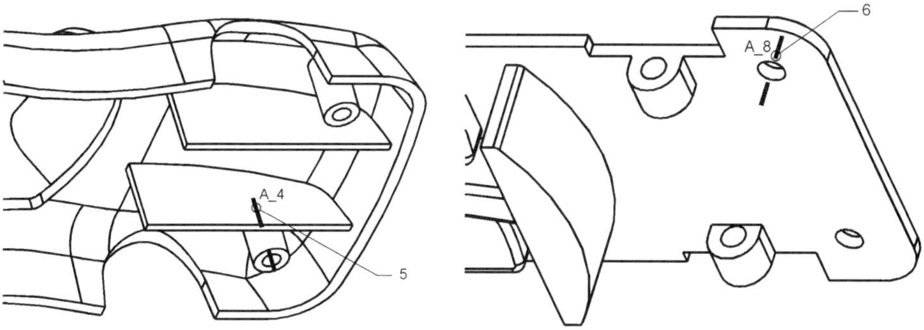

Figugre 9-13
Items to select to align a hole axis in the body with a hole axis on the chassis.

Use the axes labeled 5 on the body and 6 on the chassis. Pick **Done**, and the assembly is now complete. You aligned the holes in the body bosses with the holes in the chassis using two different constraints.

Which was better? Was there any difference? In this case, the constraint you chose was not that important. The geometry you selected on each part makes a much bigger difference.

> **DESIGN NOTE:** *When you used the inner surfaces of the body bosses to align to, you tied this assembly constraint to a draft feature. It is better practice to align axes of holes whenever possible. This way you are creating the assembly constraint between major features of your parts, not minor ones such as draft.*

It would be a good idea to place the body component on a layer called "body." Do the same for the chassis.

SAVE

More Assembly Constraint Behavior

The constraint created when you selected **Mate** is similar to the one established by the **Mate Offset** constraint used with the wheel. The difference is that the body-to-chassis mating constraint does not allow a gap to be provided. This constraint made the chassis a parent component of the body.

> **NOTE:** *The assembly constraints can be redefined if you need to change them. For example, you can redefine a* **Mate** *constraint to be a* **Mate Offset** *constraint if your design intent changes.*

The constraints you created when you selected **Align** the first time are the same as those created with **Insert** when you placed the wheel. They specified that the centerlines of the selected surfaces should always be coaxial. This constraint also made the chassis a parent of the body. The second time you used **Align**, you actually specified a rotational constraint. This second alignment simply ties the body to the chassis and controls the rotational position of the body about the first aligned hole.

> ➾ **DESIGN NOTE:** *If the chassis becomes thicker or thinner, the body will move up or down in order to maintain contact between the surfaces specified by the* **Mate** *constraint. If the first aligned hole in the chassis is moved, the body will move and maintain the coaxial constraint. If the second aligned hole in the chassis is moved, the body will pivot about the first aligned hole. The amount of pivot is determined by the position of the second hole.*

Deleting Components

Components of an assembly can be removed. From the COMPONENT menu, pick **Delete**. Pick the component you want to remove. If the component has child components associated with it, you will have to redefine their placement.

> ➾ **NOTE:** *Deleted components are permanently removed from the assembly. There is no way to undelete a component (much like deleting features in Part mode).*

> ✓ **TIP:** *You should save your assembly prior to deleting components in the event you accidentally delete the wrong one.*

Redefining Components

You can change the way a component was assembled by using the **Redefine** command on the COMPONENT menu. For this example, a different method of constraining the body will be used. The **Mate** constraint that was placed is fine. It will not need to be redone. The goal of this example will be to change the references used in the second constraint, and to use the **Orient** command for the third constraint.

Pick **Redefine** and pick the body. Pro/ENGINEER will bring up the Component Placement window, and it will begin to let you redefine the first constraint. Pick **Next** from the COMP PLACE menu to move on to the second constraint. Pick **RedoConstrnt** in order to change the references for the second constraint. Pick **AssemblyRef** and **Comp Ref** to indicate that you are changing both the assembly reference and the component reference geometry. Pick **Done**. Now, select the axes that are denoted by the labels 1 and 2 in the following figure.

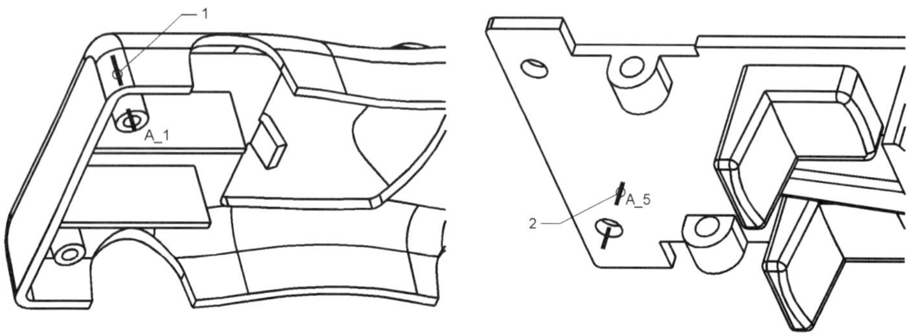

Figure 9-14
Surfaces to select to mate the body to the chassis and to align a boss of the body and a hole in the chassis.

The second constraint has now been redefined. The Component Placement window should duplicate the next figure.

Component Placement

No.	Type	Comp. Ref.	Assy. Ref.	Offset
1	Mate	Surface	Surface	0.000
2	Align	Axis	Axis	0.000
3	Align	Axis	Axis	0.000

Status: Component can be placed with present constraints.

Figure 9-15
The Component Placement window after redefining the second constraint.

Orienting with Two Datum Planes

This time, instead of specifying another set of aligned holes as you did before, you will pick two datum planes to be oriented in the same direction. This means you will need to delete the third constraint and replace it with the **Orient** constraint.

Pick **Next** from the COMP PLACE menu to move on to the third constraint. The axes in the front left corner of the car should highlight. Pick **Del Constrnt** and the third constraint will be removed from the list. Now pick **Add Constrnt**. You are now ready to choose the type of constraint you want to place.

Select **Orient** from the PLACE menu and pick DTM2 on the body. In that datum planes have a yellow and a red side, you need to tell Pro/ENGINEER which side of the datum plane you are trying to orient. From the DTM ORIENT menu, pick **Yellow**. Now pick DTM2 on the chassis. Select **Yellow** again to signify that the yellow side of this datum plane will be oriented in the same direction as the yellow side of the body's datum plane. The third constraint has now been placed. The Component Placement window should duplicate the next figure.

Component Placement

No.	Type	Comp. Ref.	Assy. Ref.	Offset
1	Mate	Surface	Surface	0.000
2	Align	Axis	Axis	0.000
3	Orient	Dtm2	Dtm2	--

Status: Component can be placed with present constraints.

Figure 9-16
*The Component Placement window after placing the **Orient** constraint.*

Pick **Done** to tell Pro/ENGINEER that you have finished redefining the body constraints.

Even More Assembly Constraint Behavior

The design intent of this method of placement and the previous method are basically the same. The main difference is apparent when the holes do not line up. Using this method will prevent the body from skewing when the chassis holes do not match the body bosses.

> ✓ **TIP:** *When you place components in your assembly that have multiple mounting locations, choose one hole as your driving alignment constraint. Use* **Orient** *to control the rotation of the component in the assembly. This will prevent the parts from becoming skewed.*

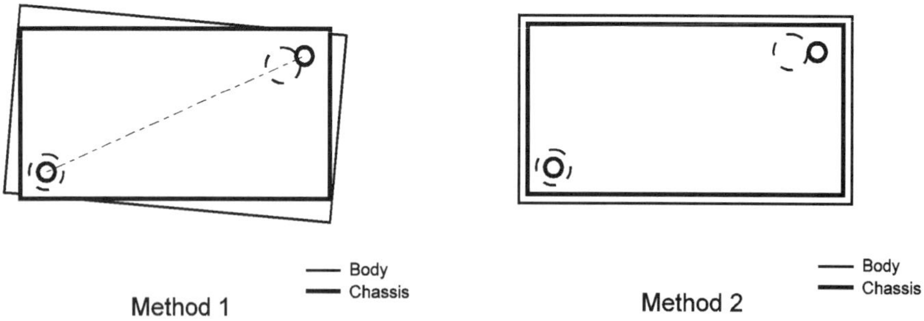

Figure 9-17
The differences between the old and new methods of constraining the body.

Handling Assembly Component Failures

Sometimes a feature on a part that was referenced in an assembly constraint is deleted. When that happens, Pro/ENGINEER cannot place the child part in the assembly during the next assembly regeneration. The TRIM MODEL menu will appear so that you can tell Pro/ENGINEER what to do. You will have the following options:

- **Trim Supp** — Suppress the failed component and continue with the regeneration of the assembly.
- **Trim Delete** — Delete the failed component and continue with the regeneration of the assembly.
- **Clip Supp** — Suppress the failed component and all remaining components.
- **Clip Delete** — Delete the failed component and all remaining components.
- **Replace** — Replace the failed component with a different one.
- **Redefine** — Redo the constraints that placed the failed component.
- **Freeze** — Assemble the failed component nonparametrically.

Most of these options should seem familiar to you because you have seen them before on the TRIM PART menu. They work the same as in Part mode, but in Assembly mode they are dealing with components rather than features.

You will have a new option called **Freeze**. By selecting **Freeze**, you can tell Pro/ENGINEER to place the component nonparametrically at the last known successful placement location.

> **NOTE:** *The* **Freeze** *command has its good and bad points. It will allow you to continue with the regeneration and let you handle the component failure later. Unfortunately, many users never go back to fix these placement problems, and as a result, their parts do not get placed properly in the assembly, which causes design mistakes.*

Modifying the Assembly

During the design process, you will surely have changes that will require you to change the size of components or add/delete component features. You can make these changes from within Assembly mode instead of entering Part mode for each of the components. In addition, you can see if the change of one of the components causes interference with another component when the change is made.

Next, when you are first creating a design, you can simply create all of the parts with the appropriate shape without being particular about the values entered for each of the dimensions. After assembling the components, you will see the interferences and misalignments. From Assembly mode, you can modify all components until they fit correctly. To modify from within Assembly mode, pick **Modify** from the ASSEMBLY menu.

Specifying One Part to Modify

If you have many components, you may want to specify only one part that will accept mouse picks. In this way, you do not have to query select through many surfaces of other parts when you are trying to pick a feature. You may also want to add a new feature to a part. Pick **Mod Part** from the ASSEM MOD menu. Then pick the part you want to work on. You can also select the part to modify by using the Model Tree. Just pick on the part name and you will see it highlighted. You can now add features, modify the part, and regenerate the part as if you were in Part mode.

To practice modifying a part's dimension in Assembly mode, change the dimension of the text feature on the bottom of the chassis. You should use the **Modify Part** command and select the chassis part. The position of the text on the bottom of the chassis can be modified by selecting **Modify Dim** from the MODIFY PART menu. Pick on the text feature to display its locating dimensions. Change the dimension `4.00` to `3.00` and then pick **Regenerate**. The part will update. Pick **Done** from the MODIFY PART menu.

Specifying Any Dimension to Modify

Now try changing the dimension back to its previous value. This time, pick **Modify** ➥ **Modify Dim**. Query select the text feature. Change the dimension `3.00` to `4.00` and then pick **Done/Return** from the ASSEM MOD menu. Pick **Regenerate** ➥ **Automatic**. The chassis text should be back to its original location.

> ➾ **NOTE:** *When a feature on a part is selected with* ***Modify Dim****, the feature's dimensions and the wheel's assembly dimensions are displayed.*

Modifying an Assembly Dimension

When you have placed components in the assembly using offset constraints such as Mate Offset or Align Off, select **Modify** ➥ **Mod Assem,** starting from the ASSEMBLY menu, to change the offset dimensions. After picking **Mod Assem**, pick one of the wheels. Pick the displayed dimension.

```
Enter value [0.08]: .8 ↵
```

Now pick **Regenerate ➥ Automatic**, and the changes will be made. Change the value back to `.08` and regenerate again.

Creating an Exploded Assembly

Did you notice that when you changed the wheel's assembly dimension to `.8` that the assembly began to look "exploded"? Pro/ENGINEER provides a method to easily create an exploded assembly. Exploded assemblies are often used in documentation when you are trying to show how an assembly is assembled. By using the constraints you chose to assemble the toy car, Pro/ENGINEER can determine how to move the components to create the exploded assembly. Because the chassis was the first component, it is used as the fixed component. All other components move with respect to the fixed component. Pick **View ➥ Cosmetic ➥ Explode** to display the exploded assembly.

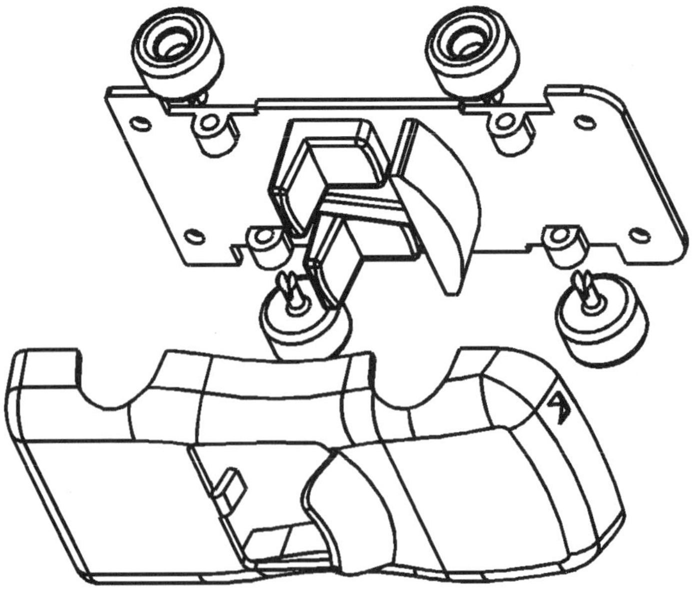

Figure 9-18
The exploded assembly.

It is easy to understand how the components are moved to create the exploded assembly. The wheels are exploded along the axis of the centerlines of the inserted surfaces. The wheels are moved in the direction that will increase the gap between the mating surfaces. The body is moved similarly. It is moved along the axis of the aligning centerline in the direction that increases the gap between the mating surfaces.

Modifying the Explosion Distance

You can change the distance the components are exploded at any time. From the ASSEMBLY menu, pick **Modify ➥ Mod Expld**. You have three options in the MOD EXPLODE menu that are used to change the exploded components.

❒ **Position** — There are several options under this command for the movement of components. The most familiar movement options in comparison to the dimensional direction of previous Pro/E revisions will be **Plane Normal | Translate**. This option allows you to select a plane normal to movement of the selected components. Simply pick and drag the component you wish to move after establishing the plane. Position also displays a status window titled Dynamic Exploded Components, which displays the **Preferences** option for component movement. Experiment with the other options in this command. There are many combinations, which make this the most powerful assembly package to date.

❒ **Expld Status** — Through this command you can specify if a component is to be exploded or remain unexploded. To accomplish this, the Model Tree is used. When **Expld Status** is picked, the Model Tree will create an additional column, labeled Explode Status. You can select the component name, which will toggle the status of that particular component from exploded to unexploded. Once you have set the status for all of the components, pick **Done** from the EXPLD STATUS menu and the exploded model will be updated. This command is particularly useful for setting the status of components to unexplode in subassemblies within top assemblies. Typically you do not want all of these subassembly components exploded.

- **Offset Lines** — Allows you to create exploded assembly offset lines, which clarify how components assemble.

Un-exploding the Assembly

You can easily un-explode the assembly. Pick **View** ➥ **Cosmetic** ➥ **Un-Explode**, and the assembly will be restored to its appearance prior to exploding. The explode locations are saved in the assembly file. Anytime you wish, you can show the exploded assembly as you last modified it by picking **Explode** from the COSMETIC menu. Use **Un-Explode** when you are through viewing the assembly as an exploded assembly.

Checking Your Assembly

Pro/ENGINEER provides many methods for you to request information about your assembly. You can refresh your memory about the constraints and parent/child relationships used during the assembly process. You can obtain diverse clearance and interference information. Pro/ENGINEER will also provide you a bill of materials (BOM) of all components used in the assembly. You can also make cross sections through the assembly. All of these functions of Pro/ENGINEER help you verify your assembly and your part designs.

Assembly Measurements

All of the measurement techniques available in Part mode are available in Assembly mode. In addition to these, there are a few more options for measuring clearances and interferences. From the

ASSEMBLY menu, pick **Info** ➡ **Measure** ➡ **Clear/Intf**. From the CLEAR/INTF menu, you will have the following choices for the type of measurement to perform:

- **Pairs** — Measure between two objects.
- **Volume Intf** — Measure interference with components enclosed in a closed quilt. The quilt option is part of the Pro Surface module.
- **Global Clr** — Measure clearance between all parts or subassemblies.
- **Global Intf** — Measure interference between all parts or subassemblies.

Most of the time you will want to measure the distance between two objects. You have control over the object types you are measuring between. The types follow:

- **Whole Subasm** — Measure to an entire subassembly.
- **Whole Part** — Measure to an entire part.
- **Surface** — Measure to a surface.
- **Cable** — Measure to a cable (requires Pro/CABLE).
- **Single Ent** — Measures to a vertex, edge, surface, or datum feature.

These commands allow you the ultimate flexibility in assessing your design.

Component Constraint Information

If you do not remember which constraints or parents were used when you assembled a component, Pro/ENGINEER provides a way to easily locate such information. From the ASSEMBLY menu, pick **Info ➥ Comp Info**. After selecting the component, Pro/ENGINEER will ask you if you want to see how the component was assembled. Press <Return> to take the default choice of `Y`. Pro/ENGINEER lists each of its constraints one at a time in the Message window while highlighting the geometry used when creating that constraint.

BOM

By selecting **BOM** from the INFO menu, you can create a text file containing a listing of all components in the assembly. The format of this text file can be customized by specifying a BOM format file in your *config.pro* file. This format file determines how the assembly information will be arranged and the information to be included in the BOM.

Assembly Cross Sections

Cross sections are created in Assembly mode just as they are in Part mode. You can make a planar cross section or an offset cross section. If you make a planar cross section, the datum plane used for the cross section must be one that was created in Assembly mode. You have independent control of the crosshatching for each individual part of the assembly.

Conclusion

There are many important characteristics of Pro/ENGINEER assemblies. The first and most obvious characteristic is their parametric nature. If the assemblies were not parametric, components would not update their positions when the parent components move.

Another attribute of assemblies is that their components are referenced. The assembly file never copies the geometry of the part into its database. This attribute allows part modifications to be automatically reflected in the assembly. The same attribute allows you to modify the wheel and have all of the wheels in the assembly update.

> ✓ **TIP:** *In that parts are referenced by assemblies, it is important to maintain your files and directories. Pro/ENGINEER does not "remember" the entire path for the part file. If you assemble parts from different directories, you need to ensure that the directories the part files reside in are in your Pro/ENGINEER search path (defined in the* config.pro *file).*

You should keep in mind that assemblies have parent/child relationships, just as parts do. Changes to a parent component can initiate a positional change in a child component. Disassembly of a parent component will cause regeneration failures of the child components.

You can also modify a part while in Assembly mode. You can change dimensions and redefine or add features. All of these capabilities make Assembly mode a powerful tool for designing part systems.

Part 3
Working with Drawings

Chapter 10

Documenting Your Design: A Simple Drawing

Making a Drawing of a Basic Part

Introduction

The next three chapters introduce basic concepts for creating a detailed drawing of your parts and the corresponding assembly. Commands and concepts covered in this chapter include the following:

- Setting the company drawing standards
- Selecting the size of the drawing sheet
- Placing the drawing views

- ❒ Moving items and changing the appearance of the drawing
- ❒ Placing the dimensions in the drawing
- ❒ Adding notes and parametric text
- ❒ Modifying part dimensions in the drawing model

In this chapter, you will create a drawing of the wheel for the toy car. When completed, the wheel will look like the next figure.

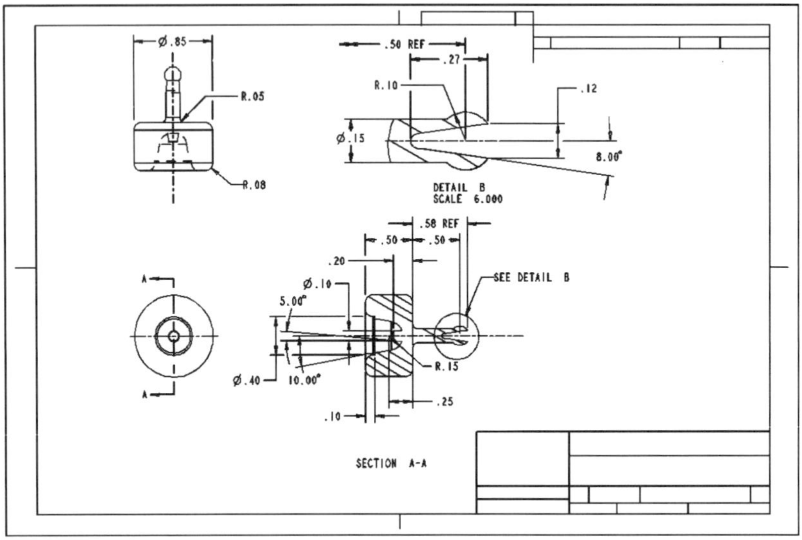

Figure 10-1
The drawing of the wheel.

Before Beginning Your First Drawing

Before you create the first drawing, you should determine the standards to be used for all of your drawings. This information will be saved in a drawing standards file that will be read by Pro/ENGINEER every time you create a new drawing. After the initial reading of the drawing standards file, changes to the drawing standards file will not be automatically reflected in existing drawings. You can, however, manually retrieve the new setup file by picking **Set Up ➥ Draw Setup ➥ Retrieve** from the DRAWING menu.

In order for Pro/ENGINEER to find your drawing standards file, you should place it in your standards directory (*/usr/standards*). In our example, the drawing standards file will be named *custom.dtl*. Place the following option in your *config.pro* file to specify the location of the drawing standards file:

```
drawing_setup_file /usr/standards/custom.dtl
```

Creating a Drawing Standards File

To create a drawing standards file, you need to use Pro/ENGINEER's Drawing mode. Select **Drawing** from the MODE menu. This will place you in Drawing mode and display the entry menu for Drawing mode. In that you are creating a drawing standards file and not a new drawing, pick **Create** from that menu and use the default file name provided by the prompt. You will not be saving the drawing file, so it does not matter what the file name is. Now select the following commands:

Set Dwg Size ➥ Landscape ➥ A ➥ Set Up ➥ Draw Setup ➥ Modify Val

This will place you into a text editor with the default drawing standards file supplied by Pro/ENGINEER. You can now edit this file to set drawing standards such as arrowhead size, text size, dual dimensions, balloon size, and so on. You may want to make several different versions of a drawing standards file. For example, one may be used for imperial units, one for metric units, and one for dual dimensioning. Only one of these files will be specified for Pro/ENGINEER to use as the default drawing standards file. After you begin a drawing session, you must then retrieve one of the other files to set the drawing to the standards specified within it. After modifying this file, exit the editing session and select **Save** from the DTL SETUP menu. Save the file as *custom.dtl* and move it to your standards directory.

➻ **NOTE:** *If a STANDARDS file does not exist in the /USR directory of your operating system, you may need to add one or have the systems administrator add one for you. You may not always have write privileges to these types of directories.*

```
Enter file name [-QUIT-]: custom ↵
```

Now select **Dbms** ➥ **Erase** ➥ **All** ➥ **Confirm**.

Beginning a Drawing of the Part

When you want to create a drawing, you need to use Pro/ENGINEER's Drawing mode. Select **Drawing** from the MODE menu. This will place you in Drawing mode and display the entry menu for Drawing mode. In that you are creating a new drawing, pick **Create** from that menu. A prompt for the new drawing's file name will appear in the message window.

```
Enter Drawing name [DRW0001]: wheel ↵
```

This will create a drawing named "wheel" in memory. When you save this drawing, Pro/ENGINEER will create a file called *wheel.drw*.

Setting the Size of the Drawing

After providing a file name, you need to specify the drawing size you will create. You can tell Pro/ENGINEER the size of a blank drawing sheet or you can tell Pro/ENGINEER to retrieve a drawing sheet that already contains the border and title block. A drawing sheet with border and a title block is called a *format* in Pro/ENGINEER. A default format is provided within Pro/ENGINEER for each drawing size. To select a default Pro/ENGINEER format, pick **Retr Format** from the GET FORMAT menu. Next, enter the desired format size.

```
Enter Format name [-QUIT-]: a ↵
```

> ➾ **NOTE:** *The format can be changed later if it is the wrong size. Select* ***Sheets*** ➥ ***Format*** ➥ ***Add/Replace*** | ***Unblank*** *from the DRAWING menu, and then select the desired format.*

Unblank specifies that the format will be displayed in the drawing. **Blank** specifies that the format will not be displayed. Not displaying the format will increase the speed of displaying the drawing entities because none of the title block or border information will be shown. A border and title block for an A-size drawing are now displayed.

Saving a Drawing

The same procedure for saving a part or assembly is used to save a drawing. Pick **Dbms** ➥ **Save** from the MAIN menu, and then specify the drawing name to save. By default, Pro/ENGINEER will save only a *changed* part or assembly whenever a drawing is saved.

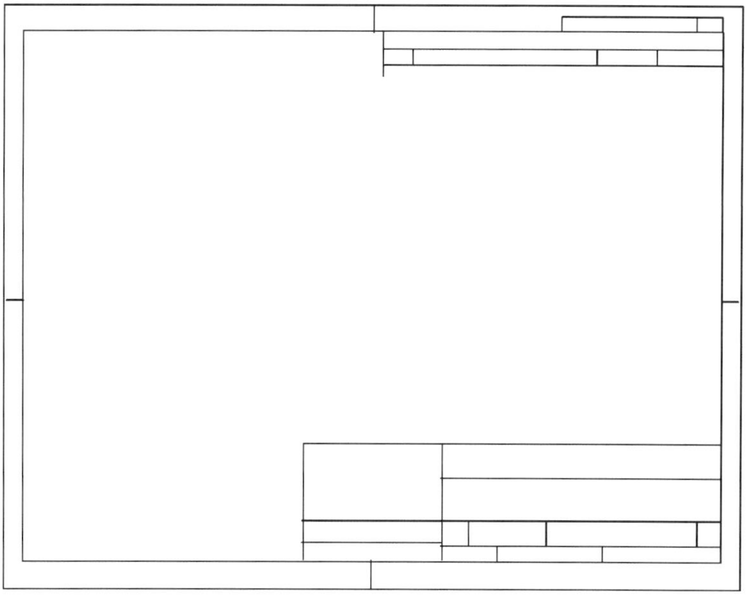

Figure 10-2
An A-size drawing sheet with a border and title block.

Placing the First View

Now you are ready to place the views of the wheel on the drawing. There are two basic styles of views that can be placed in Pro/ENGINEER. The first style is a general view. This style of view can have its own unique orientation in the drawing. It is never a child of another drawing view. A general view's location in a drawing is not affected by moving any other view in the drawing.

The other style of view is always derived from an existing view. It is the dependent style of view. The dependent view's location may be changed if its parent view is moved. You cannot delete the parent of a dependent view without first deleting the dependent view. The parent of a dependent view may itself be a dependent view. Examples of dependent-style views follow:

❒ **Projection** – This view is created by projecting the geometry of its parent view in either a horizontal or vertical direction.

❒ **Auxiliary** – This view is created by projecting the geometry of its parent view normal to a surface or along an axis.

❒ **Detailed** – A portion of an existing view can be displayed in this type of view. It will be oriented the same as its parent view, but may be displayed at a different scale.

❒ **Revolved** – A cross-sectional view can be displayed from a parent view that has a section normal to the screen. The revolved view will be projected in a direction parallel to the plane containing the section in the parent view, and it will be revolved so that it is normal to the plane containing the section.

From the DRAWING menu, pick **Views**. Because you have not specified which part or assembly will be shown in the drawing, Pro/ENGINEER will prompt you for it.

`Enter Model name [CAR.ASM]:` **`wheel`** ↵

Pick **Add View** from the VIEWS menu. When no other view exists on the drawing, only a **General** view can be selected from the VIEW TYPES menu. This is because all of the other types of views depend on an existing view for their placement.

The view types have been described previously. The next group of view options on the menu determines how much of the model will be displayed in the view.

- ❒ **Full View** – The entire model will be displayed in the view.
- ❒ **Half View** – Only the portion of the model on one side of a selected datum plane or planar surface will be displayed in the view.
- ❒ **Broken View** – For views of a large model, such as a long pipe, two points can be selected so that the middle section of the model between the two points will not be displayed. Essentially, the long pipe will be displayed as a shorter pipe because only the pipe ends are of interest for detailing in the drawing.
- ❒ **Partial View** – Only the model within a closed boundary will be displayed.

The next group of options specifies whether a cross section will be displayed in the view.

- ❒ **Section** – A cross section will be displayed in the view if the selected cross-sectional plane is parallel to the screen.
- ❒ **No Xsec** – No cross section will be displayed in the view.
- ❒ **Of Surface** – Only one surface of the model will be displayed in the view. This option is not available for detailed view types.

The next group of options specifies how the size of the view will be determined.

- **Scale –** The scale of this view can be set independent of all other views in the drawing. The scale is displayed below the view and it can be modified just as a part dimension is.

- **No Scale** – The scale of this view is set by the default scale of the drawing. The value for this scale is displayed in the lower left corner of the screen and it can be modified just as a part dimension.

- **Perspective** – A general view can be displayed as a perspective view.

Now pick **General | Full View | No Xsec | No Scale | Done**. Pro/ENGINEER will prompt you to pick the center point for the view. Pick on the drawing where you want the view to appear. The drawing should now look like the following figure.

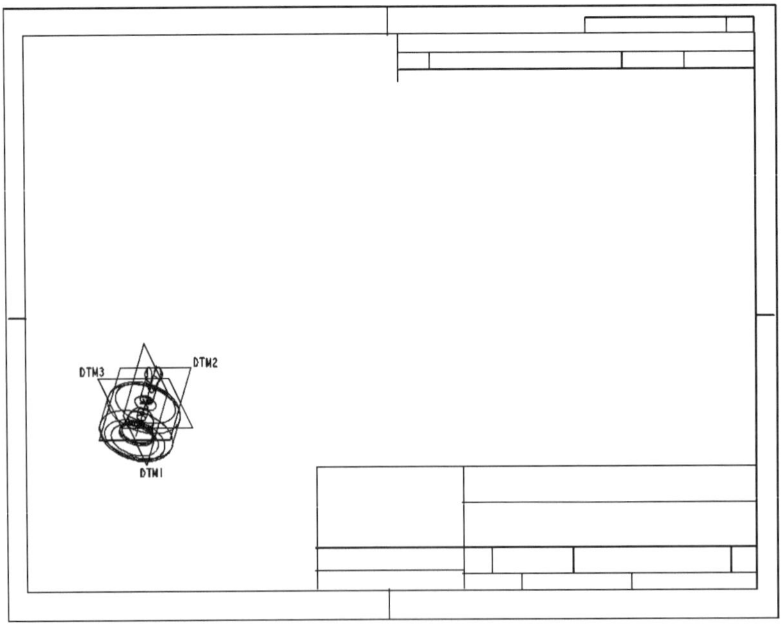

Figure 10-3
First view of the wheel as it initially appears when placed.

✓ **TIP:** *If you use named views from the part when orienting the drawing views, you may want to place the option* `display_trimetric_dwg_mode_view no` *in your* config.pro *file. By doing this, time is not taken to display the view in the default orientation just before you select the view name for reorienting.*

View Orientation

In that this is a general view, you need to orient it. Orient the wheel using the selections from the ORIENTATION menu so that the view appears as shown in the next figure. Pick **Front** and then select DTM3. Pick **Top** and select DTM1.

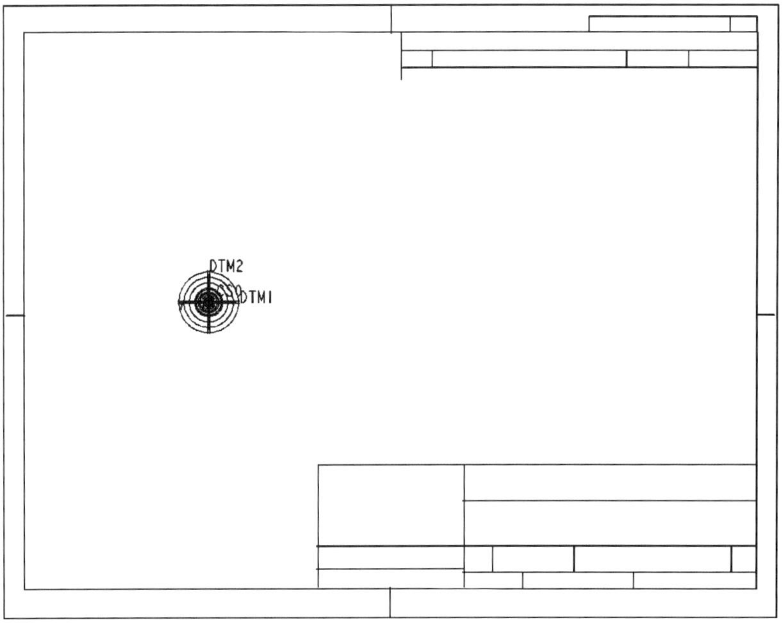

Figure 10-4
Wheel orientation for the first view.

After you are satisfied with its orientation, pick **Done/Return** from the DRAW VIEW menu.

➾ **NOTE:** *If you had named a view in the part, the named view could be selected to orient the general view, rather than orienting with two planes.*

✓ **TIP:** *Be careful when selecting planes to orient your general view. You are making the view a child of the feature that created the plane. If that plane is removed, your view no longer has the references it requires to orient correctly. Whenever possible, use default datum planes or planar surfaces from the base feature of the part as references when orienting the view.*

Placing a Projected View

Next, place a top view of the wheel. Pick

Add View ➡ Projection | Full View | No Xsec | No Scale | Done

These selections specify that you are going to add a view projected horizontally or vertically from an existing view. You will see the entire model from the chosen direction. It will not contain a cross section, and the scale of the view will be governed by the default drawing scale displayed at the lower left corner of the screen.

Pick above the first view to designate the center point for this top view. Now you have two views displayed.

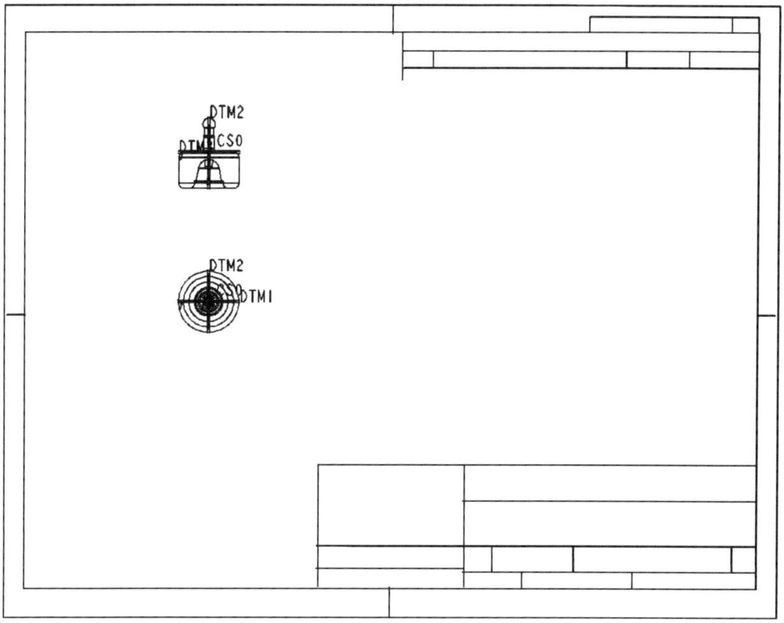

Figure 10-5
Drawing with two views displayed.

Placing a Projected Section View

Now you are ready to place the side view. This projected view will contain a cross section. It is the same type of projected view as the previous one. Pick **Add View ➥ Projection | Full View | Section | No Scale | Done**.

Specifying the Type of Cross Section

Because you specified that the view will contain a cross section, the XSEC TYPE menu is displayed. The options on this menu specify how the cross section will appear in the view. The following options determine the size of the view that will be shown as a cross-sectional view.

- **Full** – The entire view will be shown as a cross section.
- **Half** – The cross section will be shown only on one side of a plane. The other side of the plane will be displayed as a projected view with no cross section. This option is available only when **Full View** has been selected from the VIEW TYPES menu.
- **Local** – The cross section will be shown only within a closed boundary. The view outside the closed boundary will be displayed as a projected view with no cross section.
- **Full & Local** – The entire view will be shown as a cross section and a local cross section will be displayed within it.

These next options specify how the edges of the model that are behind the cross section will be displayed.

- **Total Xsec** – The cross-sectional area and the model edges behind the cross section will be displayed when this option is selected.
- **Area Xsec** – Only the cross-sectional edges will display in the view for this option.

The remaining options on the menu specify how offset cross sections will be displayed in the drawing. Offset cross sections are not defined by a plane. They are defined by a sketch projected through the model. A common use of an offset cross section is shown in the following picture. The section cuts through a pattern of bolt holes that straddle the centerline of the part.

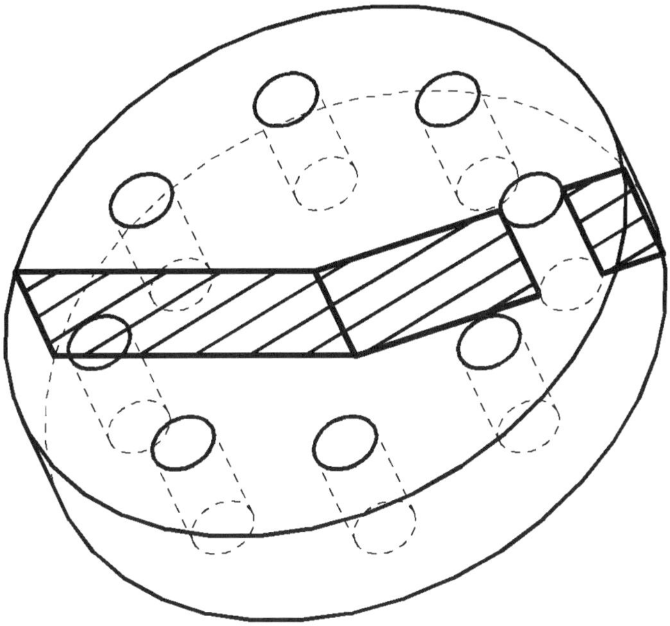

Figure 10-6
Example of an offset cross section.

The options follow:

- **Align Xsec** – The offset cross section will be unfolded about an axis and the model edges behind the cross section are not displayed.

- **Unfold Xsec** – The offset cross section of a general view is flattened and the model edges behind the cross section are not displayed.

- **Total Unfold** – The offset cross section of a general view is flattened and the model edges behind the cross section are displayed.

Pick **Full | Total Xsec | Done** from the menu. This specifies that the cross section will cover the entire view and that the edges of the model behind the cross section will be displayed also. Next, pick to the right of the first view to designate the location for the center of the new view.

Creating a Cross section

If you had created the cross section while in Part mode, you could retrieve it now by picking the cross section name from the XSEC NAMES menu. However, this cross section was not made previously, and you will have to create it. From the XSEC ENTER menu, pick **Create ➥ Planar | Done**. Provide a cross section name in response to the prompt.

```
Enter NAME for cross section [QUIT]: a ↵
```

> ➾ **NOTE:** *The name you give the cross section will be placed on the drawing as the section name. If you use the name* `a`*, the cross-sectional view will be labeled* `SECTION A-A`*.*

Now pick DTM2, from any view, as the cutting plane for the cross section. When prompted to pick a view to display the section arrows, pick in the first view you created. Your drawing now has three views placed.

> ➾ **NOTE:** *If you happen to have your datums turned off in ENVIRONMENT (which will prevent you from being able to select a datum or turn it back on in the middle of this command), you can still make the selection by using the commands **Select ➥ By Menu | Datum | Name.** This will list the datum names for you and allow you to select one. This technique also comes in handy for other commands, so log it away in your Tips bag for future use.*

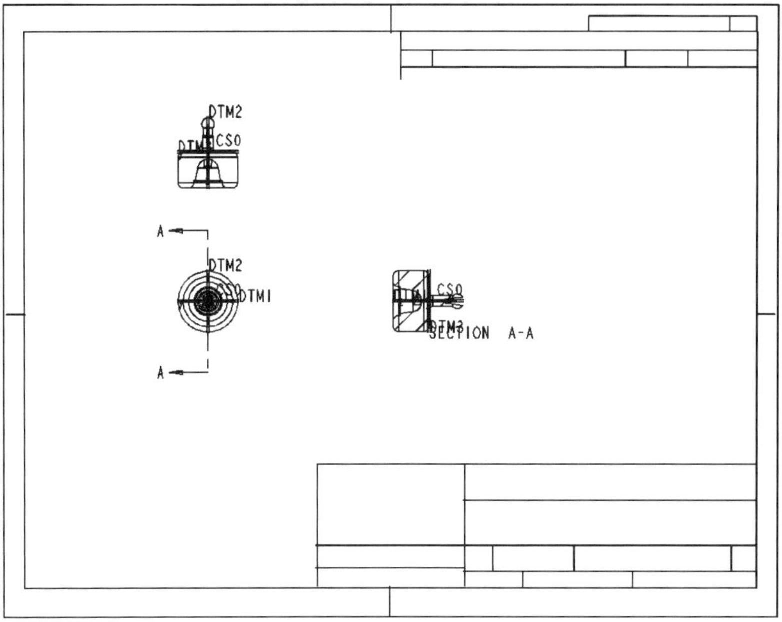

Figure 10-7
Drawing with three views displayed.

> ✗ ***WARNING:*** *Cross sections are saved in the part file instead of the drawing file. When you create a cross section in Drawing mode and save the drawing, the part file will automatically be saved. This could pose a problem if you have write privileges to the drawing file but not the part file. This could also be a problem if someone is modifying the part file at the same time you are making this change.*

Placing a Detailed View

Make a detailed view of the snap relief shown in the section view. Pick

Add View ➡ Detailed | Full View | No Xsec | Scale | Done

Pick a location in the top right corner on the drawing to place the center point of the view. For this detailed view, you specified that it would have a scale different from the other views. Enter that scale now.

```
Enter scale for view [1.000]: 4 ↵
```

Specifying the Detail to View

You will be prompted to pick a center point on an existing view. Then you can sketch a spline around that point to identify the entities you want displayed in the detailed view. The location you select for the center point must be on an entity and not out in space.

> **NOTE:** *You previously picked a position for the detail view. Because Pro/ENGINEER does not know which items or which view to detail, it needs more information. You must first identify a point on an entity in the view you wish to detail. That point will be used as the center point of the detail view.*

Zoom in on the section view. Pick near the x in the following figure to locate the center of the detail.

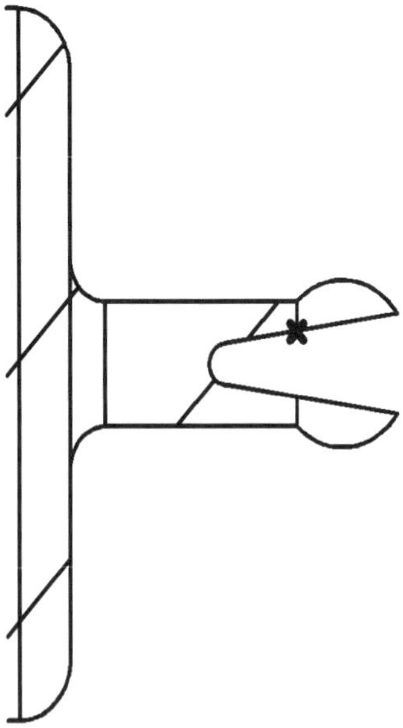

Figure 10-8
Point that will be used as the center point of the detail.

The following prompt informs you to sketch a spline to use as the boundary of your detail.

```
Sketch a spline, without intersecting other splines,
to define an outline.
```

Using the next figure as a guide, pick the points with the left mouse button, from point 1 to point 8. Then press the middle mouse button to let Pro/ENGINEER close the boundary for you.

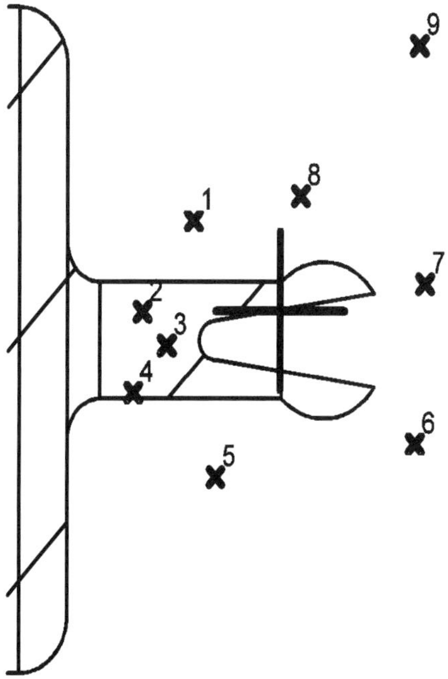

Figure 10-9
Pick points for enclosing the area to detail.

NOTE: *The center point must be enclosed within the boundary of the spline you sketch. It is a good practice not to overlap these splines when closing the loop. When overlapped splines are used, the geometry created for the break will look odd (not to normal drafting practice).*

Naming the Detailed View

You must now name the detailed view.

```
Enter a name for the detailed view: b ↵
```

Pro/ENGINEER will create a circle that will enclose your sketched spline. A note `SEE DETAIL B` will be placed with a leader pointing to the circle. You need to pick a point on the drawing where you want the note to be located. Pick point 9 from the previous figure. Fit your drawing to the screen. You now have four views in your drawing.

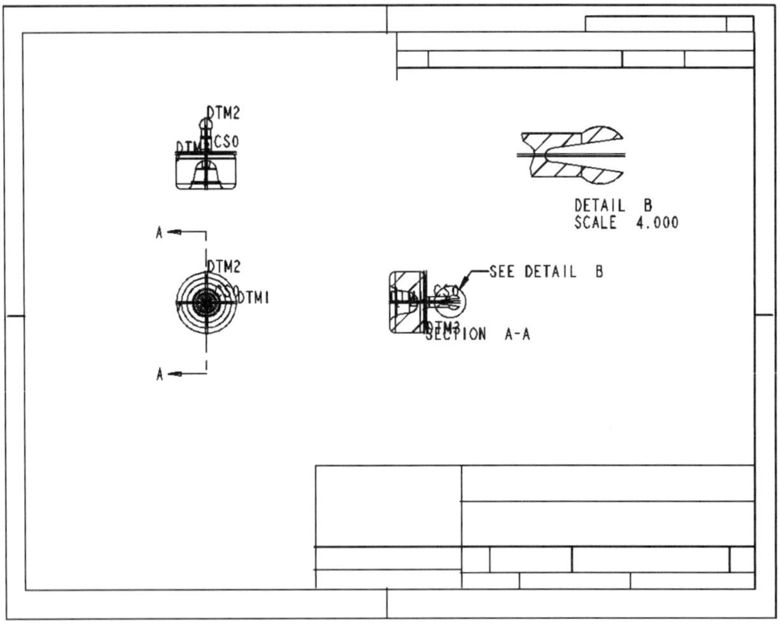

Figure 10-10
Drawing with four views.

➻ **NOTE:** *A detailed view is a child view. You must delete it before its parent can be deleted.*

Changing the Size and Position of Views

The drawing is beginning to shape up, but you still have some work to do.

- ❒ Views are not located or sized the way you want them.
- ❒ Some entities do not appear the way you want them.
- ❒ Entities are displayed that you do not want on the drawing.
- ❒ Some entities, other than dimensions and notes, are missing.

Modifying the Size of Views

Some information is displayed below the drawing. The drawing scale, the type of model (part or assembly), the model name, and the drawing sheet size are shown. The drawing scale can be modified just as a part dimension can. Pick **Done/Return** to dismiss the VIEWS menu. Now pick **Modify** from the DRAWING menu, and then pick on the value for the drawing scale. The scale will be located in the lower lefthand corner of the message window. Change it to 2.

```
Enter value for scale [1.000]: 2 ↵
```

All views that were placed with the **No Scale** option are updated to the new scale.

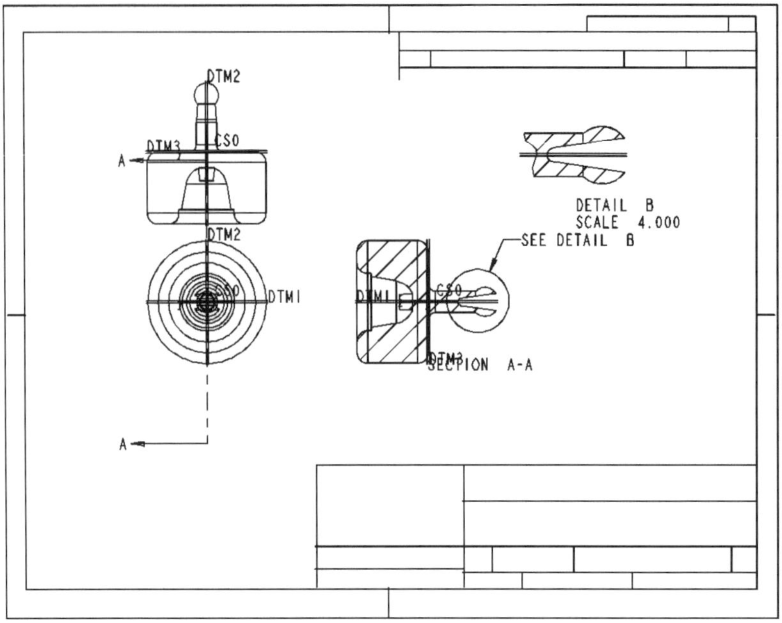

Figure 10-11
Drawing with a scale of 2.

➾ **NOTE:** *The detailed view did not change scale with the other views. That is because it was created with its own scale. You can modify the detail view scale using the same procedure used for modifying the drawing scale. Just pick on the value of the scale located below the detail view label.*

Moving Views

Your views need to be repositioned on the drawing. Select **Views ➥ Move View** from the DRAWING menu. Pick on the section view and then pick near the bottom right corner of the title block. Notice how the section view moves only horizontally.

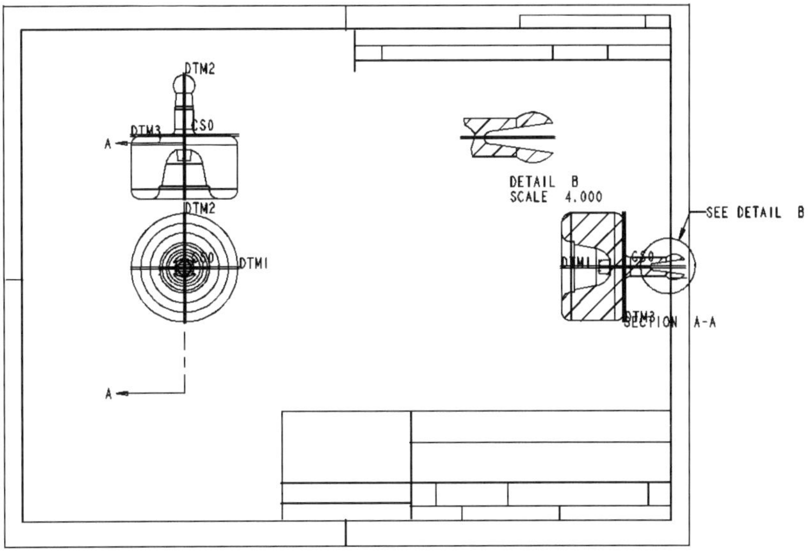

Figure 10-12
The drawing after moving the section view.

NOTE: *A projected view will always stay aligned with its parent view.*

Now move the first view. Pick on the first view and move it near the top left corner of the border. Notice how the section view moves up the same amount, but its horizontal location on the drawing remains unchanged. The top view moves horizontally to maintain its alignment with the first view but does not change its vertical location.

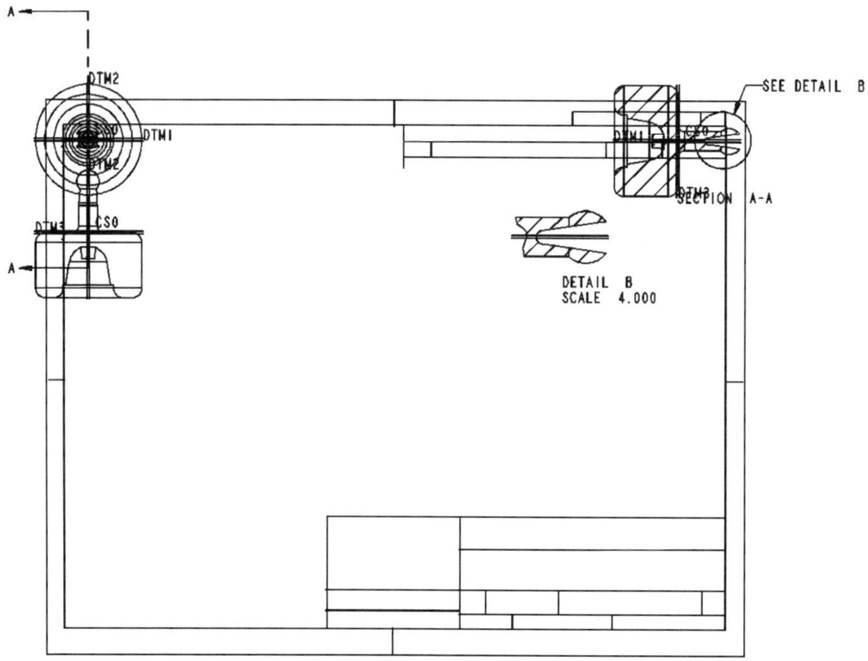

Figure 10-13
The drawing after moving the first view.

Position the views so that the section view is above the title block and the first view is inside the border. Move the top view so that it is within the border. Move the detailed view as needed. Use the following figure as a guide for the final location of the views.

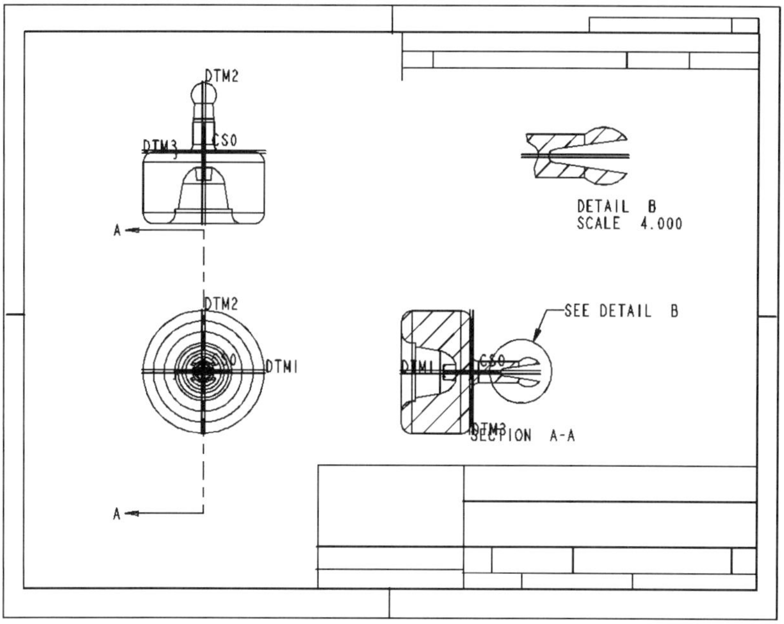

Figure 10-14
The drawing with the views relocated.

➾ ***NOTE:*** *No other views move when the detailed view is moved. It is not projected from another view, and no views are projected from it.*

Changing the Display of Entities

Entities can be turned on or off in each view.

Controlling the Display of Entities in All Views

You really do not want the datum planes, datum coordinate system, or the axes showing on the drawing. All of the options you select in this command should already be checked. To not display one of these types of entities, you must remove its check mark. From the MAIN menu, pick the following commands:

Environment ➡ Disp DtmPln | Disp Points | Disp PtnTags | Disp Axes | Disp Csys | Done-Return

The check mark should have been removed from each item you selected. Now repaint the screen and the drawing should look a little better.

> ✓ ***TIP:*** *The display of points and the point names were also turned off in the previous command even though none existed in the drawing. Get into a habit of removing all of the display settings when you work with a drawing. These environment settings can also be controlled through your* config.pro *file.*

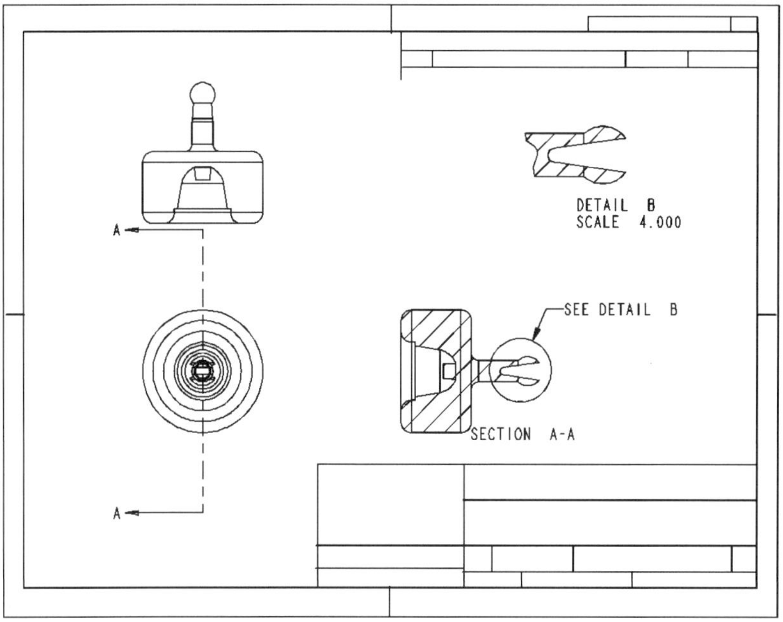

Figure 10-15
The drawing with the datum planes, datum coordinate system, and the axes not displayed.

Controlling the Display of Entities in Individual Views

You do not want the hidden line display of each view determined by the hidden line setting on the ENVIRONMENT menu. That environment setting may be changed during your next Part or Assembly mode session. From the VIEWS menu, select **Disp Mode** ➥ **View Disp**. Pick the top view, and then pick **Done Sel** to indicate that you have finished picking views to change. From the VIEW DISP menu, pick **Hidden Line** | **Done** so that hidden lines will always be shown in the top view.

All views that will have the same selections from the VIEW DISP menu can be selected at one time. Pick each of the remaining views in the drawing. Pick **Done Sel**, and then pick **No Hidden | Done** from the VIEW DISP menu. These views are now displayed with all hidden lines removed, no matter how the ENVIRONMENT menu is set for the display of hidden lines.

> ✓ **TIP:** *Each drawing view, by default, is set to display hidden lines according to the ENVIRONMENT menu setting.*

That is what **Default** specifies on the VIEW DISP menu. With such a setting, the drawing may look different each time you view it, based on the current ENVIRONMENT menu setting for the hidden line display. A better way to set the hidden line display in your drawing is to specify the desired hidden line display for each view from the VIEW DISP menu. Then you can use the ENVIRONMENT menu setting to change the hidden line display in Part or Assembly mode without causing an unwanted change in the drawing.

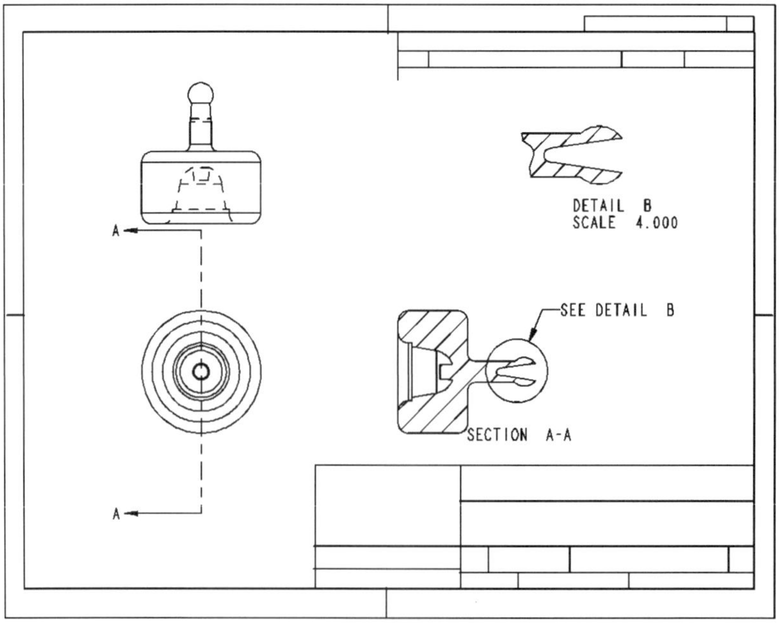

Figure 10-16
The drawing with hidden lines removed except for the top view.

The first view still has too many edges displayed. All of the tangent edges next to the rounds are displayed. By default, each view is set to display tangent edges according to the setting for tangent edges on the ENVIRONMENT menu. **Tan Default** on the VIEW DISP menu is the setting to instruct Pro/ENGINEER to use the ENVIRONMENT menu setting for displaying tangent edges. The display of these tangent edges can be specified just as you did for hidden lines. In fact, you can specify the type of tangent edge display for the view while you are specifying the type of hidden line display for that view. From the VIEWS menu, pick **Disp Mode ➡ View Disp**. Pick the first view and then select **Done Sel**. Select **No Disp Tan | Done** from the VIEW DISP menu. The display of the tangent edges in the first view is changed as shown in the following figure.

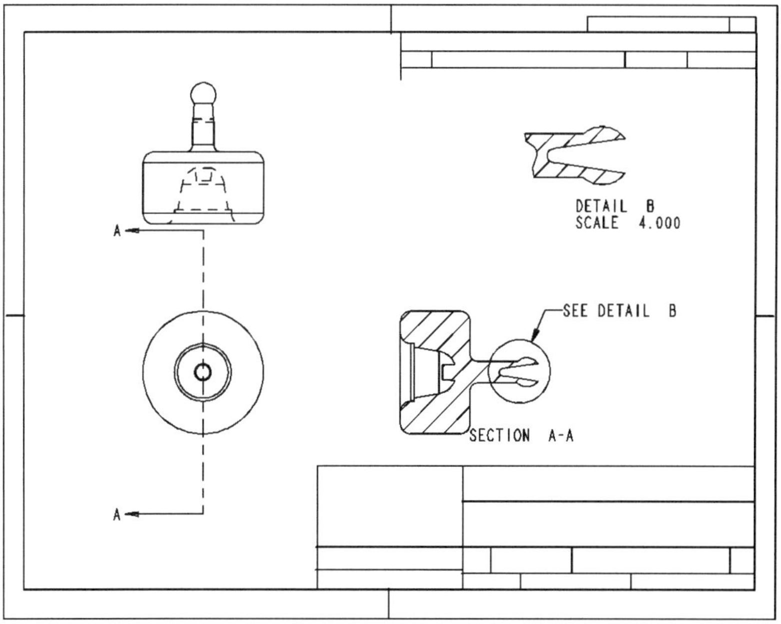

Figure 10-17
The drawing with the tangent edges removed in the front view.

Now pick the top view, and then select **Disp Tan | Done** from the VIEW DISP menu. Do the same for the section view. No view now has the tangent edge display specified by the ENVIRONMENT menu. Because the ENVIRONMENT menu is currently set to display tangent edges, none of the selected views should change.

> ➾ **NOTE 1:** *You do not need to select the detailed view because it will reflect the settings of its parent view.*

> ➾ **NOTE 2:** *In this example, you cannot pick the top view and the section view to change at the same time because they do not share the same setting for hidden line display.*

Changing the Appearance of the Crosshatching

The initial appearance of the crosshatching in the section view may not be what you want. The spacing of the lines, the angle of the lines, or the crosshatch pattern may be wrong. Pick **Modify ➡ Xhatching** from the DRAWING menu. Pick in the section view. In that there are no more sections to be modified, pick **Done Sel** from the GET SELECT menu. To change the spacing of the crosshatching, pick **Spacing**. In this case, you need to change the crosshatching spacing to be smaller, so pick **Overall | Half** from the MODIFY MODE menu. The spacing is smaller, as shown in the following figure.

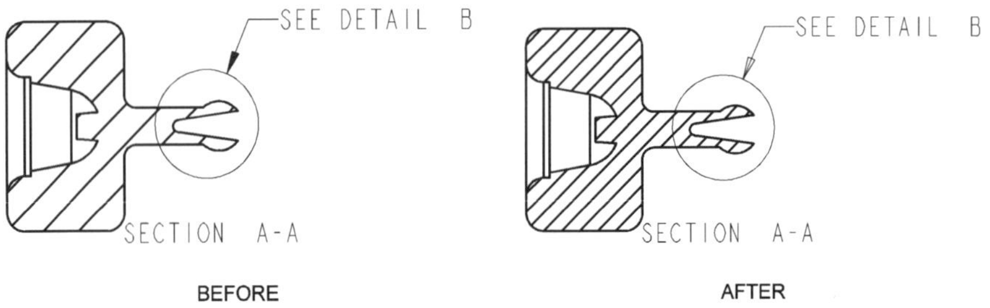

Figure 10-18
Change in the crosshatch spacing.

➾ **NOTE:** *You can select* ***Value*** *from the MODIFY MODE menu so that you can enter a value for the spacing instead of selecting* ***Half*** *or* ***Double****.*

✓ **TIP:** *Each time you pick on* ***Half****, the spacing is reduced. If you pick it too many times and the spacing gets too small, pick on* ***Double*** *to increase the spacing.*

You can also change the angle of the crosshatching. Pick **Hatch | Angle ➡ Overall | 150** from the CROSS XHATCH menu.

➾ **NOTE:** *You can select* ***Value*** *from the MODIFY MODE menu so that you can enter a value for the angle that is not listed on the menu.*

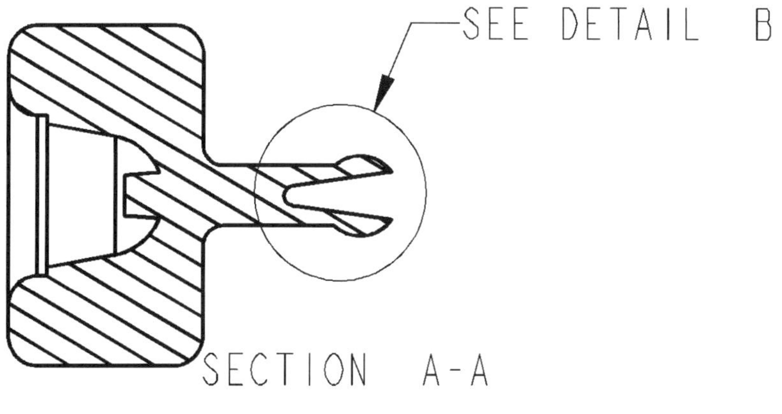

Figure 10-19
Change in the angle of the crosshatching.

You can also change the crosshatch pattern so that it matches the style for a particular material. Pick **Hatch | Retrieve ➡ List | Plastic** from the CROSS XHATCH menu to change to the pattern to represent a plastic material.

➾ **NOTE:** *Whenever you retrieve a crosshatch pattern, the crosshatch angle is reset to the default angle and the crosshatch spacing* *is reset to the default spacing. Select* ***Hatch*** | ***Angle*** ➡ ***Overall*** | ***150*** *to change the angle back to 150 degrees.*

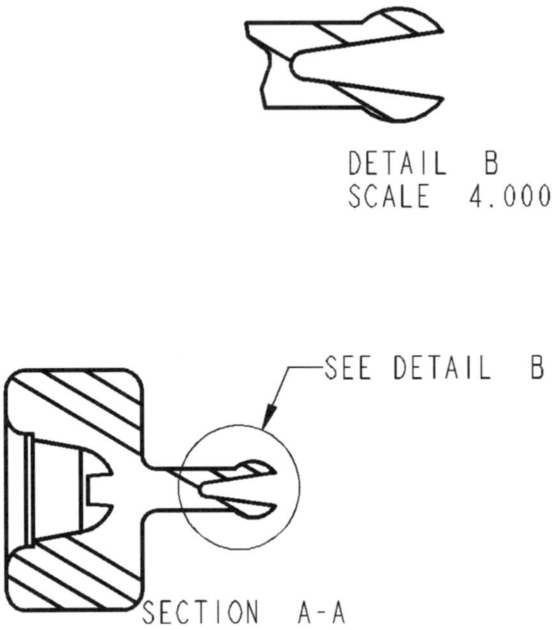

Figure 10-20

New crosshatch pattern with the angle set to 150°.

➾ ***NOTE:*** *You can create a new crosshatch pattern and save it for future use by selecting* ***Hatch*** | ***Save****, and then providing a name for the new crosshatch style.*

After making all of your crosshatch modifications for the selected section view, pick **Done** from the CROSS XHATCH menu. All child views of the selected section view will update.

➾ ***DESIGN NOTE:*** *The detailed view's crosshatching changed as the crosshatching in the section view changed. That occurred because the section view is its "parent view." To break this hatching relationship, use* ***Modify*** ➡ ***Xhatching****. Pick the detailed view and*

Done Sel *from the GET SELECT menu. Next, select* ***Det Indep*** *from the CROSS XHATCH menu. You can select* ***From Parent*** *to restore the crosshatching parent/child relationship.*

Displaying Selected Centerlines

A cross hair should be displayed in the first view to represent the axis of the wheel, and centerlines should be used in all of the other views. Pick **Detail ➡ Show ➡ Axis ➡ By Feature** from the DRAWING menu. In the first view, query select until you get feature 5, the first protrusion of the wheel. Now you have the cross hair displayed in the first view and a centerline displayed in the projected views. The centerline did not appear in the detailed view because the detailed view does not show any portion of feature 5. To place the centerline in the detailed view, pick **Feat & View** from the SHOW ITEM menu. In response to the prompt to select the view in which to display the centerline, pick in the detailed view. Next, pick on the axle in the detailed view. All of the centerlines are now displayed.

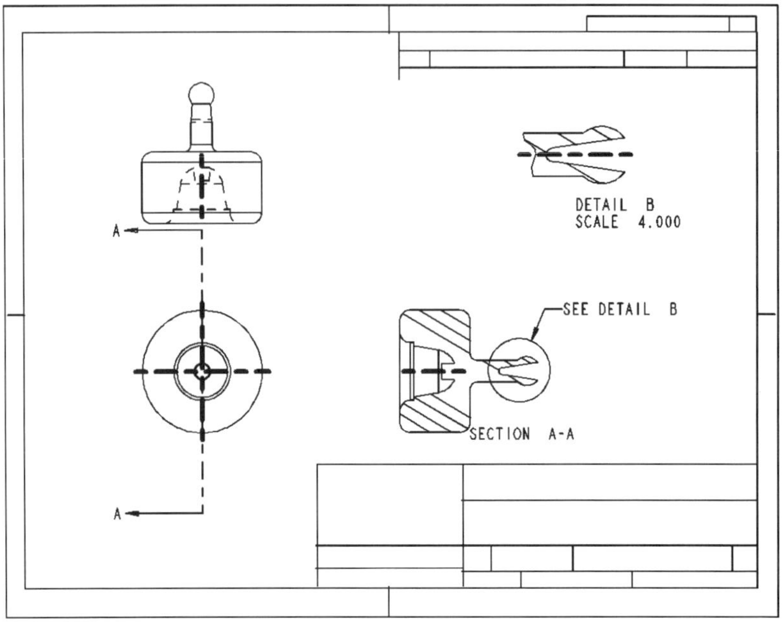

Figure 10-21
The drawing with a cross hair and centerlines displayed.

Modifying the Length of the Centerline

The centerlines in the projected views are fit to the size of feature 5. You should extend them so that they cross through the axle. Pick **Done/Return** to exit from the SHOW ITEM menu and then pick **Move** from the DETAIL menu. Pick on the centerline in the section view, and a circle will display on one of its endpoints. That is the end of the line that will now move. Move the cursor to the new location for that endpoint and press the left mouse button. Now pick **Clip**, and pick the centerline again. The circle is displayed on the other end of the centerline this time. Change its location in the same fashion as before. Carry out the same operation for the centerline in the top view so that the drawing appears as follows.

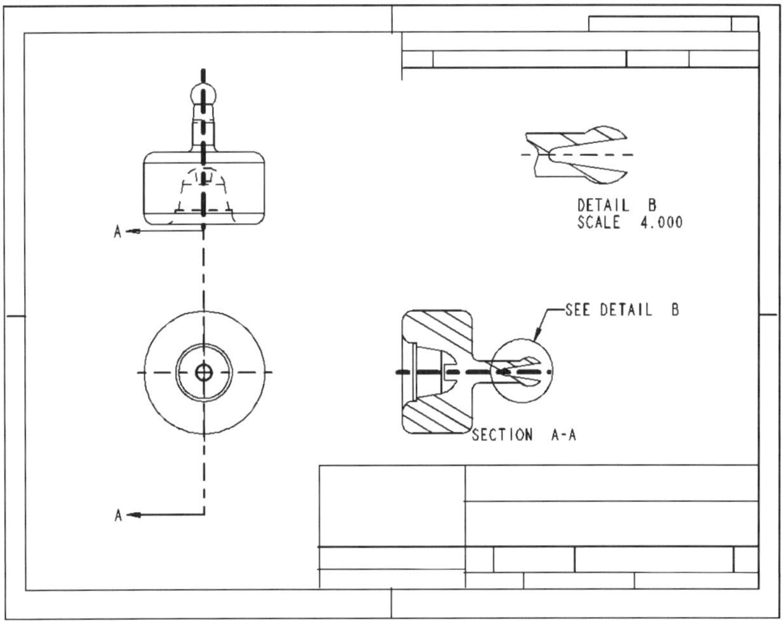

Figure 10-22
The drawing with centerline endpoints moved.

Changing Drawing Formats

It is apparent that an A-size drawing will not be big enough for these views. You need to change to a B-size drawing format. Pick **Sheets ➥ Format ➥ Add/Replace | Unblank** from the DRAWING menu. Now select the new drawing size.

```
Enter Format name [A]: b ↵
```

The drawing is now changed to a B size. Pro/ENGINEER will automatically change the default drawing scale by a factor equal to the ratio of the new format size to the old format size. Modify the drawing scale to be 2 again. Also modify the scale of the detailed view. Change it to 6. Move the centerline endpoints as required. Move the views so that the drawing looks like the following figure.

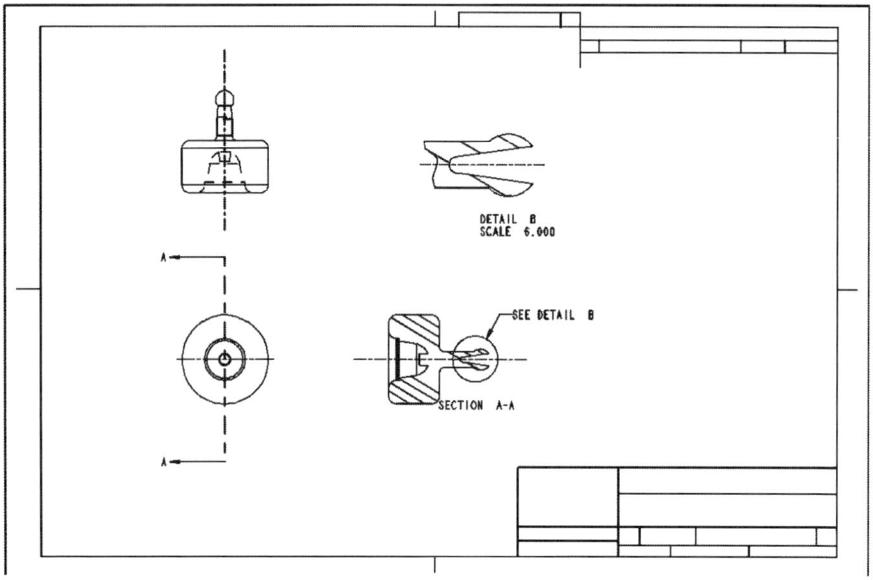

Figure 10-23
The drawing on a B-size format.

Now the drawing views are complete, and you need to show the dimensions and add the notes.

Placing Dimensions on the Drawing

The part already contains all dimensions required to fully constrain it. Instead of creating dimensions for the drawing, just display the part dimensions.

Showing the Dimensions

From the DETAIL menu, pick **Show ➡ Dimension ➡ Show All.** Now all of the dimensions for the part are displayed.

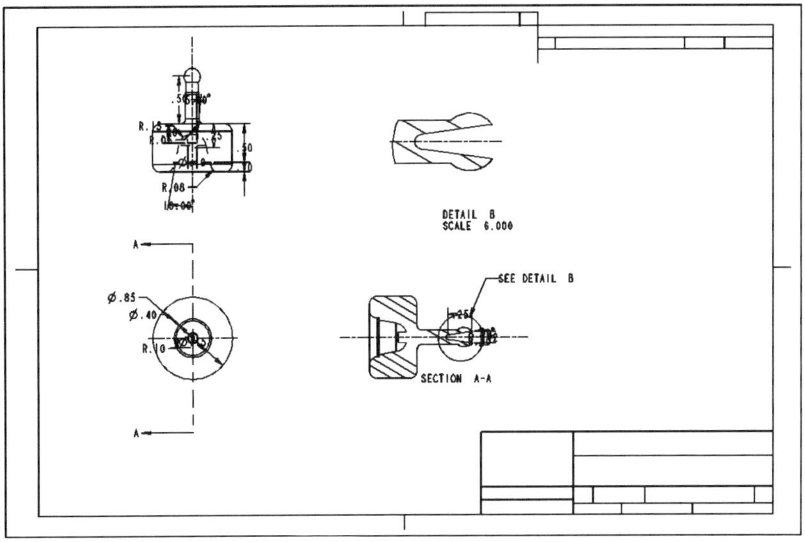

Figure 10-24
The drawing with the dimensions shown.

> ➾ ***NOTE:*** *As you see in the drawing, no dimensions are shown in the detailed view. The majority of the dimensions are shown in the top view. When a dimension is valid in several different views, Pro/ENGINEER will choose the view that was placed earliest. In this case, the top view was placed before the section view. If the section view had been placed before the top view, most of the dimensions would have appeared in the section view.*

Well, the drawing looks like a mess now, but you know that all of the dimensions required to fully constrain the part are displayed. They are just not located properly. The first step is to move the dimensions from on top of each other so that you can tell what they are.

Cleaning Up Dimensions

There is a way to perform a preliminary cleanup of dimensions. From the DRAWING menu, pick **Clean Dims**. Pro/ENGINEER will ask you for the spacing from the geometry to the first dimension line and for the spacing between dimension lines.

```
Enter offset for the first dimension line [0.5]: ↵
Enter distance between dimensions [0.375]: .75 ↵
```

Now pick in the top view. Pro/ENGINEER will move the linear dimensions so that they are not on top of each other.

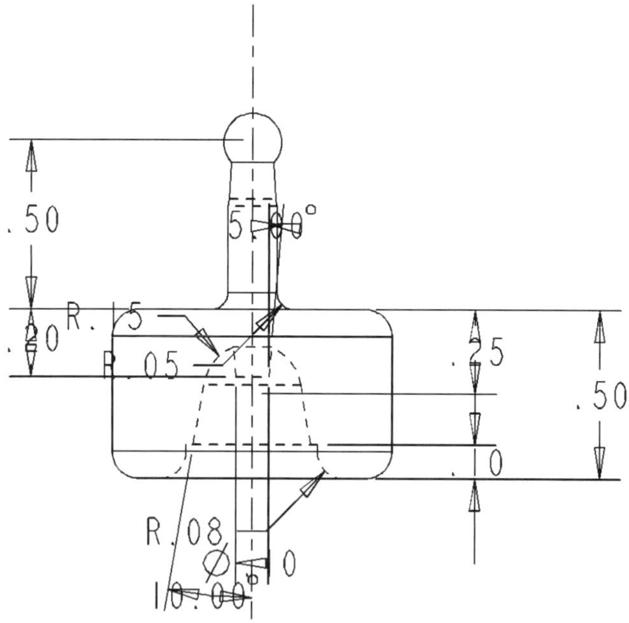

Figure 10-25
Top view after using the Clean Dims *command.*

Pro/ENGINEER will ask you if you want to keep this change. Press <Return> to keep the changes.

```
Do you want to restore previous dimensions? [N]: ↵
```

Switching Dimensions Between Views

Some of the dimensions would look better if they were shown in another view. The next step is to switch the dimensions to their ideal view. Zoom in on the section view and the detailed view. Select **Switch View** from the DETAIL menu. Switch all dimensions for the snap relief in the section view to the detailed view. Those dimensions are the two 8.00-degree snap relief angles, the .25 relief depth, and the .12 relief width. To make the switch, pick the text of the dimension you want to switch and then pick in the detailed view.

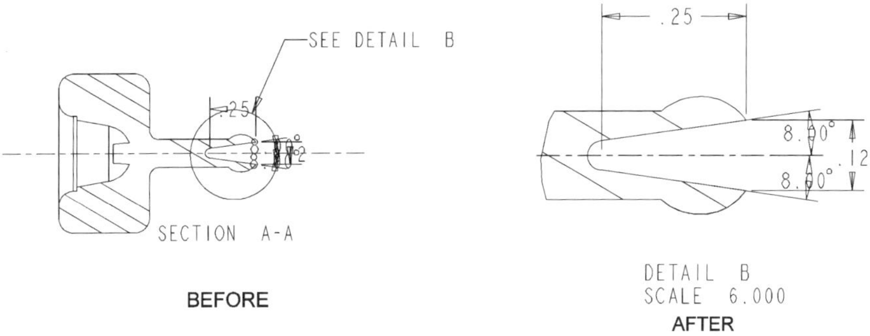

Figure 10-26
The snap relief dimensions switched to the detailed view.

Now switch the .10 radius for the snap and the .15 axle diameter from the front view (the first one you placed) to the detailed view.

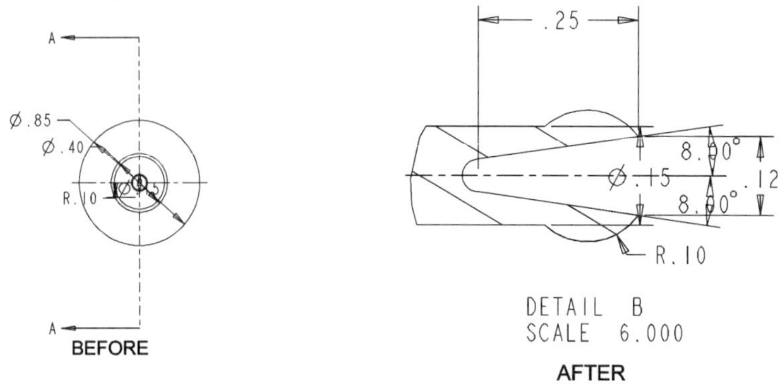

Figure 10-27

The snap radius and axle diameter dimensions switched to the detailed view.

Switch all of the other dimensions in the top view, except the .08 radius and the .05 radius, to the section view.

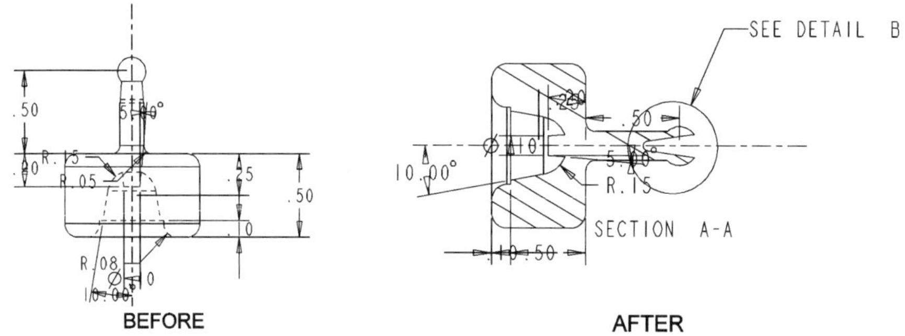

Figure 10-28

The top view dimensions switched to the section view.

Now switch the .85 diameter from the first view to the top view, and switch the .40 diameter to the section view.

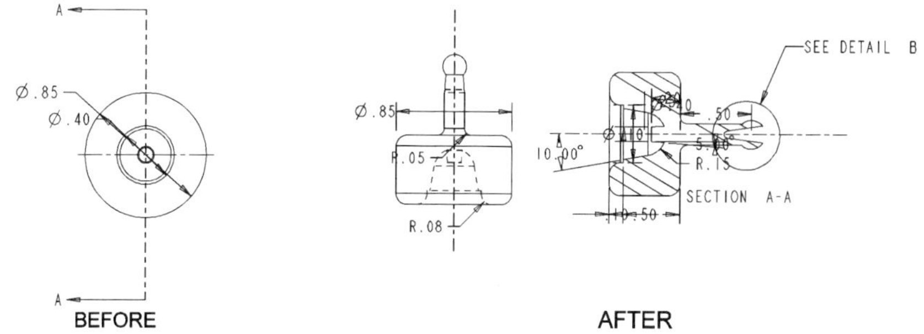

Figure 10-29
The remaining dimensions switched between views.

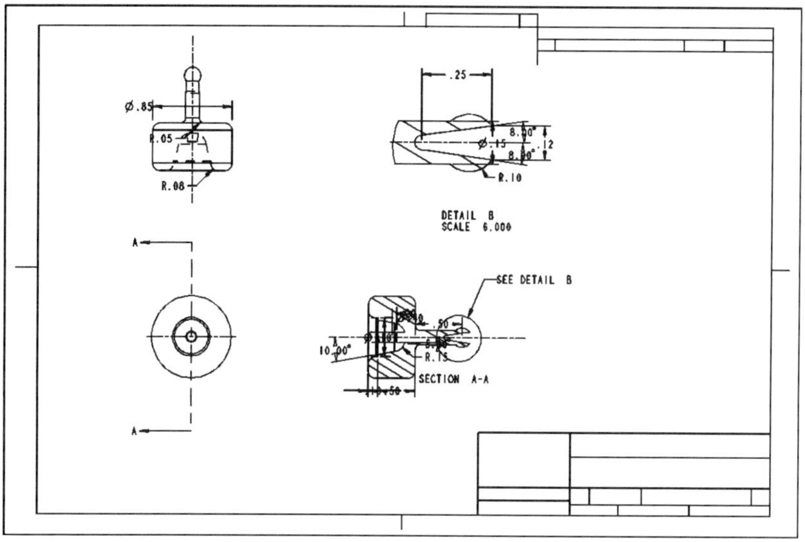

Figure 10-30
The drawing after all dimensions have been switched to the appropriate view.

Changing Dimension Appearance

Moving Dimensions

Now you will get started moving the dimensions so that you can read them easily.

> ✓ **TIP:** *You can select **Clean Dims** again to begin moving the dimensions after they have been switched between views.*

Select **Clean Dims** from the DRAWING menu, and let Pro/ENGINEER rearrange the dimensions in the top, section, and detailed views using the same offset dimensions as before. Pick **Move** from the DETAIL menu. Now pick the dimension text you want to move. Pick the .50 dimension for the length of the axle. Move the cursor down and press the left mouse button again. The dimension relocates to the new cursor position. Notice how the dimension dynamically drags and follows your cursor. Now select the .50 dimension again and move the cursor to the location for the dimension shown on the following figure and press the left mouse button again.

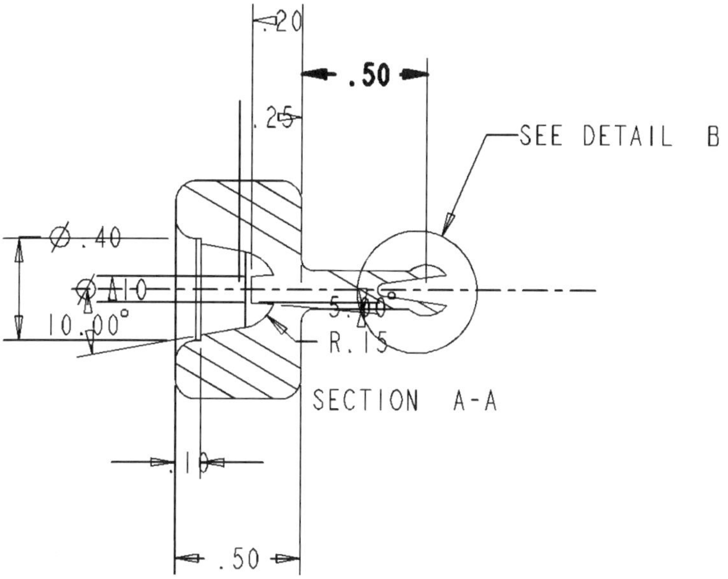

Figure 10-31
Position for moving the .50 axle length dimension in the section view.

Using the **Move** command, pick the .05 radial dimension in the top view. Move the cursor to different positions around the view. The angle of the leader line and the arrow tip remain fixed.

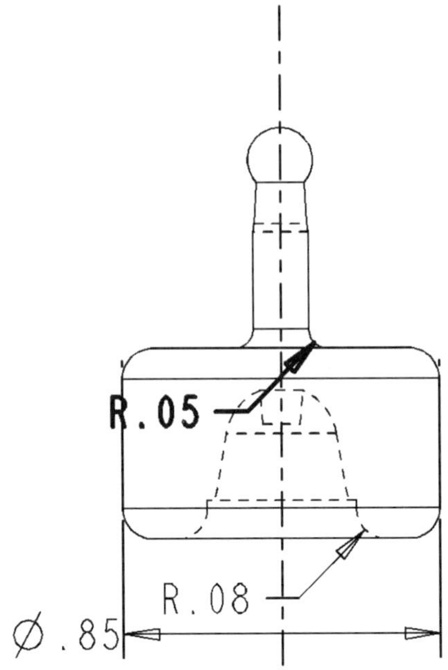

Figure 10-32
The radial dimension attachment location.

Now pick **Move Attach**. This time when you try to move the dimension, the arrow position and the angle change, but you cannot change the length of the leader line. Use a combination of **Move** and **Move Attach** to position the dimension as seen in the next figure.

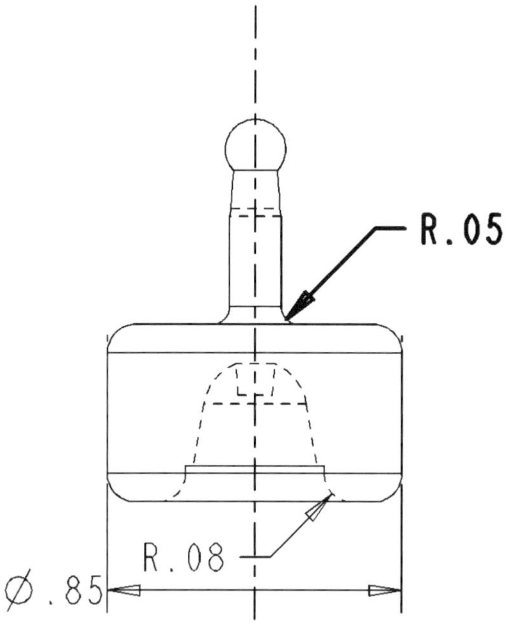

Figure 10-33
The radial dimension with the arrow on the other side of the arc.

To get the text of the dimension in the correct location, pick **Move Text**. When you move the dimension text this time, the elbow of the leader stays fixed and only the text moves horizontally.

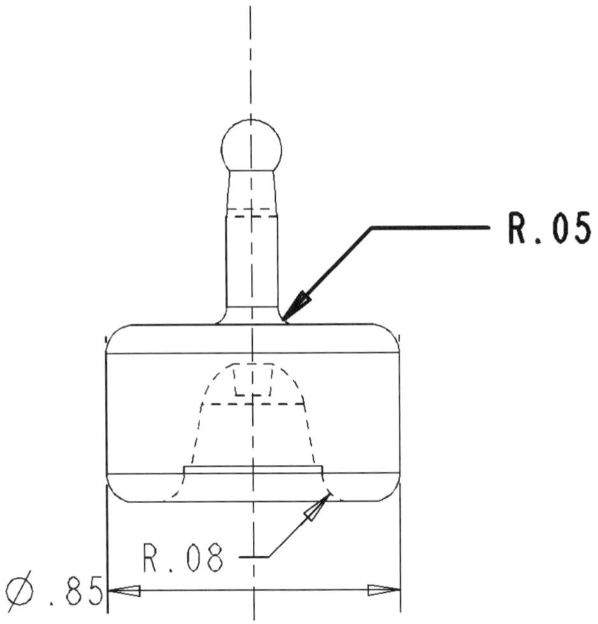

Figure 10-34
The radial dimension in its final position.

The .08 radius in the top view may be attached to an arc that is a hidden line, as is the case in the previous figure. The dimension is also valid for the visible arcs on the tire because you rounded all of those edges using the same radius value in a feature in Part mode. There are two ways you can attach the radius to one of the visible arcs. You can retrieve the wheel part again and redefine the round feature. You will need to delete the edge you want the dimension attached to and then select it again.

✓ ***TIP:*** *The last edge in the selection set of edges to round is the one the radius dimension will be attached to in the drawing.*

The other way to change the attachment location is the easiest. Just make a parametric note with a leader that attaches to the desired arc. Select the following commands:

Create ➡ Note ➡ Leader | Enter | Horizontal | Normal Ldr | Default | Make Note ➡ Arrow Head

Using the following figure, pick on one of the visible arcs to attach the note to and pick a location for the elbow of the leader line to appear. The variable name for the radial dimension in the example is `d29`. Now enter the text for the note.

```
Enter NOTE: R&d29
Enter NOTE: ↵
```

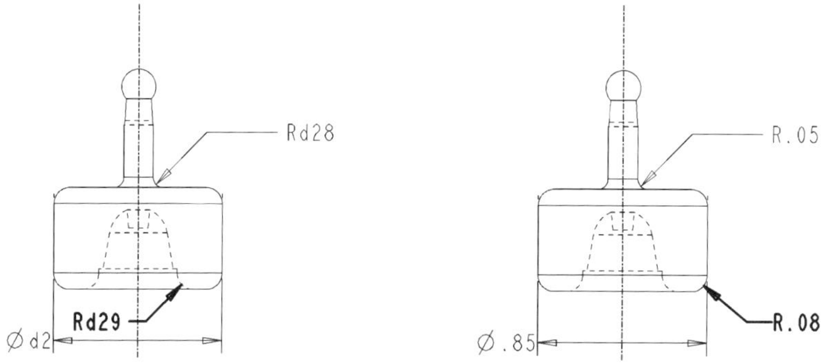

Figure 10-35
The radial dimension with variable name displayed and the equivalent parametric dimension for the radial dimension.

✓ ***TIP:*** *To move the note using **Move Attach**, you must select the new location by selecting on the entity it is attached to. This is different than the way you use **Move Attach** to move a dimension.*

Now, using these move commands you just learned, move all of the dimensions until the drawing looks like the following figure.

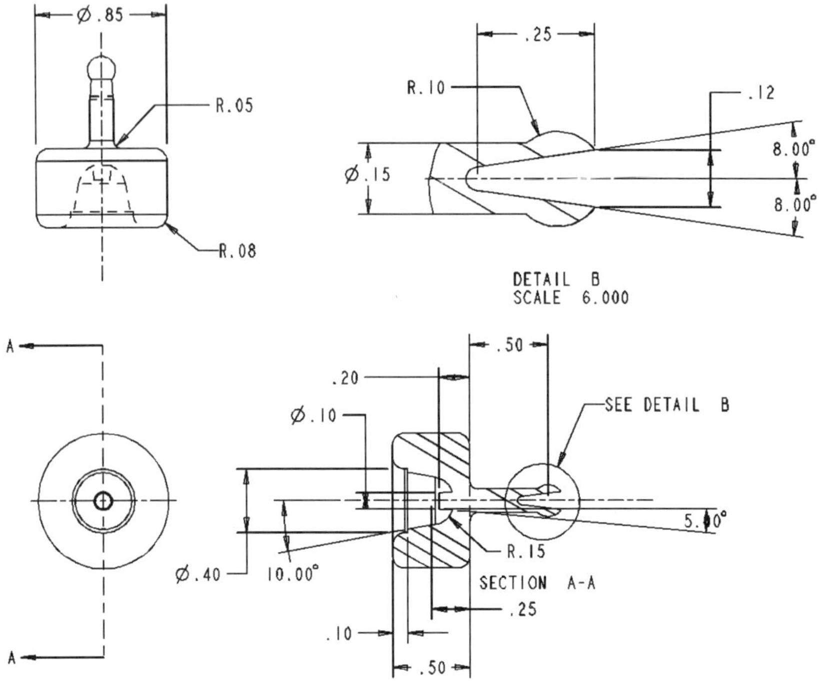

Figure 10-36
The current drawing.

Now move the 5-degree angle so that the dimension extension lines flip as shown in the following figures. To display the angle this way, pick on the angle text to be moved, and then pick on the opposite side of the angle's vertex.

> ✓ **TIP:** *If you cannot locate the text exactly where you want it without the arrows displaying on the wrong side of the angle vertex, place the text at a location where the arrows appear correctly. Then pick the text again and place it at the desired location.*

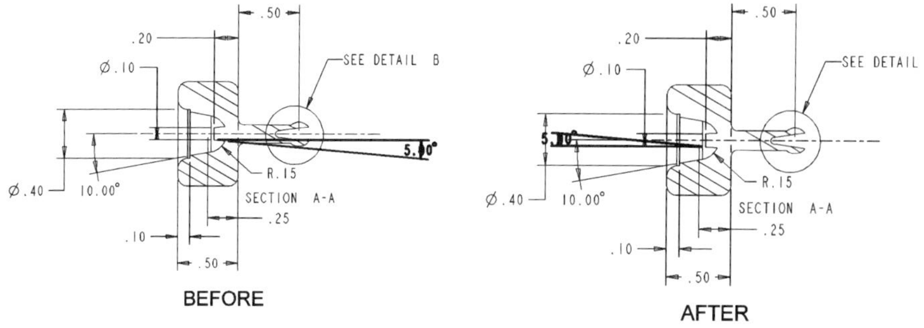

Figure 10-37
Dimension line flipped for the 5-degree angle.

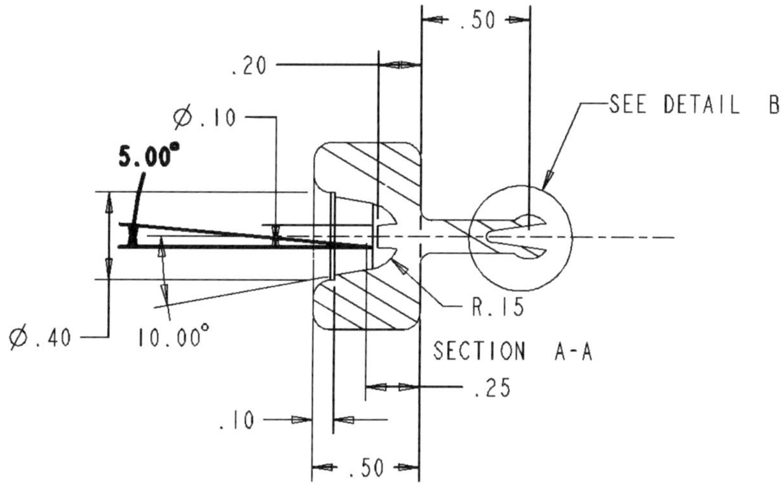

Figure 10-38
The section view with the angle dimension relocated.

Moving View Labels

The section view label overlaps some of the dimensions. You need to move the text to a better position. View labels are moved just the way dimensions are. Use the **Move** command to move the section view name so that it is centered below its view.

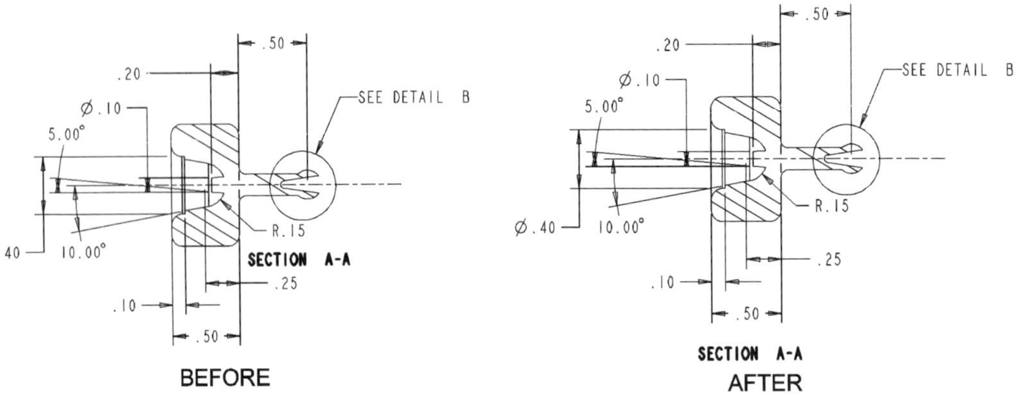

Figure 10-39
The drawing after moving the view label.

The section arrow labels can be moved in the same way as the other text. Select **Move** and then pick on the A of one of the section arrow labels in the first view. Now pick a new location that is closer to the section plane and closer to the geometry in the view.

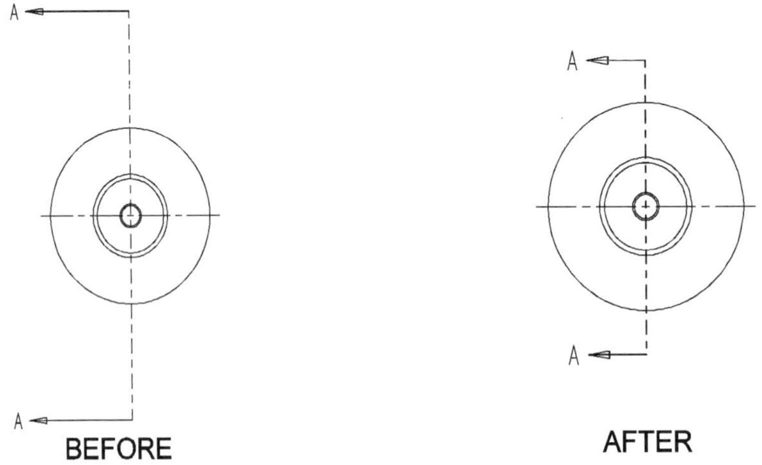

Figure 10-40
Section arrow labels moved.

➾ ***NOTE:*** *The opposite arrow moves closer to the section plane at the same time as the selected arrow does.*

Aligning Dimensions

You can align the arrows of some of the dimensions. Pick **Align** and then pick the .50 dimension for the axle length and the .50 dimension for the tire width. Press the middle mouse button to indicate that no other dimensions will be aligned with these. Next, pick a location on the drawing to align these dimensions through. Notice that these dimensions dynamically move just like when you used the move commands.

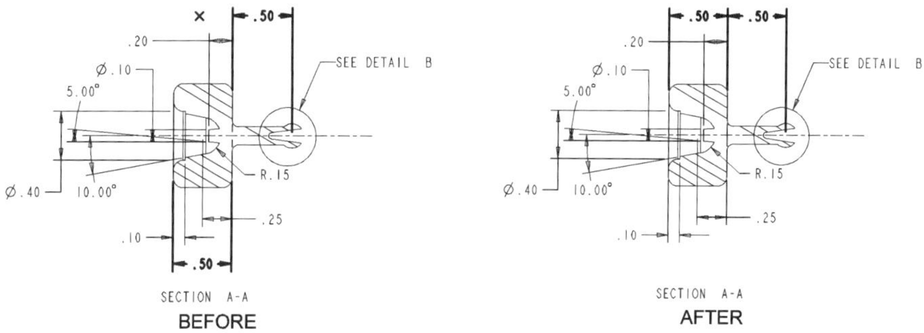

Figure 10-41
Aligned dimensions.

Erasing Dimensions

Sometimes dimensions required to fully constrain a part are not necessary on the drawing. As an example, the two 8.00-degree dimensions in the detailed view are not both required. Pick **Erase ➡ One Item | Dimension** and then pick the upper 8.00-degree dimension. That dimension will now disappear when you pick **Done Sel** from the GET SELECT menu.

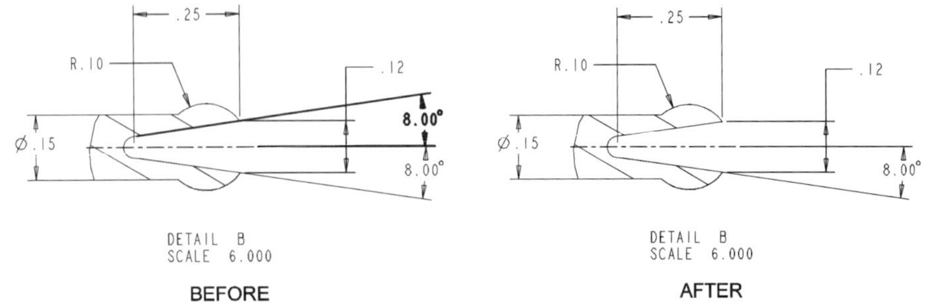

Figure 10-42
Dimension erased from the detailed view.

➾ ***NOTE 1:*** *Dimensions used to constrain a part cannot be deleted. They can only be removed from display by being erased.*

➾ ***NOTE 2:*** *The method to display a dimension after it has been erased is the same for showing a dimension for the first time. If you want all erased dimensions shown, pick* ***Detail*** ➥ ***Show*** ➥ ***Dimension*** ➥ ***Standard*** ➥ ***Show All****. If you only want one erased dimension shown, use* ***Feat & View*** *instead of* ***Show All****. Then, in the view you want the dimension to appear, pick the feature that used the dimension.*

A dimension used to constrain a feature in the part is frequently not the one you want to see on the drawing. The .25 depth for the snap relief groove is an example of this. Based on what you learned in Chapter 8, you could go back to Part mode and redefine the dimensioning scheme of the feature. To illustrate how to create new dimensions in Drawing mode, replace the unwanted dimension with a different created dimension. Erase the .25 dimension just as you did the 8.00-degree dimension.

Adding a Driven Dimension

To create the dimension for the groove depth, pick **Detail** ➥ **Create** ➥ **Dimension** ➥ **Standard** ➥ **New Ref** ➥ **On Entity**. Now pick on point 1 and then point 2, as shown in the following figure.

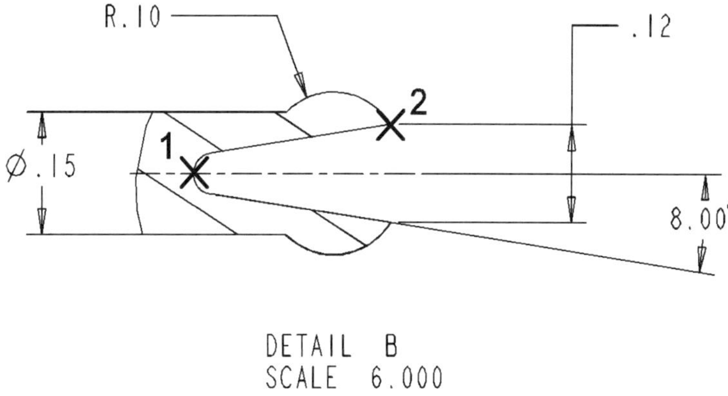

Figure 10-43
Pick location for creating the driven dimension.

Place the dimension text using the middle mouse button so that the detailed view will look like the following figure. Select **Tangent** from the ARC PNT TYPE menu. Next, pick **Horizontal** to orient the dimension as a horizontal dimension.

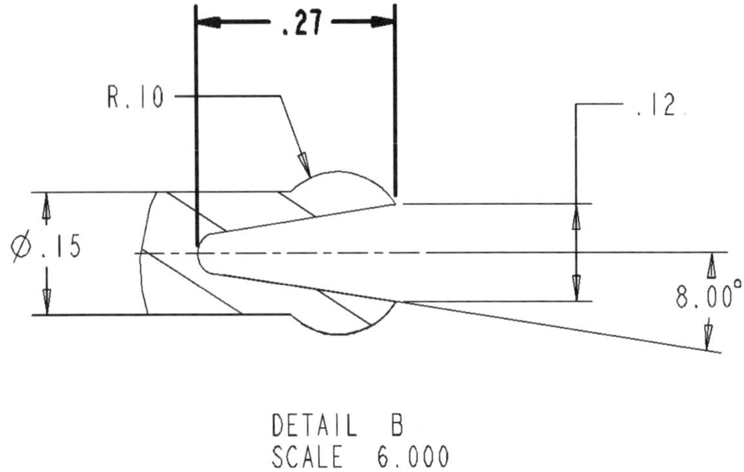

Figure 10-44
The detailed view with a driven dimension for the groove depth.

NOTE: *This new dimension for the groove depth is different from the original in more ways than its value. The dimension you just created is a* driven *dimension. It reflects the size of the model only. Later you will see that you can modify a dimension on a drawing, and the part will change. However, you cannot modify the value of a driven dimension. Driven dimensions only change their value when the part changes.*

Placing Reference Dimensions

Sometimes you may want to display a dimension in more than one view. To do this, you need to display one of the dimensions as a reference dimension. Make a reference dimension of the .50 axle length dimension and display it in the detailed view. This dimension is from the side of the tire to the center of the snap. Pick **Create ➥ Ref Dim ➥ Standard ➥ New Ref ➥ On Entity**. Now pick on point 1 and then point 2, as shown in the following figure.

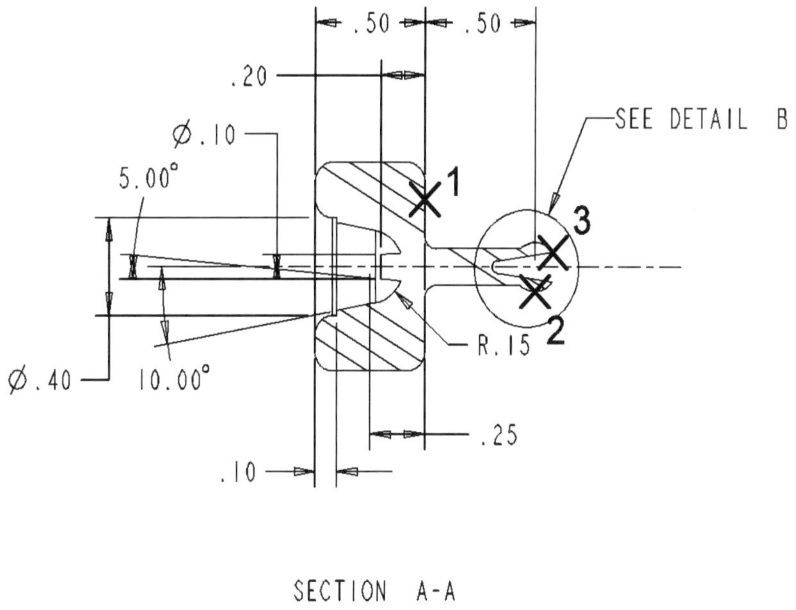

Figure 10-45

Pick locations for creating the reference dimension.

Next, place the dimension text and then pick **Center** from the ARC PNT TYPE menu.

> ➾ ***NOTE:*** *By default, Pro/ENGINEER appends the text REF after a reference dimension. If you want a reference dimension to be denoted by placing it in parentheses, add a setting to your* config.pro *file. The setting is* `parenthesize_ref_dim yes`.

Place a reference dimension in the section view for the axle length to the end of the snap. Pick point 3 and then point 1 from the previous figure, and place the text. Next, switch the .50 reference dimension to the detailed view. Move the reference dimensions so that the drawing looks like the following figure.

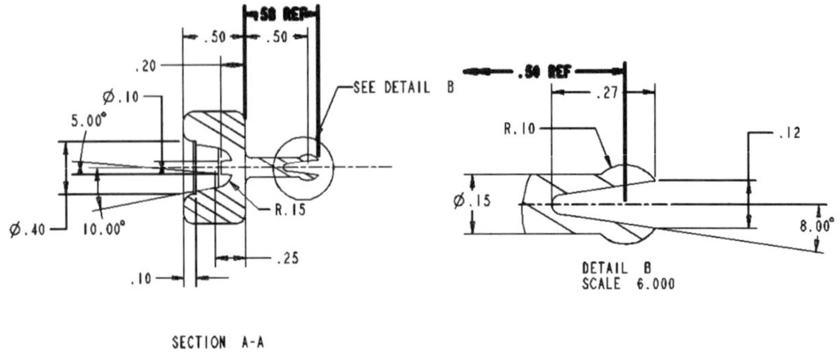

Figure 10-46
Drawing the reference dimensions.

➾ ***NOTE:*** *Reference dimensions behave similarly to driven dimensions. They cannot be modified: they merely reflect the size of the geometry.*

Flipping the Direction of Arrows

Some of the arrows need to point in the opposite direction to improve their appearance. Pick **Flip Arrows** and then pick each dimension text you want the arrows flipped on. Pick the .10 radius and the 8.00-degree angle in the detailed view. Pick the .58 reference dimension and the .15 radius in the section view. Now that you have the idea, flip the arrows on all of the dimensions until the drawing appears as in the next figure.

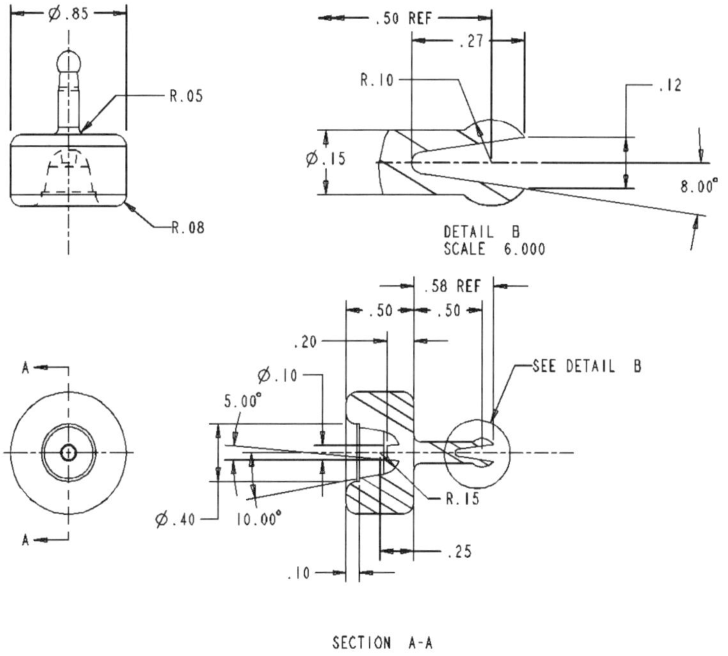

Figure 10-47
The drawing with arrows flipped.

Clipping Dimension Witness Lines

As you see on the drawing, many of the dimensions do not have a gap between the witness lines and the geometry. Pick **Clip** from the DETAIL menu. In this example, pick only the .50 dimension for the wheel thickness in the section view. Press the middle mouse button after selecting the dimension to indicate that no other dimensions will be selected to be clipped. Next, pick on the witness line of the dimension at the location to begin the gap.

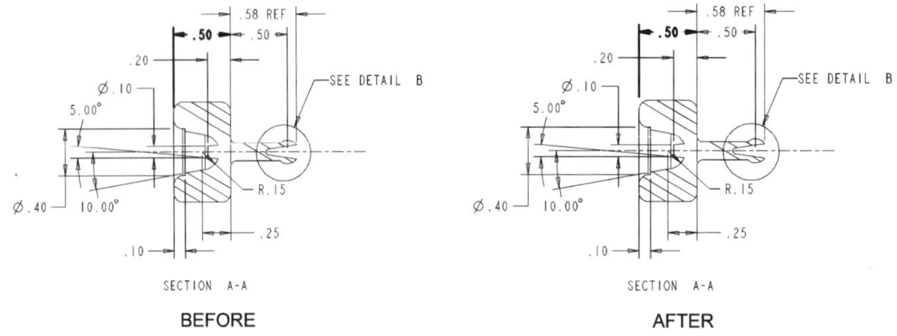

Figure 10-48
Clipping the witness line of a dimension.

> ✓ **TIP:** *It is not necessary to clip the dimension to provide a gap between the witness line and the part. Whenever you plot the drawing, the gap will be created, even though there is no gap when viewing the drawing on your monitor. The default selection* **Options ➥ Quality 3** *from the PLOT EDIT menu provides this function.*

Occasionally, you may need to move the double arrowheads of clipped dimensions, such as the .50 REF dimension in the detailed view. To move the double arrowheads, select **Clip** and then pick the dimension text. Move the mouse over the double arrowheads and press the middle mouse button. Now pick on the location you want the arrowheads to be moved to.

> ✓ **TIP:** *If you want to clip the witness line instead of moving the double arrowheads, pick the dimension text and then move the mouse to the opposite side of the witness line from the double arrowhead. Move the mouse, without crossing over the witness line, and pick the desired witness line clip location.*

Breaking Dimension Witness Lines

Notice how the text of the .10-radius dimension passes over the witness line of the .27 dimension in the detailed view. Place a gap in the witness line so that it does not pass through the text. Pick **Break ➥ Add** from the DETAIL menu. Next, pick the .27 dimension. Pick **Simple**. Now pick on the witness line above the .10-radius leader line and then pick on the witness line below the leader line. The gap is now placed in the witness line. Continue breaking the other dimensions until your drawing looks like the following drawing. Move your views as needed now that you are finished detailing the drawing.

> **NOTE:** *Press the middle mouse button whenever you have finished adding breaks to a dimension and want to select another one.*

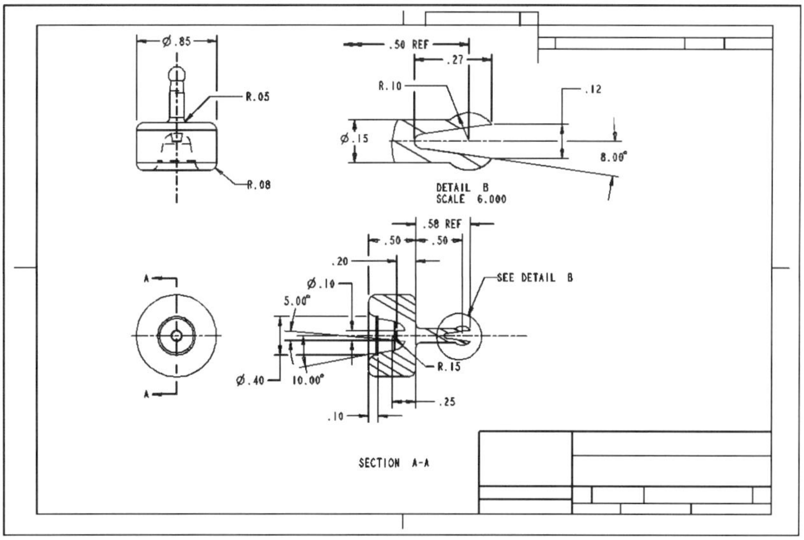

Figure 10-49
Final drawing of the wheel.

SAVE

Changing the Part from the Drawing

Pro/ENGINEER's drawings exhibit *bi-directional associativity*. This means the drawing is associated with a part so that when the part changes, the drawing changes. In addition, if a dimension on the drawing is changed, the part will change. You have already seen that you can modify a part from Assembly mode. You can also modify the part from Drawing mode.

Modifying a Dimension Value in a Drawing

Change the width of the tire. Pick **Modify** ➥ **Value** from the DRAWING menu. Pick on the .50 tire width dimension. Change it to .60.

```
Enter depth [0.50]: .6 ↵
```

Now pick **Regenerate** ➥ **Model** from the DRAWING menu, and the drawing changes to the new size.

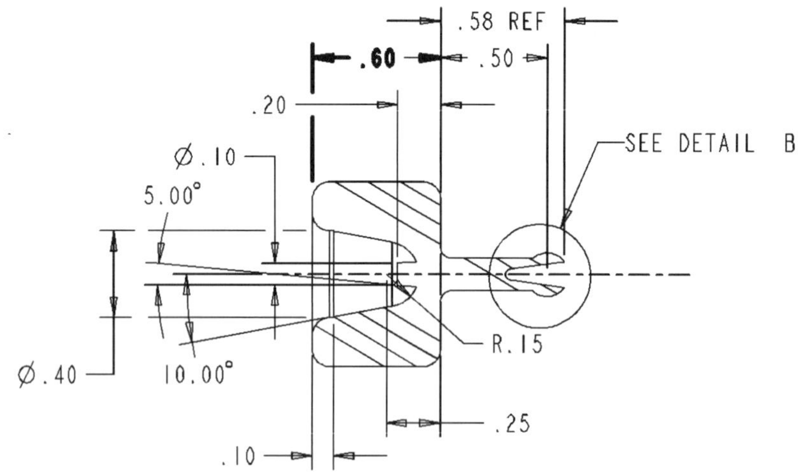

Figure 10-50
Wheel drawing with a wider line.

Modifying a Dimension Value in a Part

Quit the drawing window, and retrieve the wheel part now. Select **Modify**, and pick feature 5. You will see that the dimension for the tire width has changed to .60. Pick the .60 dimension and change it back to .50. Regenerate the model and then quit the window containing the part. Retrieve the drawing.

The tire width has changed back to .50.

Conclusion

After completing the wheel drawing, you should have a good understanding of the basic drawing commands. The following topics were covered:

- ❐ Placing general, projection, section, and detailed views
- ❐ Orienting views
- ❐ Moving views
- ❐ Changing the view scale
- ❐ Modifying view display characteristics
- ❐ Creating and modifying cross sections
- ❐ Changing drawing formats
- ❐ Showing dimensions
- ❐ Moving dimensions
- ❐ Erasing dimensions

- ❒ Creating reference and driven dimensions
- ❒ Modifying the part from the drawing

Remember, drawing views have the same behavior as views in Part mode. The orientation of the general view is dependent on the features selected to establish the orientation. A change in one of these parent features can affect the orientation of the general view. In addition, projected views are children of the view they are projected from. You cannot delete the parent view unless you delete all child views first.

If you model the part using the proper dimensioning scheme, you do not have to create dimensions in the drawing. The dimensions were already provided when you built the part. All you need to do is show these dimensions and arrange them as required. If you build the part and do not like your dimensioning scheme, you can always make driven dimensions in the drawing that update as the model dimensions change.

Now try a more advanced level of drawing in Chapter 11. Try to develop the same philosophy in regard to documentation (drawings) developed for model construction. The next chapter will enhance your documentation expertise and take you up another level of complexity for drawing creation.

Chapter 11

Documenting Your Design: A Larger Drawing

Drawing an Involved Part and an Assembly

Introduction

The chassis part is going to require a more complex drawing than the wheel drawing. You will need to create more difficult cross sections and use different dimension types and formats. You will also practice making an exploded view assembly drawing. In this chapter, you will create a drawing of the chassis that will look like the following figure when complete. You will also create a drawing of the toy car assembly.

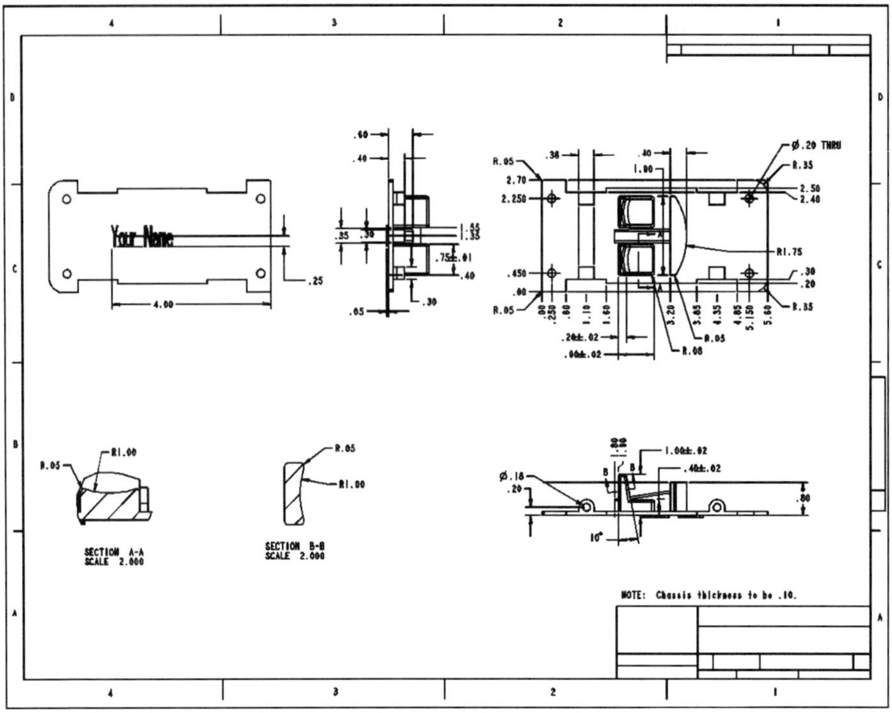

Figure 11-1
The drawing of the chassis.

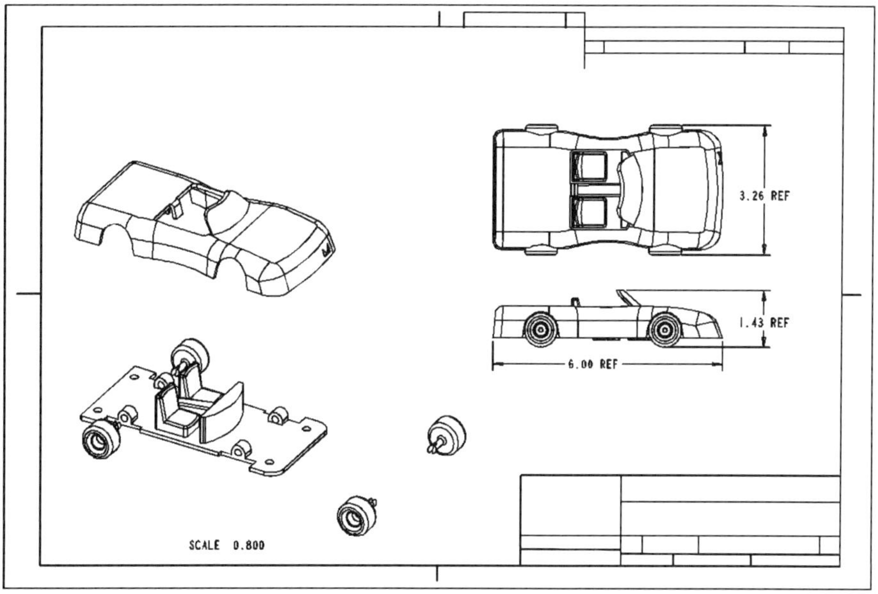

Figure 11-2
The drawing of the toy car assembly.

Starting the Drawing

Select **Quit Window** from the MAIN menu so that the wheel drawing is removed from the graphics window. Create the drawing for the chassis just as you did for the wheel except that this drawing should be called *chassis*. Use a C-size format. Place a top view first, then a view from the passenger side of the chassis, a view from the back side, and a view from the bottom so that the drawing looks like the following. Review the procedures shown in the drawing in the previous chapter if you have trouble placing the views.

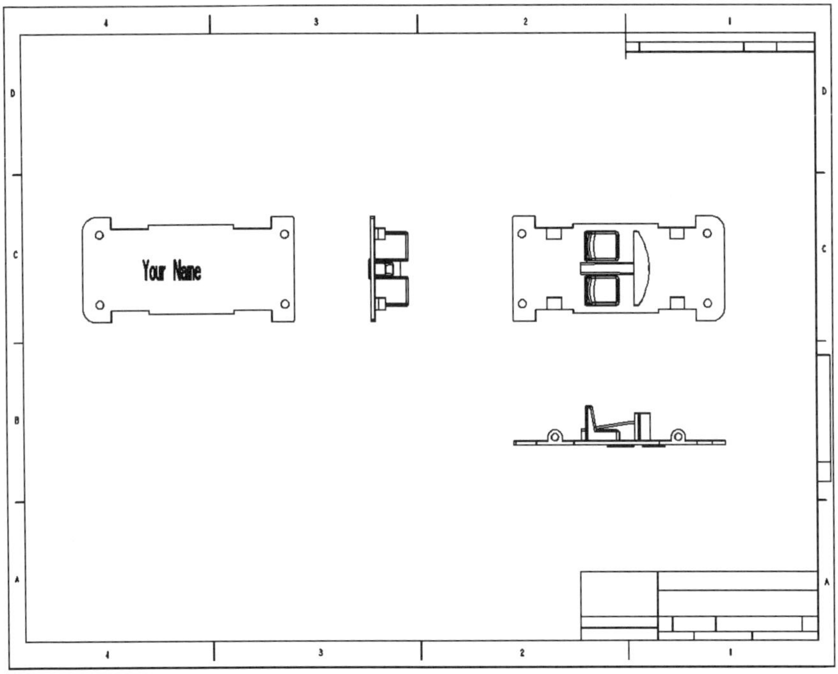

Figure 11-3
Four views in the chassis drawing.

The seat bottoms and backs require section views to detail their curved surfaces. To begin, you should create a partial view of the seat bottom.

Creating a Partial Section View

Make the section view of the seat bottom as a general view. To create the partial, general view, pick the following commands:

Views ➡ Add View ➡ General | Partial View | Section | Scale | Done

> ✓ **TIP:** *If you make all of your section views with general orientation, you will have the flexibility to move them to any place on the page. This can be important when you are trying to arrange views on a sheet to maximize sheet usage and improve drawing clarity.*

The partial view will only show the seat bottom, but the section will pass completely through the chassis. Pick **Full | Total Xsec | Done**. Locate the position on the drawing sheet for the view, and set the scale of the view to 2. Orient the view by facing DTM3 toward the top and DTM1 toward the back. From the XSEC ENTER menu, create the cross section by picking **Create ➡ Planar | Done**. Name the section *a* (lowercase "a"). Make the datum plane for the cross section by offsetting the back surface of the seat back forward by .5. Pick **Make Datum ➡ Offset**. Pick the back surface of the passenger seat back as the plane to offset from. Pick **Enter Value** and provide a value of `-.5` at the prompt, in that the displayed arrow points toward the back of the chassis. Select **Done** to complete the cross section creation.

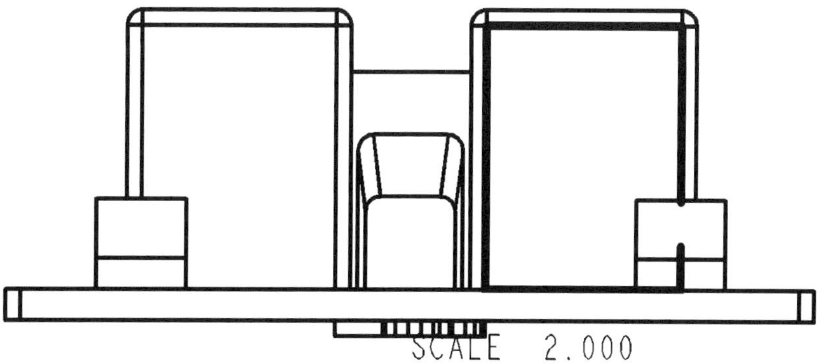

Figure 11-4
Surface to pick for offsetting the datum plane.

Place the section arrows in the top view. Using the following figure, place the center point for the outer boundary at the location of point 1.

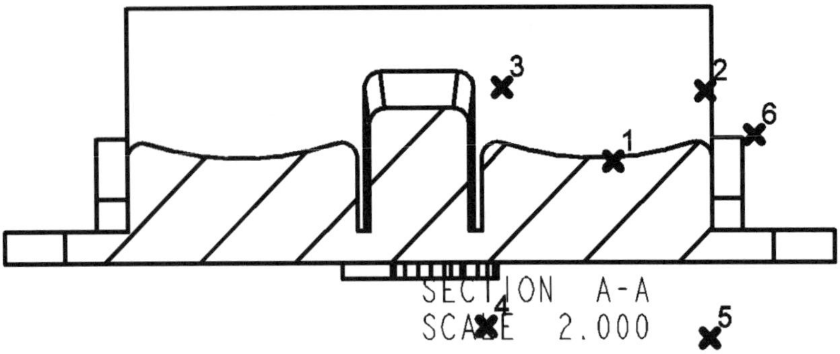

Figure 11-5

Pick points for denoting the spline for the partial view.

Sketch the boundary spline by selecting the points in order from point 2 through point 6. Do not forget to use the middle mouse button after placing the last point. `SECTION A-A` is now placed.

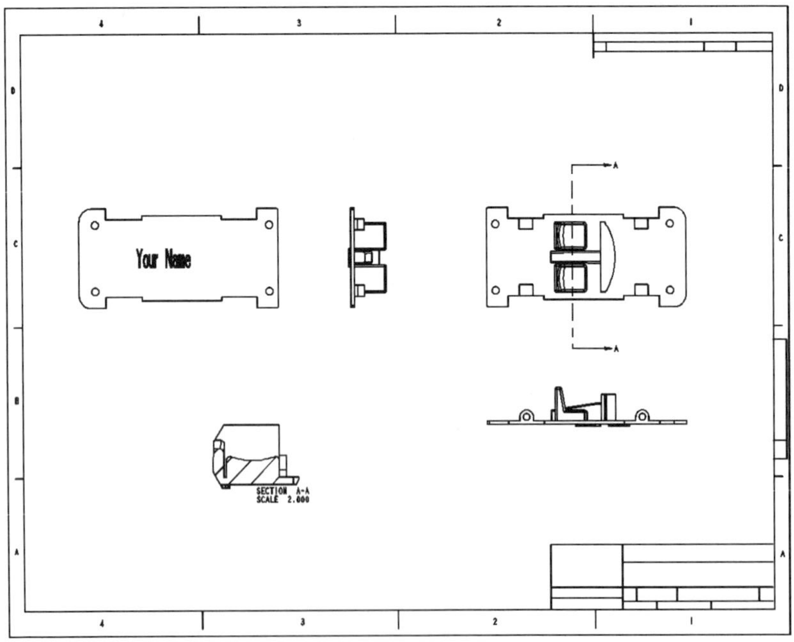

Figure 11-6

The drawing with a partial section view of the seat bottom.

The last view will be the partial section view of the seat back. This section view is needed so that you can provide the radius dimension for the front surface of the seat back. Therefore, it needs to be normal to the axis of that curved surface. Currently there is no plane oriented that way. In order to create the plane, you need to suppress the feature that created the curved surface on the seat back. You cannot suppress the feature in Drawing mode.

Temporarily stop working on the drawing, go back to the chassis part, and place the datum plane needed. Pick **Quit Window** ➥ **Mode** ➥ **Part** ➥ **Retrieve** and then accept the default part name.

```
Enter Part name [CHASSIS]: ↵
```

Insert a datum plane feature before placing the curved surfaces on the passenger-side seat back. Pick **Feature** ➥ **Insert Mode** ➥ **Activate**. Pick **Query Sel** and then pick on the front side of that seat back. The message window will display the feature number of that curved surface, feature 19. Because you must enter the feature number of the feature to insert after, pick **Sel By Menu** ➥ **Number** and then enter `18`.

```
Enter feature regeneration number (1-31) [QUIT]: 18 ↵
```

All of the features placed after feature 18 are now suppressed. You will be creating a datum plane, so you need to turn on the display of datum planes. Pick **Environment** and then select **Disp DtmPln** to toggle on the check mark. Now pick **Done-Return** and the datum planes will be displayed.

Place a datum plane through the top front edge of the passenger-side seat back that will also be normal to the front side of the seat back. From the FEATURE menu, pick the following commands:

Create ➥ Datum ➥ Plane

Using the following figure as a guide, pick **Through** and then the top, front edge of the seat back. Next, pick **Normal** and the front surface of the seat back.

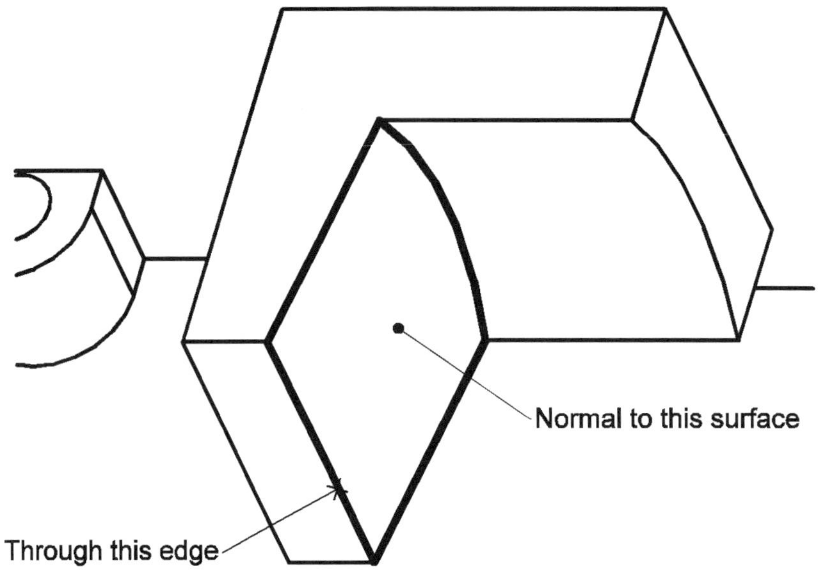

Figure 11-7
Selections to create the datum plane.

Finally, pick **Done** and the datum plane is created.

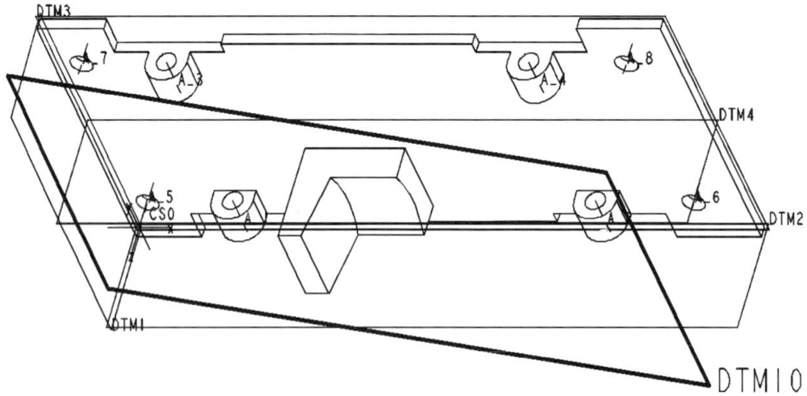

Figure 11-8
Datum plane inserted after feature 18.

Next, create another datum plane offset downward from the previous one by .375. Pick the following commands:

Create ➥ Datum ➥ Plane ➥ Offset

Then select the most recently created datum plane. You want to enter an offset distance for this new datum plane, so pick **Enter Value**. In that the arrow points in the desired direction, toward the seat bottom, enter `.375` at the prompt. Pick **Done** and the second datum plane is created.

Because these datum planes do not need to be as large as the part, fit them to the size of the top surface of the seat back. From the FEAT menu, select **Redefine** and pick the first datum plane. From the REDEFINE menu, select **Attributes | Done ➥ Fit Surface | Done**. Now select the top surface of the seat back. Do the same for the second datum plane.

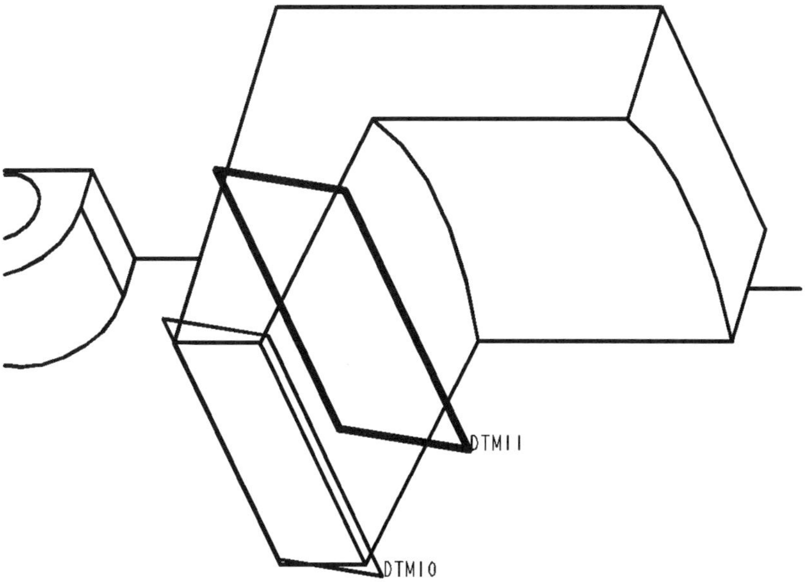

Figure 11-9
Second datum plane inserted after feature 18.

To resume all of the features that were suppressed by the **Insert Mode** command, pick **Insert Mode** ➥ **Cancel** from the FEAT menu. Be careful not to use the **Resume** command from the FEATURE menu because this will resume all suppressed features, not just the ones suppressed during the insert stage.

```
Resume features that were suppressed when activating
insert mode? [Y]: ↵
```

You can create cross sections in Part mode. Pick **X-section** ➥ **Create** ➥ **Planar** | **Done** from the PART menu. In that you have already created a cross section named `A`, this cross section will be named `B`. When prompted to select a plane, pick the last datum plane you just created. The cross section will then be created.

SAVE

Quit the window and retrieve the drawing of the chassis you were just working on. This partial section view will be created the same as the last, except that you do not need to create the cross section. Pick **Views** ➥ **Add View** ➥ **General** | **Partial View** | **Section** | **Scale** | **Done** and then **Full** | **Total Xsec** | **Done**. Place this view at a scale of 2 and orient the view so that the second datum plane you created is oriented to face front. DTM2 will face bottom. The cross section will now be displayed, and the section arrows will be added in the side view of the chassis. **Pick Retrieve** ➥ **B**. Pick in the side view of the chassis for displaying the section arrows. From the following figure, pick the point to be used as the center point for the outer boundary.

Figure 11-10
Point to use as the center point for the boundary of the partial view.

Pick each point in order from the following figure so that the spline for the outer boundary is created. Press the middle mouse button after picking the last point shown in the figure.

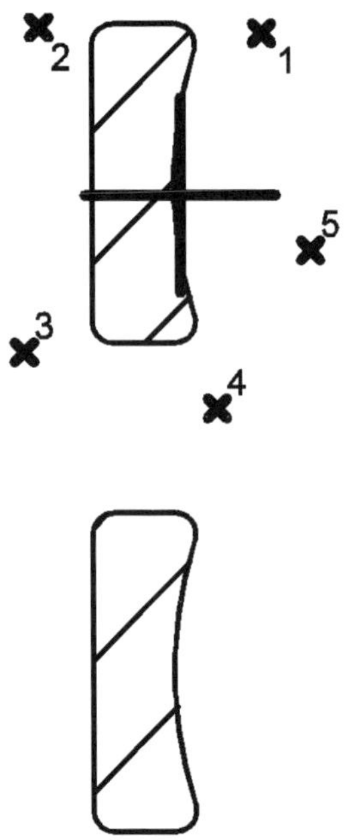

Figure 11-11
Points to create the spline for the boundary of the partial view.

Turn off the display of the datum planes, points, axes, and coordinate systems from the ENVIRONMENT menu and repaint the screen. Specify that no hidden lines will be displayed in any of the views, but tangent edges will. Select **Disp Mode** from the VIEWS menu like you did for the drawing of the wheel. Select each view and then **Done Sel**. Finally, pick **No Hidden | Tan Default | Done**. Now, all of the views required for the chassis drawing are created.

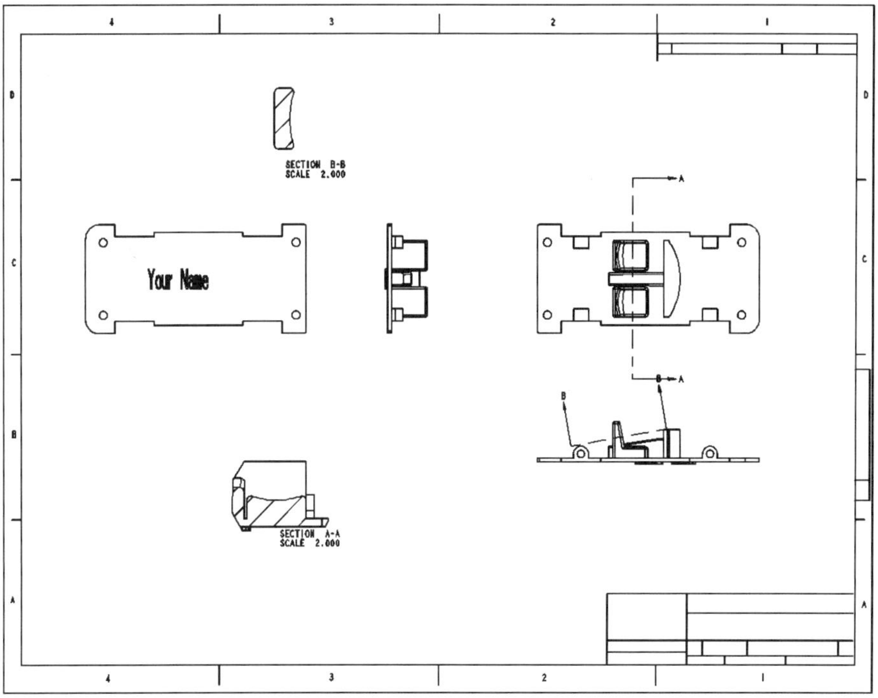

Figure 11-12
Drawing with all views displayed.

Adding Axes

Place axes in the form of cross hairs on the holes in the top view. Pick **Show ➥ Axis ➥ Feat & View** from the DETAIL menu. Now, pick in the top view and then pick one of the body mounting holes. A cross hair will appear on that hole. Continue picking the top view and the other body mounting holes until all four holes have a cross hair. The top view should look like the following figure when you are finished.

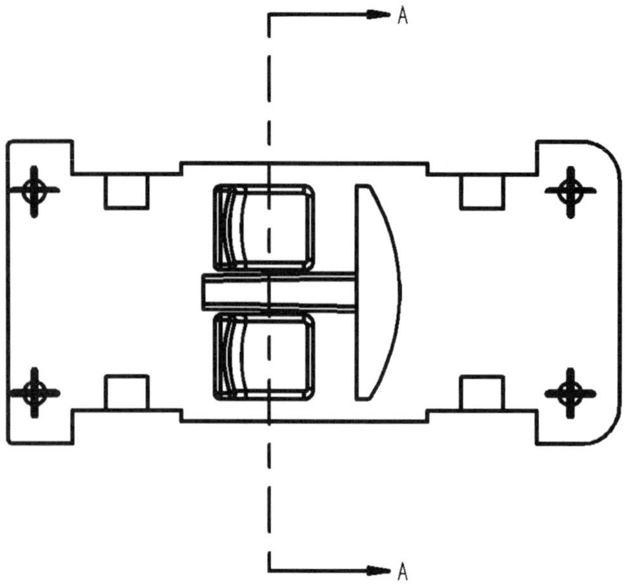

Figure 11-13
Top view with cross hairs displayed on the body mounting holes.

Ordinate Dimensioning

After you have the views positioned as in Figure 11-12, you are ready to place the dimensions on the drawing. This time some of the dimensions will be ordinate dimensions. The dimensions do not have to be created as ordinate dimensions from within Part mode for you to be able to show them as ordinate dimensions in Drawing mode.

When you change them to ordinate dimensions in Drawing mode, they are also changed to ordinate dimensions in Part mode.

> ➾ **NOTE:** *A change in dimension type between linear and ordinate dimensions requires changing the part file. When the drawing file is saved, the part file will automatically be saved.*

Establishing the Baseline

You should show the ordinate dimensions first. However, because there are no ordinate dimensions in the part, you must establish a baseline for the ordinate dimensions. In order to do this, you need to use a dimension already visible that has a witness line on the desired baseline.

> ➾ **NOTE:** *When no ordinate dimensions exist in the part, a linear dimension must first be displayed before any dimensions can be converted to ordinate dimensions. A witness line of this displayed dimension must be selected to establish a baseline for the ordinate dimensions. You can then display all dimensions as ordinate dimensions if they have a witness line coincident with that baseline.*

Show the two dimensions you need to establish the baselines for your ordinate dimensions. Pick **Detail** ➥ **Show** ➥ **Dimension** ➥ **Standard** ➥ **By Feature** and then pick the first protrusion feature of the chassis (feature 5). Now you have the linear dimensions 5.60 and 2.70 displayed. Pick **Ordinate** ➥ **Make Bases** from the LIN ORD menu. Pick on the dimension text of the 5.60 dimension and then pick on its witness line at the back of the chassis. That edge is now established as a baseline and the dimension is converted to an ordinate dimension.

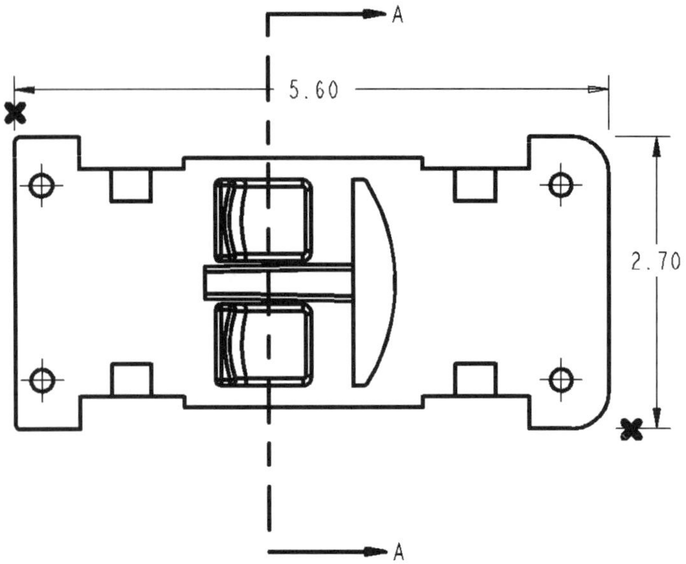

Figure 11-14
The selections to establish the ordinate dimension baseline.

Now create a baseline for the direction that is orthogonal to the first one. Pick **Make Bases** and the 2.70 dimension. Pick the witness line on the right front side of the chassis. That edge is now established as a baseline.

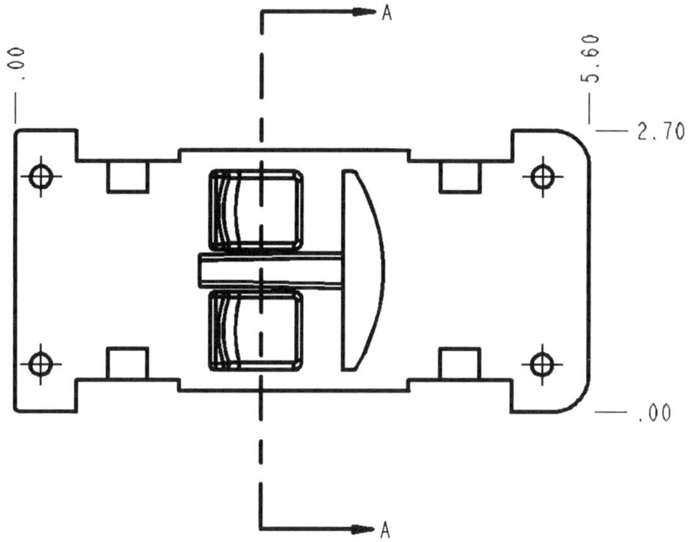

Figure 11-15
The drawing with the baseline ordinate dimensions displayed.

Displaying the Ordinate Dimensions

You can now display all dimensions that reference the baseline edges as ordinate dimensions. Pick **Show Dims** ➥ **Show All** from the ORD BASES menu. Only dimensions that can be shown as ordinate dimensions are displayed.

> **NOTE:** *If you select* **Done/Return** *from the LIN ORD menu before showing all ordinate dimensions, Pro/ENGINEER will require you to specify the baseline edges before the ordinate dimensions can be displayed. Pick* **Bases** *from the ORD BASES menu and pick the two baselines. Then pick* ***Show Dims*** ➥ ***Show All*** *to display all ordinate dimensions that use these baselines.*

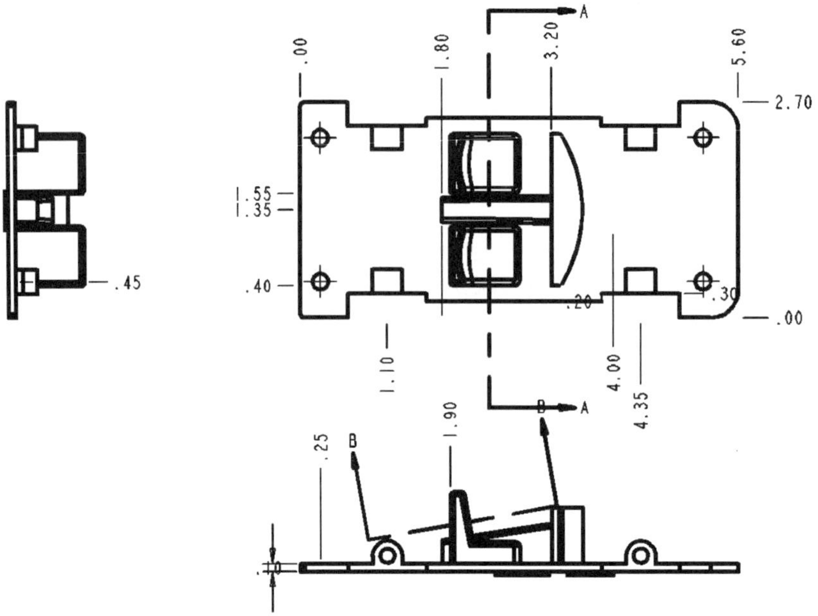

Figure 11-16
Drawing with the ordinate dimensions displayed.

Switch the dimensions to other views, move them, and align them until the drawing looks like the following figure.

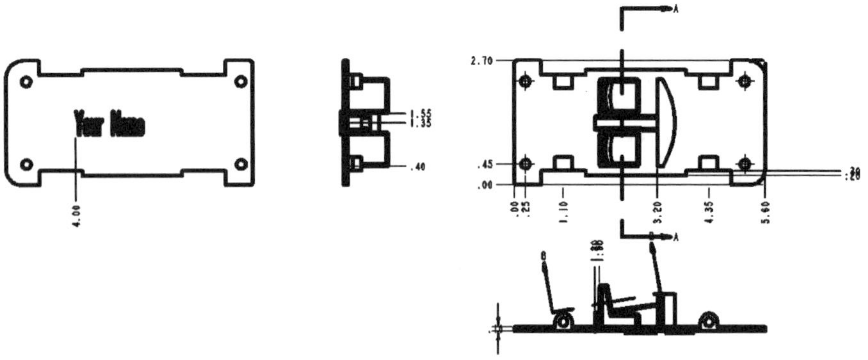

Ffigure 11-17
Drawing with repositioned ordinate dimensions.

Adding Jogs to Dimensions

Ordinate dimensions are occasionally so close to each other that they overlap. This is the case with the .20 and .30 dimensions in the top view and the 1.80 and 1.90 dimensions in the side view. You should put a jog in their witness lines so that they no longer overlap.

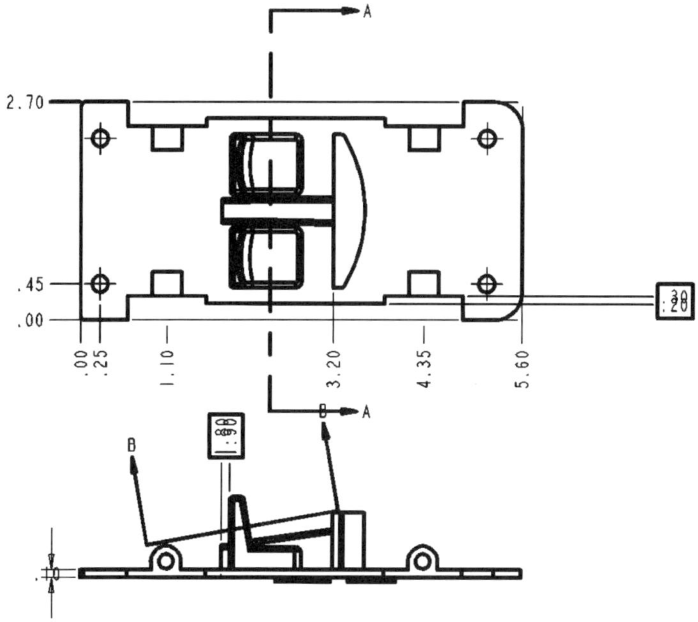

Figure 11-18
Dimensions that overlap.

Pick **Make Jog** from the DETAIL menu. Next, pick on the 1.90 dimension text. Next pick beside the witness line at the location you want the jog to begin. Now pick at the spot you want the next jog vertex placed. You can continue placing more jog vertices using the left mouse button, or use the middle mouse button to implement the jog. Do the same for the .20 and the .30 dimension so that your drawing looks like the following figure.

➾ **NOTE 1:** *To delete a jog, select* **Delete** *from the DETAIL menu and pick a vertex that created a jog. Picking the vertex farthest from the dimension will not remove a jog because it is only used to initiate the jog.*

➾ **NOTE 2:** *A jog vertex can be moved by selecting* **Move** *from the DETAIL menu, picking the vertex to move, and then picking its new position.*

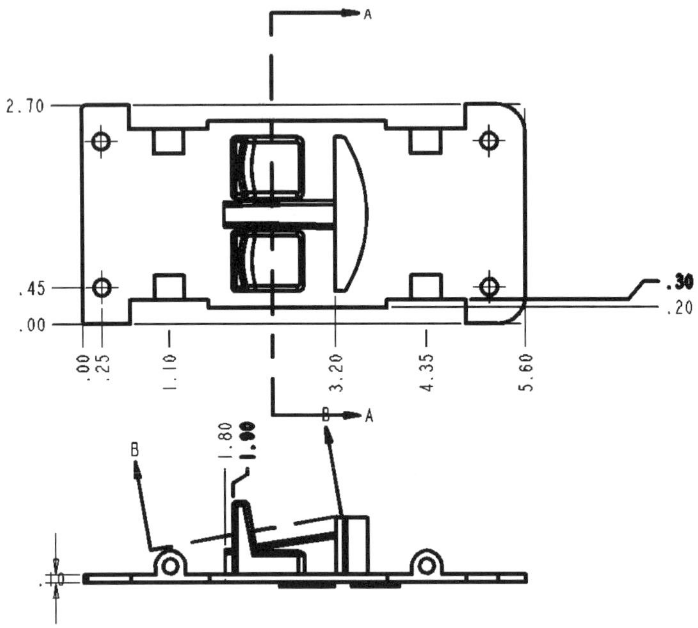

Figure 11-19
Jogs on dimension witness lines.

Converting Dimensions Between Ordinate and Linear

Display the remaining dimensions that could not be displayed as ordinate dimensions.

➾ ***NOTE:*** *Only standard dimensions with a witness line coincident to a baseline can be converted to ordinate dimensions.*

Pick **Show ➥ Dimensions ➥ Standard ➥ Show All**. Change one of the dimensions shown as an ordinate dimension back to a linear dimension. From the DETAIL menu, pick the following commands:

Modify ➥ Dim Params ➥ Dim Type ➥ Ordinate Dim ➥ Ord to Lin

Select the 4.00 dimension that locates the text on the bottom of the chassis. It will change to a linear dimension.

Linear dimensions can be modified to ordinate dimensions by a similar process. The only requirement is that one of the witness lines of the linear dimension must be coincident with a baseline. Pick **Set Base** from the MOD DIM TYPE menu and then pick an existing baseline. Next, pick **Lin to Ord | No Jogs** and the dimension to be changed. The ordinate dimension is then displayed. If the **One Jog** had been picked instead of the **No Jogs** you would have been able to create a jog in the leader for that dimension during the conversion process. This will save you from going back and creating jogs later. There are several linear dimensions on this drawing that should be replaced with ordinate dimensions, but they do not meet the requirement of having a witness line coincident with a baseline.

Erase these dimensions and replace them with driven dimensions that reference the baseline. Pick **Erase ➥ One Item | Dimension** from the DETAIL menu. Select the 4.90 and 1.80 dimensions for the spacing of the holes. Next, select the two .50 dimensions and the 1.00 dimension for the cutouts for the wheel clearance. Pick **Done/Return**, and these selected dimensions are erased.

You are ready now to create the ordinate dimensions to replace the ones you just erased. From the DETAIL menu, pick **Create ➥ Dimension ➥ Ordinate ➥ Create Dims**. Now pick on the baseline dimension text `(.00)` for the back edge of the top view. Select the five edges shown in the following figure, and then press the middle mouse button to display the ordinate dimensions.

> ➾ **NOTE 1:** *When selecting the body mounting hole, pick on the vertical cross hair to get an ordinate dimension to the center of the hole.*

> ➾ **NOTE 2:** *In that you probably want all dimensions you have been creating to be in alignment, the easiest way to accomplish this is to have your GRID SNAP set to On in your environment. This way, you can snap them to the same grid point easily.*

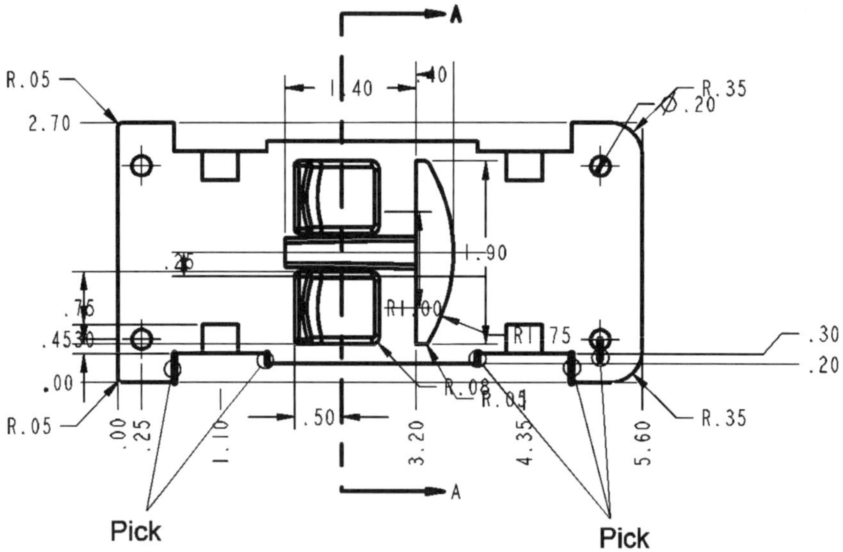

Figure 11-20

The edges to select for creating ordinate dimensions along the length of the chassis.

Now pick the baseline dimension text `(.00)` parallel to the passenger side of the chassis in the top view. Use the following figure as a guide for selecting the three edges to create the ordinate dimensions.

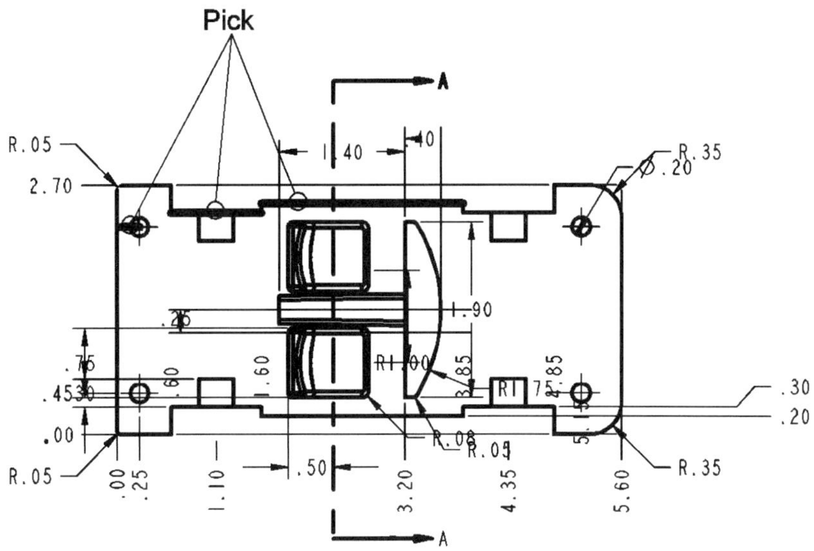

Figure 11-21

The edges to select to create ordinate dimensions along the width of the chassis.

The top view of the drawing now contains the ordinate dimensions that replaced the linear dimensions you erased.

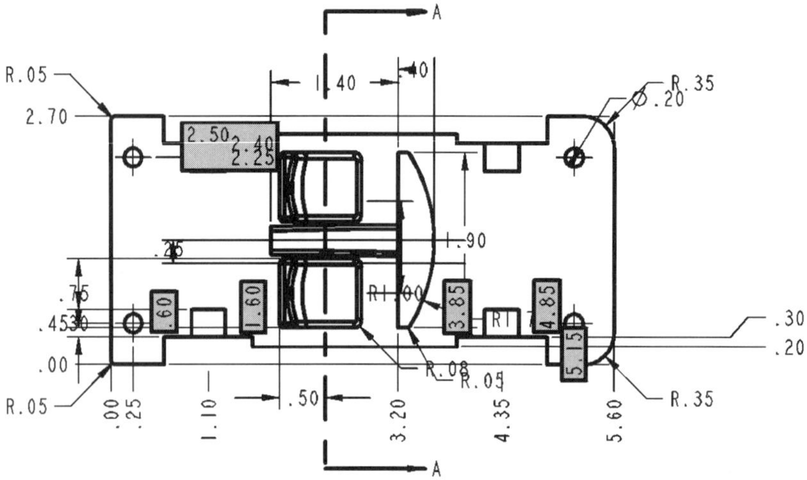

Figure 11-22
The driven ordinate dimensions in the top view.

> ✓ ***TIP:*** *If you have a linear dimension you wish to convert to an ordinate dimension, but it does not currently have a witness line coincident with a baseline, you must either modify the scheme of dimensioning so that it references the baseline or erase it and create a driven dimension that references the baseline.*

To change the dimension scheme, pick **Modify** ➡ **Dim Params** ➡ **Scheme** from the DETAIL menu. Pick the feature containing the dimension to be changed. Now you are able to change the way the feature was dimensioned so that ordinate dimensioning can be used. If you create a driven ordinate dimension, you cannot later modify its value to change the part. Reposition all dimensions so that you can read them. Move the views as required. Erase the .50 dimension in the top view and the .375 dimension in the side view that were for locating the position of the cross sections. Erase the 1.40 dimension in the top view as well. When you are through, the drawing should look like the following figure.

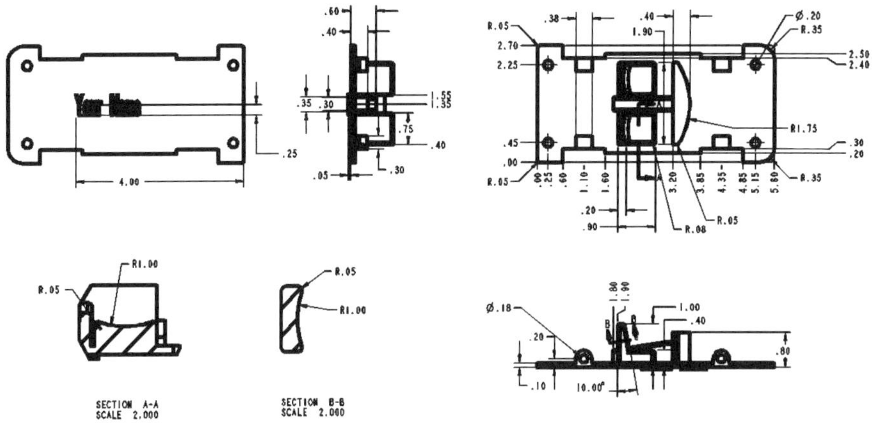

Figure 11-23
The drawing with repositioned dimensions.

Adding Text to a Dimension

Append some text to the end of the .20 diameter hole dimension in the top view of the chassis. From the DRAWING or DETAIL menu, pick **Modify** ➥ **Text** ➥ **Text Line** and then pick the dimension. Now enter the text `THRU` at the end of the line of text.

```
Edit line [0]>>{0:^An^B}{1:@D} THRU ↵
```

The dimension is now changed.

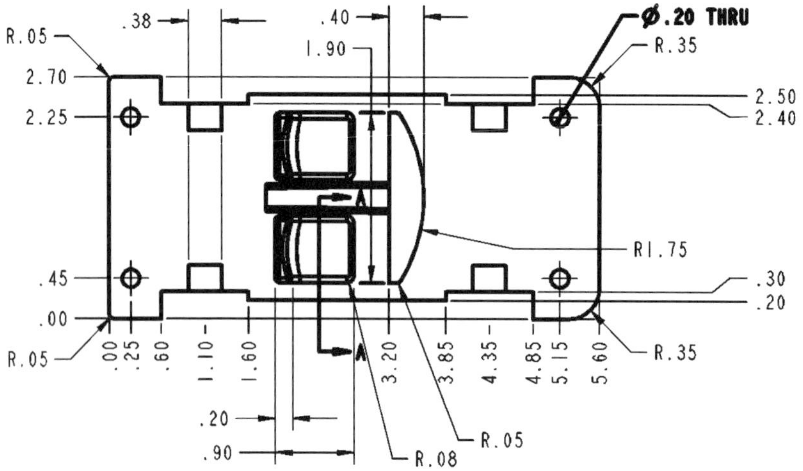

Figure 11-24
Dimension with text appended.

> ➾ ***NOTE:*** *If you want to add multiple lines of text to the dimension, pick* ***Modify*** ➥ ***Text*** ➥ ***Full Note****. The text will be displayed in a text editor that will allow you to add more lines of text to the dimension.*

Adding Parametric Notes

A parametric note can be placed on the drawing. It will be parametric because a dimension will be removed from the drawing, and its value will be placed in the note. The value continues to behave as if it were still a dimension. You can pick it to modify just as you can any other dimension.

> ✓ ***TIP:*** *Use* ***Switch Dim*** *from the DRAWING menu to toggle the dimensions back and forth between their numeric values and their variable names. Remember the appropriate variable names so that you can include them in the note.*

It may help to zoom in on the side view so that you can more easily see the variable name for the chassis thickness dimension. Select the following commands from the DETAIL menu:

Create ➥ Note ➥ No Leader | Enter | Horizontal | Standard | Default | Make Note

Pick the location on the drawing to place the note, and then enter the note. Your dimension variable name may be different than the one used in the example. Just locate the dimension for the chassis thickness and use its variable name.

`Enter NOTE [QUIT]:` **`NOTE: Chassis thickness to be &d0.`** ↵

`Enter NOTE [QUIT]:` ↵

The dimension for the thickness of the chassis plate is removed from the view of the side of the chassis and placed in the note.

> ➾ **NOTE 1:** *A dimension can be shown only once in a drawing. If you want it displayed in more than one place on the drawing, create a reference dimension for the additional occurrences.*

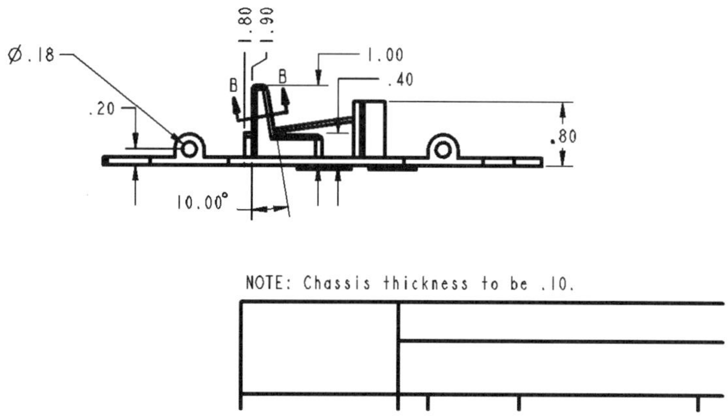

Figure 11-25
Parametric note for the chassis plate thickness.

➾ **NOTE 2:** *Parametric text is created by placing a parameter name or dimension within the text. These parametric values will always begin with an ampersand (&). A dimension could appear as* `&d3` *in the text string, whereas a parameter name could appear as* `&test`. *If the name following the ampersand has not been defined as a parameter, the ampersand and the name will be displayed as text.*

Displaying the Tolerance on Selected Dimensions

All part dimensions have tolerances associated with them as soon as they are created. If you have not checked the ENVIRONMENT menu setting for the tolerances, the tolerances are not displayed in Part mode. However, if **Display Tol** is checked in the ENVIRONMENT menu and tolerances are not displayed with your dimensions in Part mode, the default format for displaying your tolerances is *nominal*. The display of tolerances in Drawing mode is not controlled by the tolerance setting on the ENVIRONMENT menu. In fact, that option on the menu is grayed out when you are in Drawing mode. To turn the tolerance display on for Drawing mode, pick **Set Up ➡ Draw Setup ➡ Modify Val** from the DRAWING menu. In the editor, change `tol_display` to `yes`.

➾ **NOTE:** *Tolerances are displayed in your drawing after you have turned tolerances on from within the SETUP menu. If you want the tolerances in your drawings to always be on by default, you must save the drawing standards information in the current drawing file after it has been modified, and specify that Pro/ENGINEER should copy that file into each drawing as you create new drawings. To specify such a default, add a configuration setting to your* config.pro *file that specifies which drawing standards file to use. For this example, if you used a file name of* custom *for your drawing setup file and placed it in your company standards directory (*/usr/standards*), your configuration setting in the* config.pro *file would be* `drawing_setup_file /usr/standards/custom.dtl`.

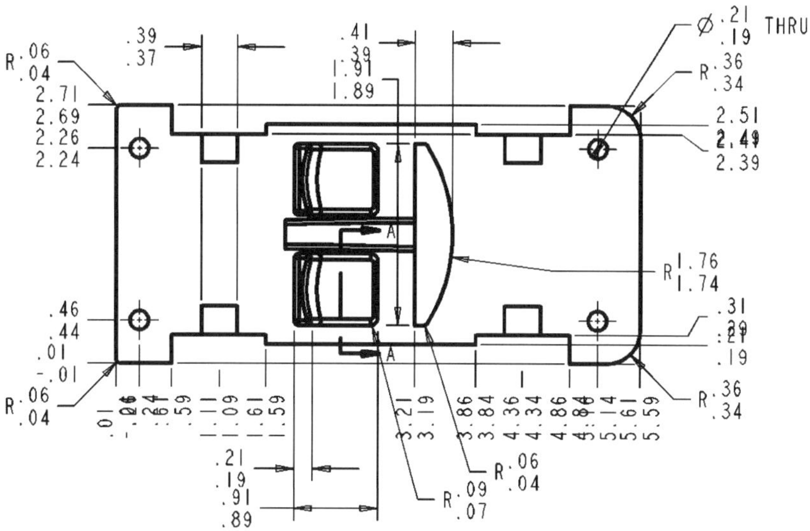

Figure 11-26
Drawing with tolerances shown for all dimensions.

Changing the Format of the Displayed Tolerance

When the chassis was created in Part mode, the default way of showing toleranced dimensions was as limits. When you turn tolerances on in the drawing, they are displayed as limits. It is easy to change to another type of tolerance display format. The following list shows the available types of dimension formats.

- **Nominal** — The dimension is shown without a tolerance.
- **Limits** — The dimension is displayed as a maximum value and a minimum value.
- **Plus-Minus** — The dimension is displayed with a maximum tolerance value and a minimum tolerance value.

- ❒ **+–Symmetric** — The dimension is shown with one tolerance value for both the maximum tolerance and the minimum tolerance.

Make all of the toleranced dimensions shown as nominal. Pick the following commands:

Modify ➥ Dimension ➥ Pick Many ➥ Pick Box

Draw a fence around all toleranced dimensions. After you select **Done Sel** from the GET SELECT menu, pick Tolerance mode, then **Nominal** from the dialog window, and **OK** from the dialog window. All dimensions are changed to show only the nominal size.

> ➾ **NOTE 1:** ***Pick Many*** *can be used with several other commands, such as* ***Delete, Erase,*** *and* ***Align****.*

> ➾ **NOTE 2:** *If the majority of your dimensions usually have their tolerances specified the same way, you can add a configuration setting in your* config.pro *file to set your default tolerance format. To make tolerances display as nominal, place* `tol_mode nominal` *in your* config.pro *file.*

Change the tolerance format for the dimensions of the seat. Change the tolerance format for the .90 and .20 dimensions in the top view, the .40 and 1.00 dimensions in the side view, and the .75 dimension in the back view. Make their tolerances displayed as a symmetrical plus/minus tolerance. Pick **Modify ➥ Dimension** and select the five dimensions. After you pick **Done Sel**, pick Tolerance mode, and **+– Symmetric** from the dialog window. Then pick **OK** from the dialog window. The tolerances are changed. Changing the format for displaying the toleranced dimensions is not difficult.

> ➾ **NOTE:** *A change to the dimension format causes the part to be changed. When you save the drawing, the part will automatically be saved.*

Specifying the Number of Digits to Display on Selected Dimensions

If you look at the default tolerances, three decimal dimensions indicate a tolerance of plus or minus .001. This tight tolerance is required for locating the mounting holes. Change the number of decimal places for the .45, 2.25, .25, and 5.15 dimensions in the top view to three digits. Pick **Modify ➥ Num Digits** and then specify the number of digits.

`Number of decimal places for value [2]:` **`3`** ↵

Next, pick each of the dimensions you want to change to three decimal places. Change the number of digits in the angle to zero.

> ➾ ***NOTE 1:*** *If most of the dimensions on your drawing are three digits, set the options in your* config.pro *file to make three digits your default. Add* `default_dec_places 3` *and* `sketcher_dec_places 3` *to the file.*

> ➾ ***NOTE 2:*** *If your tolerances are determined by the number of digits displayed in a dimension, use the tolerance block in your title block to specify dimensional tolerances. In addition, specify the tolerance to use for each number of digits you specify in your tolerance block in your* config.pro *file. To specify a tolerance of .005 for a three-digit dimension, place* `linear_tol 3 .005` *in your* config.pro *file.*

> ➾ ***NOTE 3:*** *A change to the number of digits for a dimension causes the part to be changed. When you save the drawing, the part will also be saved.*

Modifying the Value of a Tolerance

Changing the value of a tolerance is like changing the value of the dimension. Change the tolerance of the seat dimensions from .01 to .02. Pick **Modify ➥ Value** and then pick the tolerance of one of the dimensions. Enter the new value after picking the dimension.

```
Enter value [0.01]: .02 ↵
```

Repeat this procedure for each of the seat dimensions. This action completes the drawing for the chassis. This function can also be performed in the dialog window, when the initial type of dimension tolerance is specified.

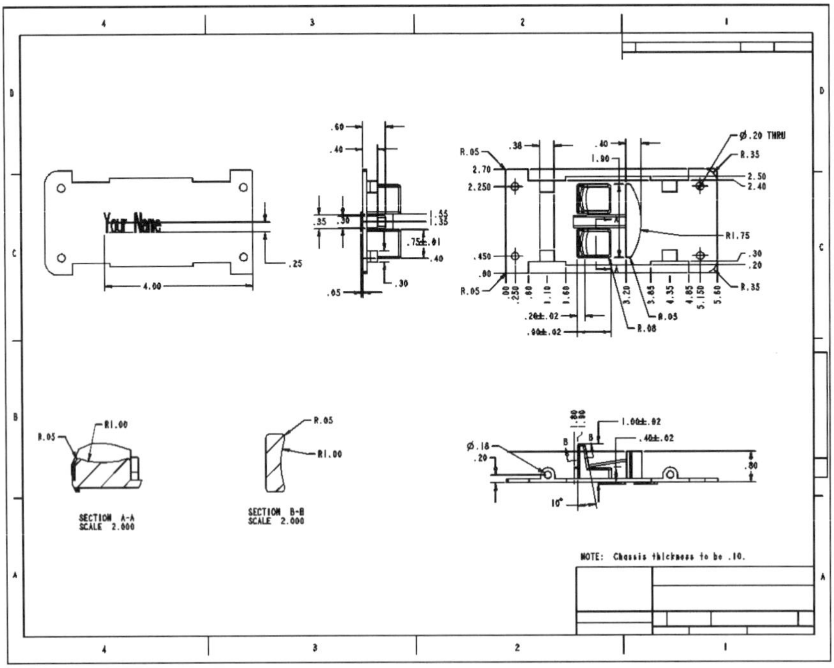

Figure 11-27
The completed chassis drawing.

Making a Drawing of an Assembly

Drawings can be made of assemblies as well as parts. Assembly drawings are associated with the assembly just as the part drawing is associated with the part. If you change a component in the assembly, the change is automatically reflected in the assembly drawing. Begin creating the drawing just as you did for the others. Use a drawing name of "car." Put it on a B-size format.

Placing the Views

Place the exploded view first. Place the view just as you did with the other drawings, but select the **Exploded** option. Provide a scale for this view and let any other views use the default drawing scale. Pick the following commands:

Views ➥ Add View ➥ General | Full View | No Xsec | Exploded | Scale | Done

Pick the location for the view, and then enter a scale of `.8`.

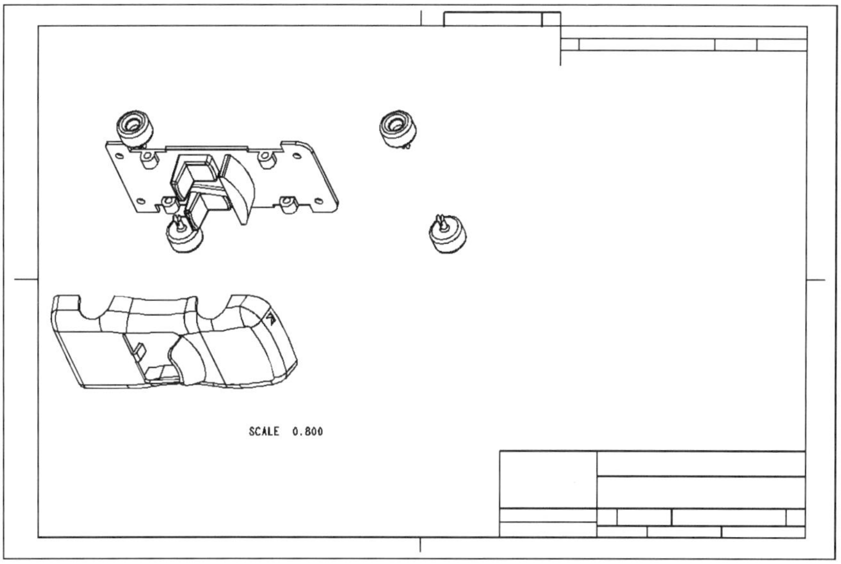

Figure 11-28
The exploded view as first displayed.

The orientation of the exploded view is not very well suited for the drawing. Pick **Top** for DTM3 in the chassis and **Right** for DTM1 in the chassis. You should get a side view of the exploded assembly.

> ✓ ***TIP:*** *When making views of an assembly, use the datum planes of the base component as the referenced planes if you did not create default assembly datum planes.*

Now pick **Angles ➥ Vert**, and enter the rotation angle.

```
Enter ROTATION ANGLE about the VERTICAL AXIS: -30 ↵
```

Next, pick **Horiz** and enter another rotation angle.

```
Enter ROTATION ANGLE about HORIZONTAL axis: 30 ↵
```

Pick **Done/Accept** and the exploded view is then reoriented.

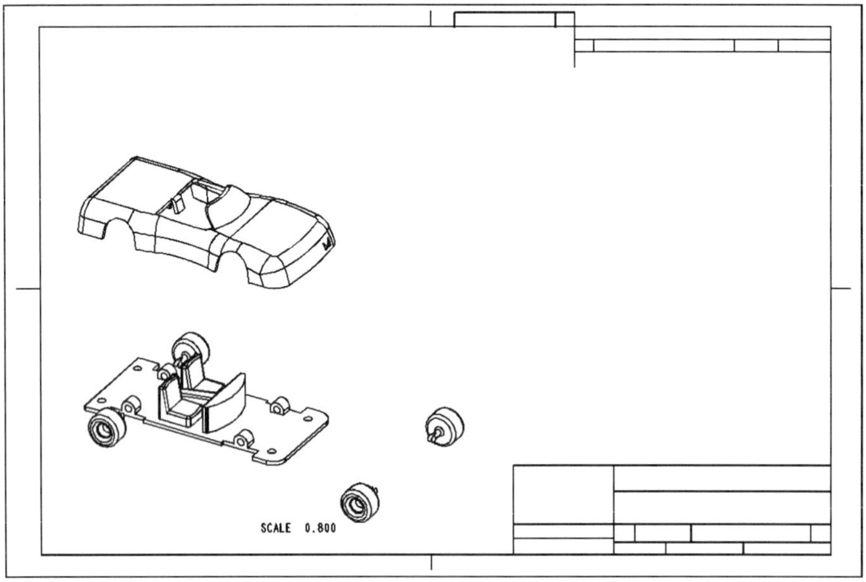

Figure 11-29
The exploded view after reorienting.

The views for displaying the overall dimensions of the toy car should now be placed. Place a general top view of the assembly with **No Scale**, and then project a side view.

> ✓ ***TIP:*** *There is a chance that you could get a message such as the following when placing the projected view:* `Conflict in parent view exists. Select parent view for making the projection.` *Such a message indicates that Pro/ENGINEER does not know which view to project from. In this example, pick on the top view of the car. Modify the scale of the drawing to* `.75`. *Remember to set the display mode for each of your views so that the ENVIRONMENT menu setting does not govern the display of the hidden lines.*

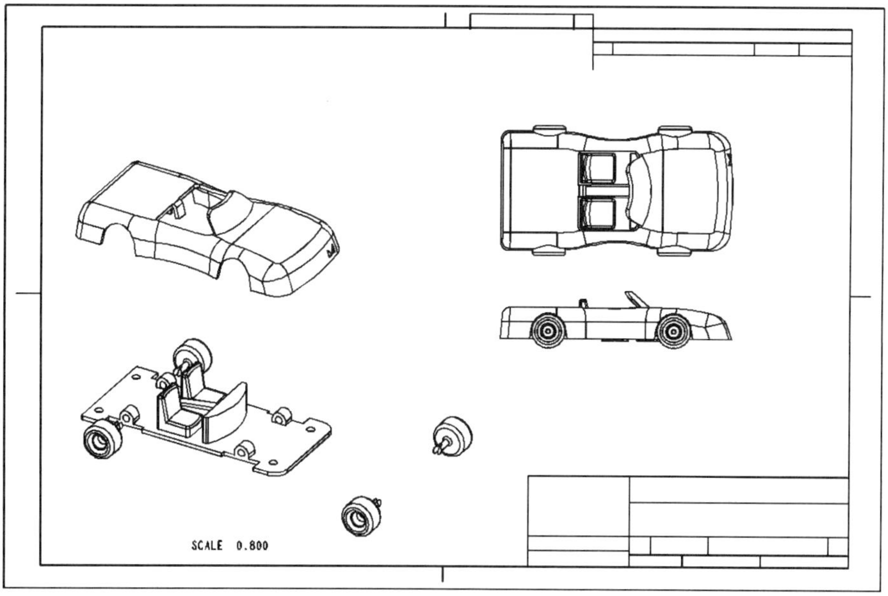

Figure 11-30
The assembly drawing with all views displayed.

Placing Reference Dimensions for the Overall Dimensions of the Assembly

Reference dimensions are needed to provide the overall length, height, and width of the toy car. From the DETAIL menu, pick **Create ➡ Ref Dim ➡ Standard ➡ On Entity**. In the top view of the car, pick the edges of the two front tires and then place the dimension. In the side view of the car, pick the top edge of the windshield and the bottom edge of the front tire. Pick the dimension text location, and then designate that the dimension is tangent to the tire and vertical. Finally, pick the vertex farthest toward the front of the body and the vertex farthest toward the back of the body. Pick the location for the dimension text and place a horizontal dimension. Now the assembly is dimensioned appropriately.

Making a Drawing with Multiple Models

Drawings can be made of multiple models, which include assemblies as well as parts. The two can be combined on one drawing that may include one assembly and several different parts. It is very simple to add models to drawings. Pick **Dwg Models ➡ Add Model** from the VIEWS menu and enter the model name to be added (if you enter an asterisk in the prompt for the model name, this will give you a listing of your current directory that you can pick from). This technique is extremely useful if you don't remember the exact model name or don't have a hard copy listing of the directory handy. Regardless of the number of models in a drawing, the command structure remains the same as for single-model drawings. You can also combine different view types, such as exploded and general, just as in the single-model drawing.

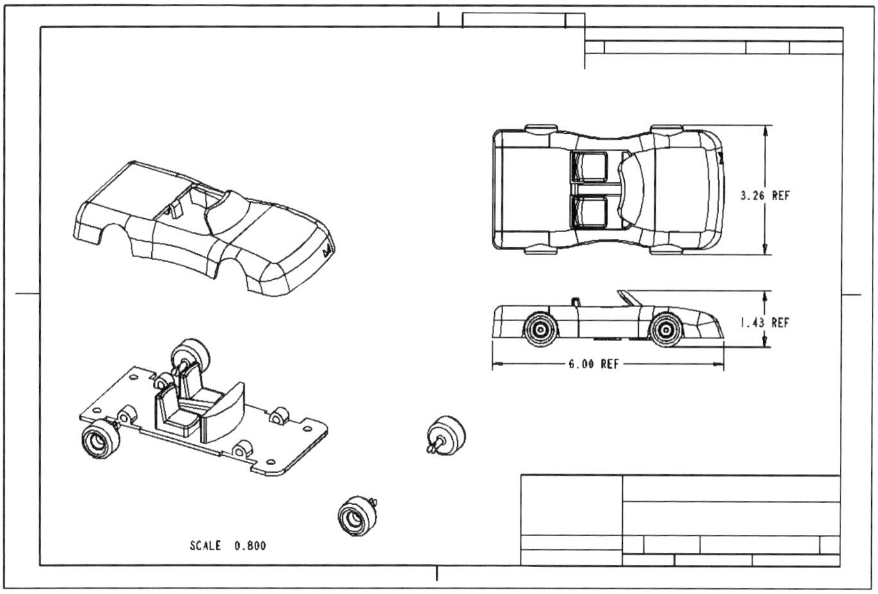

Figure 11-31
The final assembly drawing.

Conclusion

You should know how to perform the following tasks after creating the chassis and assembly drawings.

- ❒ Making partial views
- ❒ Making exploded views
- ❒ Making datum planes to create cross sections
- ❒ Modifying dimension tolerances
- ❒ Creating notes
- ❒ Using ordinate dimensions
- ❒ Creating driven and reference dimensions

There are many important characteristics of Pro/ENGINEER drawings. Like assemblies, the drawing references the part or assembly model. The geometry is not copied into the drawing file. It is important to note that some changes to the drawing are not saved in the drawing file, but in the model file. In that Pro/ENGINEER has bidirectional associativity, the value of a dimension in the drawing can be changed, and the referenced part will also change.

Drawing Methodology

You probably noticed some similarities in the drawing process between the wheel and chassis drawings. You should follow this sequence every time you create a drawing:

STEP 1 — Decide on the drawing format.

STEP 2 — Place the main view first.

STEP 3 — Place the projected views.

STEP 4 — Create the cross sections and place the section views.

STEP 5 — Show the dimensions.

STEP 6 — Clean up the dimensions.

STEP 7 — Add required notes.

Sticking to this simple process will reduce your efforts when trying to create drawings. Sometimes you will not know exactly which views are needed for a large part's drawing. If this is the case, an iterative process that repeats steps 4 to 6 would be appropriate. You should now possess the basic skills to handle any type of drawing that would be required in an engineering environment. In the next chapter you will take an in-depth look at drawing formats.

Chapter 12
Documenting Your Design: Understanding Drawing Formats

How to Add Information to the Drawing

Introduction

In this concluding chapter on the basics of drawing, you will practice the following:

- Creating a custom border and title block
- Placing a table for a tabulated drawing
- Creating a BOM table for an assembly drawing
- Adding balloons for the assembly components

Creating a Custom Title Block in a New Format

As you probably noticed, the title blocks for the drawings in the previous chapters were left blank. You could have placed text at the appropriate locations to fill in the existing title block, but the text would not be located with respect to any lines in the title block. It would be free-floating text.

Making a Title Block Using a Drawing Table

There is another way to manage the title block. Some information can even be placed automatically in your title block. Take the B-size format and change its title block to make it more usable. The new title block will look like the old one when no text is placed in it, but it will have more functionality.

Pick **Quit Window** from the MAIN menu. This time, select **Format** from the MODE menu. Format mode is another mode of Pro/ENGINEER that is used specifically for making drawing formats. Retrieve the format named `B`. Select **Dbms ➥ Save As** and copy the format `B` to the name `btitle`.

```
Enter Object To Save [B.FRM]: ↵
Enter TO: btitle ↵
```

Now pick Quit Window again, and this time, retrieve the format `BTITLE`. Next, delete the old title block. Pick **Detail ➥ Delete ➥Pick Many ➥ Pick Box**, and place a fence around the title block, as shown in the following illustration.

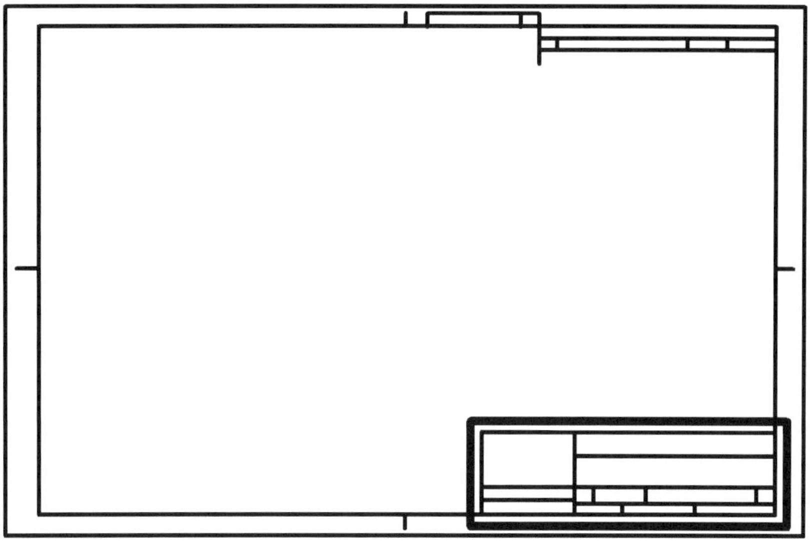

Figure 12-1
The fence around the title block before deleting it.

After placing the fence, pick **Done Sel**. After the old title block is deleted, it is time to create the new one.

Creating the Cells for the Title Block

The new title block will be created by placing a table containing cells similar to a spreadsheet. Adjacent cells in the table can be merged to form a bigger cell. This is the method to use so that the table matches the appearance of the desired title block. The available options when creating the table follow:

- **Descending** — Specifies that the table will extend downward from its origin.
- **Ascending** — Specifies that the table will extend upward from its origin.

- ❒ **Rightward** — Specifies that the table will extend to the right of its origin.
- ❒ **Leftward** — Specifies that the table will extend to the left of its origin.
- ❒ **By Num Chars** — Specifies the size of a cell in the table by the number of characters it can contain.
- ❒ **By Length** — Specifies the size of a cell in the table by an entered length.

From the FORMAT menu, pick the following commands:

Table ➡ Create ➡ Ascending | Leftward | By Length ➡ Vertex

Pick the lower right corner of the drawing border as the origin of the table. It is now time to create the cells of the title block. Enter a width of `.4` for the first column from the origin of the table.

```
Enter the width of the first column in drawing units
(INCH) [QUIT]: .4 ↵
```

Continue providing column widths until you get to the left border of the title block. Provide the values 1.35, 1.50, 1.00, and 2.00 as you are prompted for additional widths. Now that you have provided the width for the last column, enter ↵ to accept the default entry DONE. Now you are prompted for the height of each row in the title block.

```
Enter the height of the first row in drawing units
(INCH) [QUIT]: .25 ↵
```

Provide the values .10, .25, .65, and .50 for the rest of the rows as you are prompted for them. Press <Enter> again to indicate that all of the rows have been specified. The next figure shows the modified title block.

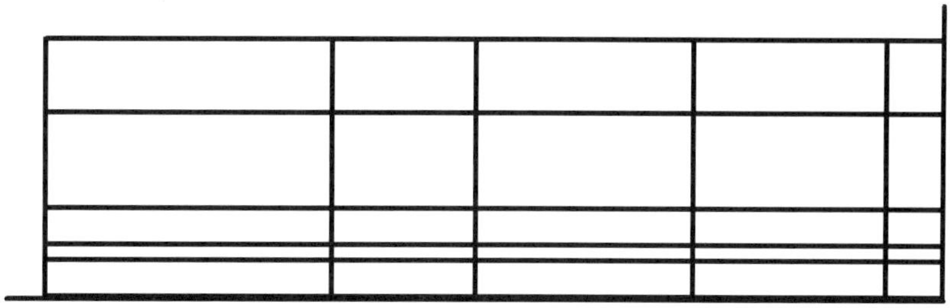

Figure 12-2
The new title block after specifying the column and row size.

All lines of the desired title block are in the current title block, as well as a few extra.

Merging Cells of the Title Block in the Format

You need to combine some of the cells and make larger ones. This procedure is called merging cells. Pick **Modify Table ➥ Merge ➥ Rows & Cols**. To help you pick the correct cells while merging, use the following figure of the title block.

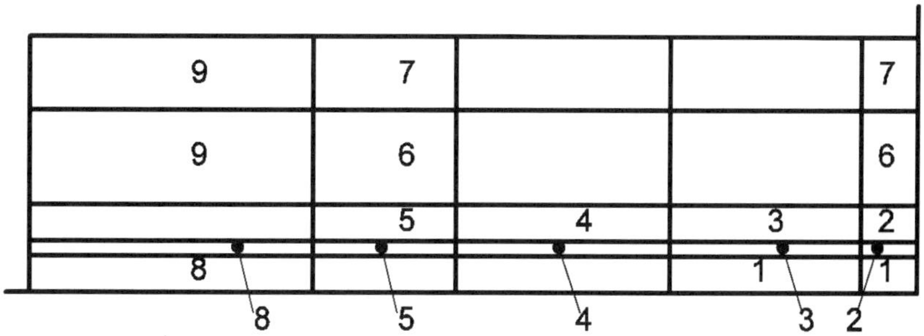

Figure 12-3
The title block with cell picks designated for merging the cells.

To create the first merged cell, pick either one of the two cells marked with a `1` when you are prompted to pick a table cell for one corner. Then pick the other cell marked with a `1` when prompted to pick another table cell. The first merged cell is created.

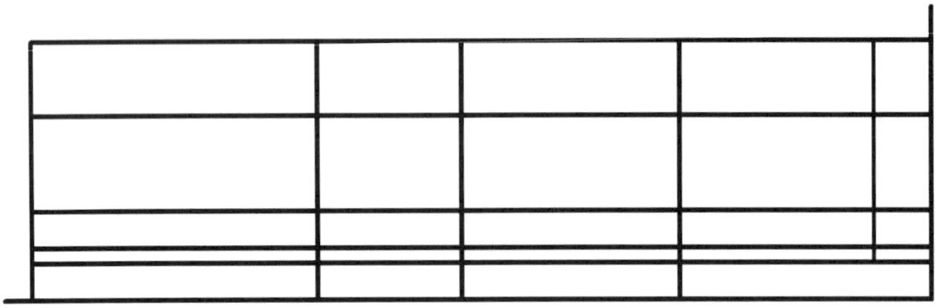

Figure 12-4
The title block with the first merged cell.

Now create the other eight merged cells by using the previous figure to aid you in the selection of the correct cells.

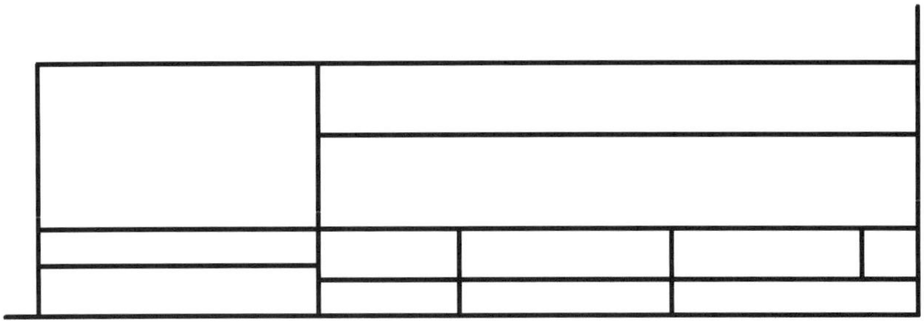

Figure 12-5
The title block with the merged cells.

After you create the nine merged cells, all lines of the title block are displayed correctly. All that is needed is to add the text.

Adding Nonparametric Text to the Title Block

Add the text shown in the following figure.

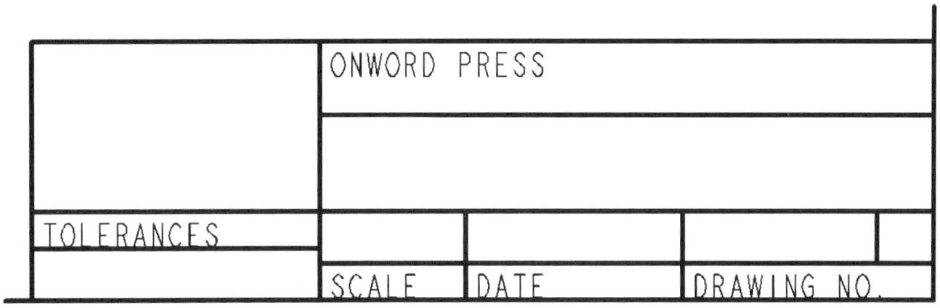

Figure 12-6
The title block with text in selected cells.

The text displayed in the figure will appear the same on every drawing. Pick **Enter Text** from the TABLE menu and then select one of the cells to place the text in. For instance, pick the cell that contains the text "SCALE." Then enter the text at the prompt.

```
Enter text [QUIT]: SCALE ↵
Enter next line of text for this cell [QUIT]: ↵
```

The text is now shown in the cell. Pick each of the other cells shown with text and enter that text into the title block.

Adding Drawing Labels to the Title Block

Now you are ready to add the drawing labels to the title block. These are Pro/ENGINEER-supplied variables that will automatically display specific information in the drawing as the drawing changes. You will place drawing labels in the title block to specify the default drawing scale, the date the drawing was created, the drawing name, the name of the model, and the default tolerances for the dimensions. The following list shows the text you must enter in the cell to make the drawing label display the indicated information.

- scale — &scale
- date — &todays_date
- drawing name — &dwg_name
- tolerance for a 2 decimal place linear dimension — &linear_tol_0_00
- tolerance for a 1 decimal place angular dimension — &angular_tol_0_0

To aid in selecting the correct cell for placing the drawing labels, numbers are shown in the cells in the following figure. The numbers will be referenced in each label placement procedure.

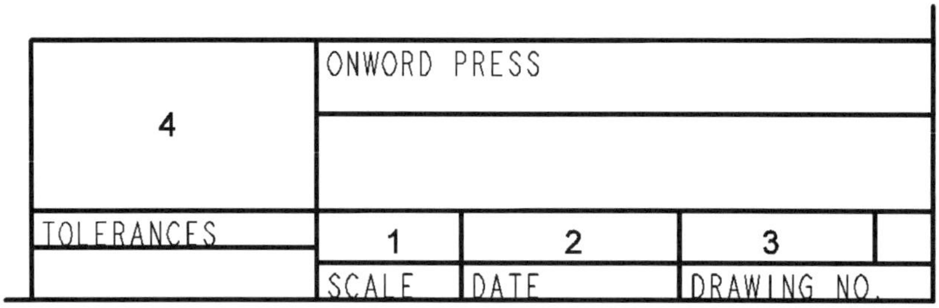

Figure 12-7
The title block with the cells numbered for text placement.

Pick **Enter Text** as you did before. If only one line of text is to be provided, enter ↵ when prompted for the second line of text. Pick cell 1 and enter "&scale" for the text.

> **NOTE:** *To display the scale as a ratio, modify the drawing setup file. Set the value for* `view_scale_format` *to* `ratio_colon`.

Pick cell 2 and enter `&todays_date`. For cell 3, enter `&dwg_name`.

> ✓ **TIP:** *You can use part numbers or drawing numbers as the drawing file name. The number will appear in the cell containing the entry* `&dwg_name`.

The default tolerance information will be placed in cell 4. Text will be used along with the drawing label and more than one line will be entered. For the first line of text, enter `X.X   &linear_tol_0_0`. There are three spaces entered before the ampersand on this line. The second line will be `X.XX  &linear_tol_0_00`, with two spaces before the ampersand. The third line will be `X.XXX &linear_tol_0_000`. For the final line, enter `X.X&angular_tol_0_0`, with two spaces before the ampersand.

> **NOTE:** *To enter the degree symbol or the ampersand (&) symbol in the text, pick it from the displayed PALETTE window.*

Now you almost have all of the desired text placed in the title block.

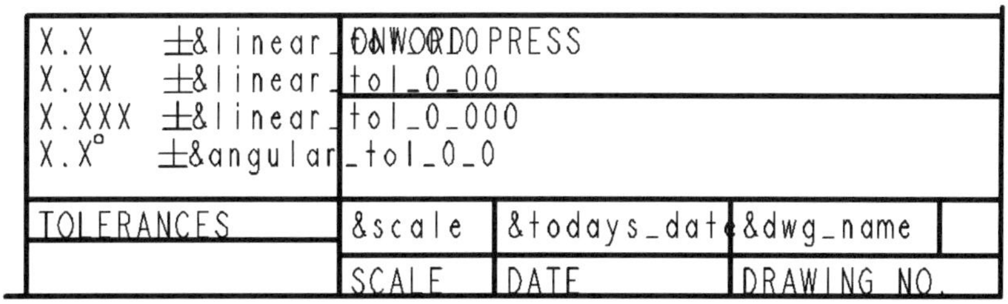

Figure 12-8
The title block with the drawing labels displayed.

Adding Your Own Parameters to the Title Block

You can place a parameter in the title block just as you did the drawing labels. Place a parameter you create called *rev_level* in the cell next to the drawing label *dwg_name*. Pick the cell and enter `&rev_level`. Next, place a parameter for the drawing description called *descr* in the large cell below the cell containing "ONWORD PRESS." Pick the cell and enter `&descr`.

> **NOTE:** *When you retrieve a format that contains parametric text, the value for the parameter in the model will be placed in the drawing at the specified location. If the value of the parameter changes, the displayed text changes. If the parameter has not already been created in the model at the time the format is retrieved, or the format is retrieved prior to specifying which part or assembly will be shown in the drawing, the parametric text will be deleted from the drawing and replaced with text that you are prompted for. This text is not parametric.*

Changing the Size of the Text

From the FORMAT menu, pick **Modify ➥ Text ➥ Text Style**. Select all of the nonparametric text except for `ONWORD PRESS`. Select **Done Sel**, or press the middle mouse button, pick Use **Default Next to Height** in the dialog window, pick within the window, use the backspace to remove the default value, and then enter a new text height of `.125`. Pick **Apply** from the dialog window, and then **Done/Return** from the MODIFY TEXT menu and the text height is changed.

```
Enter new value for text height [0.1562]: .125 ↵
```

Use the same method to change the text height of *ONWORD PRESS* to `.25`.

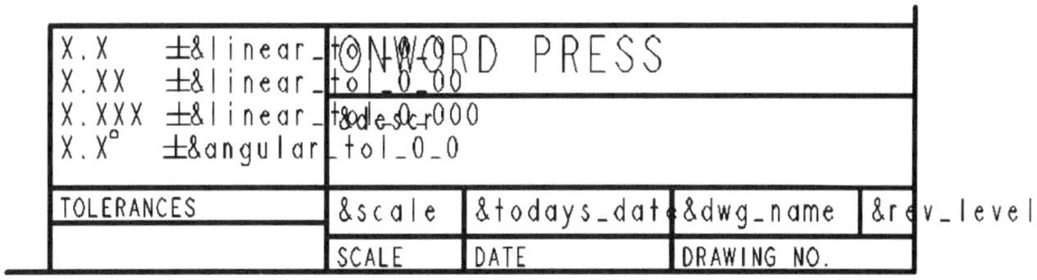

Figure 12-9
The format with the new text size.

Justifying Text

All of the text has been added to the format, but it is not located correctly within the cells. Center the text within each cell, except for the cell containing the tolerance values. Methods of justifying the text appear below.

- **Left** — Justifies the text to the left of the cell.
- **Center** — Justifies the text horizontally to the center of the cell.
- **Right** — Justifies the text to the right of the cell.
- **Top** — Justifies the text to the top of the cell.
- **Middle** — Justifies the text vertically to the middle of the cell.
- **Bottom** — Justifies the text to the bottom of the cell.

Pick **Modify** ➥ **Text** ➥ **Justify** from the FORMAT menu. Pick the text in a cell and then pick **Center | Middle | Done/Return**. Repeat this for the text in each of the cells except the one containing the tolerance values. Set the justification for the tolerance values to **Left | Middle | Done/Return**. After you finish, the title block should look like the following figure.

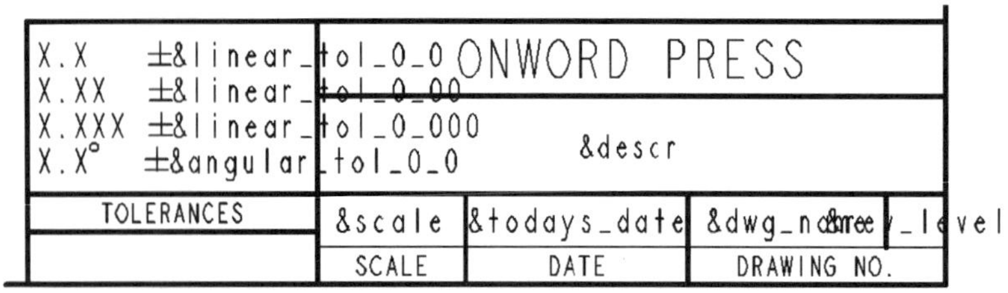

Figure 12-10
The format with the new title block.

You have completed the process of adding the title block to the format. Whenever this format is placed in a drawing, much of the information will be placed automatically.

✓ ***TIP:*** *You should place all format files in a subdirectory of your standards directory (*/usr/standards*). Place the following option in your* config.pro *file to specify the location of these format files:* `pro_format_dir/usr/standards/formats`.

Placing the New Format in a Drawing

Replace the format used in the drawing of the wheel with the new one. Retrieve the drawing of the wheel. From the DRAWING menu, pick **Sheets** ➥ **Format** ➥ **Add/Replace** ➥ **Unblank**. Now enter the name of the new format to use.

`Enter Format name [B]:` **`btitle`** ↵

Because the parameters "rev_level" and "descr" have not been defined in the wheel part, you are prompted to enter text that will replace the parameter name.

`Enter text for the parameter rev_level [NONE]:` **`A`** ↵

`Enter text for the parameter descr [NONE]:` **`WHEEL`** ↵

> ➾ ***NOTE:*** *From this point forward, whenever the revision level of the drawing changes, you have to edit the text on the drawing to update it. Even if you now add a parameter named* `rev_level` *to your part, the revision level on the drawing will not change automatically because the parameter no longer exists in the title block. It has been converted to text. The only way to maintain the* `rev_level` *parameter in the title block is to define the parameter in the part prior to replacing the format in the drawing. This also applies to adding the parameter* `descr` *to the part file.*

Now the new format is placed, and information is automatically updated in the title block.

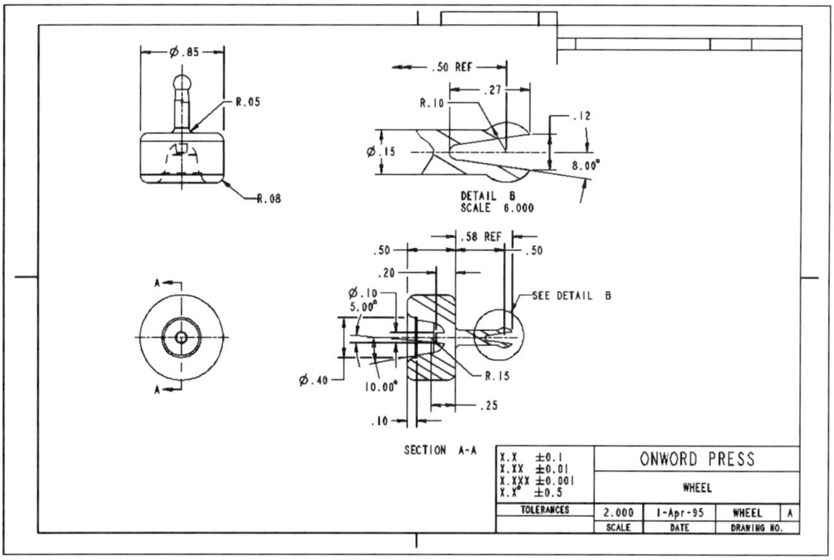

Figure 12-11
The drawing of the wheel with the text in the title block.

Adding a Parameter to the Part

Add a parameter for the revision level and a description to the part, and then retrieve the format again. First, retrieve the wheel part. The parameter for the description can be a single line of text similar to when you create the parameter for the revision level, or you can prepare a multi-line note. The procedure for setting a parameter as a note follows. Pick **Set Up ➥ Notes ➥ Add** from the PART menu.

```
Enter note (or a <CR> to finish): WHEEL, ↵
Enter note (or a <CR> to finish): PLASTIC ↵
Enter note (or a <CR> to finish): ↵
```

You have now created the note. Remember the ID number that shows in the message window to identify this note. In our example, the ID number is `0`. You will use it when you create the description parameter. From the PART menu, pick **Relations ➥ Add Param ➥ Note**.

```
Enter Parameter name: descr ↵
Enter value for descr [0]: ↵
```

You have now defined a parameter, `descr`, whose value is set to the text in the model note. Now you need to define the parameter "rev_level."

> ✓ ***TIP:*** *One way to add a parameter to a part is by entering it as a relation. In this fashion, you can later modify the parameter by editing the relation. The value of the parameter can also be changed if it is displayed on a drawing. However, to modify it that way, you must either delete the relation if the parameter was created by entering a relation, or you would have to create the parameter by using* ***Add Param*** *from the RELATIONS menu.*

Select **Add** from the RELATIONS menu. Enter the relation `rev_level="B"`.

SAVE

Placing the Format in the Drawing Again

Retrieve the drawing of the wheel. From the DRAWING menu, pick **Sheets** ➥ **Format** ➥ **Add/Replace** ➥ **Unblank**. Now enter the name of the format, and remove the highlighted table from the old format when prompted to do so.

```
Enter Format name [BTITLE]: ↵
```

```
Highlighted table is copied from old format. Remove
it from drawing? [N]: y ↵
```

In that the parameters "rev_level" and "descr" have already been defined in the part, the format is placed on the drawing without prompting for information. If you change the parameter value of `rev_level` in the part by editing the relation, the drawing will display the new value. You can change the revision level on the drawing, and the parameter `rev_level` will change in the part (but it will not be a permanent change unless there is no relation present for the parameter in the part).

> ➾ **NOTE 1:** *You can modify the value of the parameter* `rev_level` *without removing the relation from the part file. However, the next time the part is regenerated, the parameter value will change back to the value determined by the relation. That is the reason the relation must not appear in the part file if you want to be able to change the value of the parameter from the drawing.*

➾ ***NOTE 2:*** *To modify the text in the model note referenced by the parameter* `descr`*, you must change to Part mode and select* ***Set Up*** ➥ ***Notes*** ➥ ***Edit*** *from the PART menu. You will then enter the ID number for the note you want to modify.*

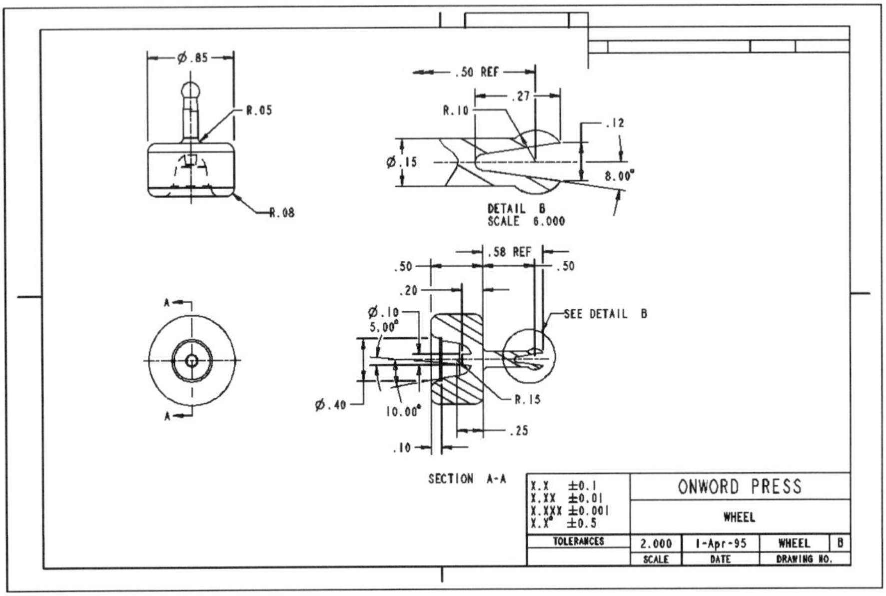

Figure 12-12
The drawing of the wheel.

SAVE

Creating a Custom Company Logo on a Drawing Format

You can also add your company's logo to the title block. First, you will need a .dxf file of the logo image to import into the format. Activate your format (new or existing), pick **Interface** from the FOR-

MAT menu, and then pick **Import** from the INTERFACE menu and **DXF** from the INTFIMPORT menu. Now enter your DXF file name at the prompt, then a Return. In that you probably do not want the logo to fill the format when it comes in, answer N for no, then return to the SCALE TO FIT FORMAT prompt. The next prompt is for the location where the logo will appear when it is read into the format. For this prompt, answer Y for yes, then <Enter>. In that you will be moving this image anyway, it really doesn't matter where it initially appears.

To move the logo to its new location, pick **Tools** from the DETAIL menu and use **Translate** from the TOOLS menu. Because an imported DXF file comes in as segmented geometry, you will need to use **Pick Many** from the GET SELECT menu and place a window around the entire logo. Once all of the geometry has been selected for translation, pick **Done Sel**, and then pick a point on the logo that will be referenced where the logo will appear when the new location is selected.

Your logo may require some cleanup: additional lines may have been created during the DXF conversion, which sometimes occurs when a file originates from a software package different from Pro. To remove any of these unwanted lines, use the **Delete** command from the DETAIL menu.

Sizing Your Logo

Your logo may come in either too large or too small for the space in which it will reside on the format. It can, however, be resized by using the **Rescale** menu pick from the TOOLS menu. Remember that the logo is in segments and therefore that the **Pick Many** command will also need to be used for this process. Once all entities have been higlighted and you have picked **Done Sel**, a prompt for the selection of an Origin Point for Scaling will appear. In this case, you want to select the lower left corner of the logo and then enter a scale value at the prompt. If you want the logo to be reduced, enter a decimal value such as .5 for the scale. If you want the logo to be enlarged, enter a whole number value, such as 2. Once the desired size and the final location have been achieved, it is time to save your new format and start implementing it into your drawings.

Creating a Custom BOM on a Drawing Format

Automatic updating of title block text is great stuff. What about assembly drawings that should contain bill of materials (BOM) information? Pro/ENGINEER addresses BOMs as well. Create a BOM table that will be used in the assembly drawing. Information is available for each component used in the assembly, and it can be automatically placed in the BOM table.

> **NOTE:** *The rest of this chapter discusses features that require the Pro/REPORT module.*

> ✓ **TIP:** *It is best to place the BOM table in a drawing format that will be used for assembly drawings. Although a BOM table can be created in a drawing, you will have to make one in every assembly drawing. In addition, a BOM table created in a drawing format can be aligned to the drawing format border. It cannot be aligned to a drawing border in Drawing mode because the border entities cannot be selected.*

Setting Up the BOM Table on the Format

Make a copy of the B-size format in which you placed the parametric title block, and add the BOM table to it. Quit the window you are in and retrieve the format named *btitle*. Pick **Dbms** ➥ **Save As** to copy the format to a new name.

```
Enter Object To Save [BTITLE.FRM]: ↵
Enter TO: bassy ↵
```

Retrieve the format "bassy." When you create a table and place it with respect to the top left corner of the drawing border, it will start from the corner of the border and extend downward. Pick the following commands:

Table ➡ Create ➡ Descending | Rightward | By Length ➡ Vertex

Next, pick on the top left corner of the drawing border. Make the column widths .5, 5, and .75, and the row heights .5 and .375.

```
Enter the width of the first column in drawing units
(INCH) [QUIT]: .5 ↵
```

Enter 5 and then .75 before accepting the `DONE` default.

```
Enter the height of the first row in drawing units
(INCH) [QUIT]: .5 ↵
```

Enter .375 before accepting the `DONE` default.

The table now contains two rows and three columns. The top row will contain a title in each cell describing the contents of the column. The next row will contain the parametric information to be repeated for each component in the assembly. You must now designate the bottom row of cells as a region of cells that will be repeated. Pick **Repeat Region ➡ Add ➡ Simple**, pick the bottom left cell, and then pick the bottom right cell. The bottom row is now outlined by a magenta fence that identifies it as a repeat region.

Figure 12-13
The BOM table in the drawing format.

Your remaining task is to place the text in the appropriate cells. Before you begin, specify how the text in the columns will be justified. In the title block example, you placed the text and then individually justified each cell's text. This time, if you provide the justification information for each column, the text will be created with the correct justification at the start.

Providing Justification Information for Table Columns

To provide justification information, pick the following commands:

Mod Rows/Cols ➥ Justify | Column ➥ Center | Middle

Next, pick in each of the three columns. They will be outlined in red after you select them. Any text placed in those columns will now be centered both horizontally and vertically.

Adding the Column Title Text

The text in the top row will be the titles for each of the columns. To place the text, pick **Enter Text ➥ Keyboard** and then pick the cell to place the text in.

> **NOTE:** *In that you previously specified a repeat region for this table, you can pick report symbols from a menu to place in the*

*repeat region's cells. For cells outside the repeat region, you must select **Keyboard** and type them.*

Enter the text for the cell. Repeat for each of the top cells of the BOM table until the table looks like the following figure.

Figure 12-14
The BOM table with the text inserted for the column descriptions.

Adding Report Parameters

Pro/ENGINEER supplies you with some parameters that provide information that cannot be placed in a user-created parameter in a part or assembly. These *report* parameters can be used in the cells of the repeat region of your BOM table. To place them, pick **Enter Text ➡ Report Sym** and then pick the first cell on the bottom row. The first column of the BOM will contain the index number on each row of the BOM. From the REPORT SYM menu, pick **rpt... ➡ index**. The second column will contain the name for the model. Pick the second cell on the bottom row. From the REPORTS SYM menu, pick **asm... ➡ mbr... ➡ name**.

> **NOTE:** *In this example, the use of the report parameter, "asm.mbr.name," assumes that you have used a file name for each part that will serve as a description for the BOM. The third column of information will contain the quantity of each component used in the assembly. Pick the third cell, and then select **rpt... ➡ qty**. Now the BOM table has all of the correct text and parameters placed, and most of the text is correctly justified.*

Modifying the Justification of a Text Entry

The description to be used for each part should be left-justified instead of centered. Pick **Modify ➥ Text ➥ Justify** from the FORMAT menu. Pick the text in the bottom middle cell. Pick **Left | Middle | Done-Return** to set its justification. The BOM table is now complete.

Figure 12-15
The BOM table with the text inserted.

Displaying the BOM on the Assembly Drawing

Now you are ready to use the format that contains the BOM table. Retrieve the car assembly drawing. Replace the existing format with the new B-size format that contains the BOM table (called BASSY). In that your assembly does not contain a parameter named `rev_level` or `descr`, you are prompted to provide the text. Key in `A↵` for `rev_level` and `CAR↵` for `descr`. The first thing you notice about the BOM table is that no quantities are displayed. Next, each wheel part is listed on a separate row of the table. You can fix this.

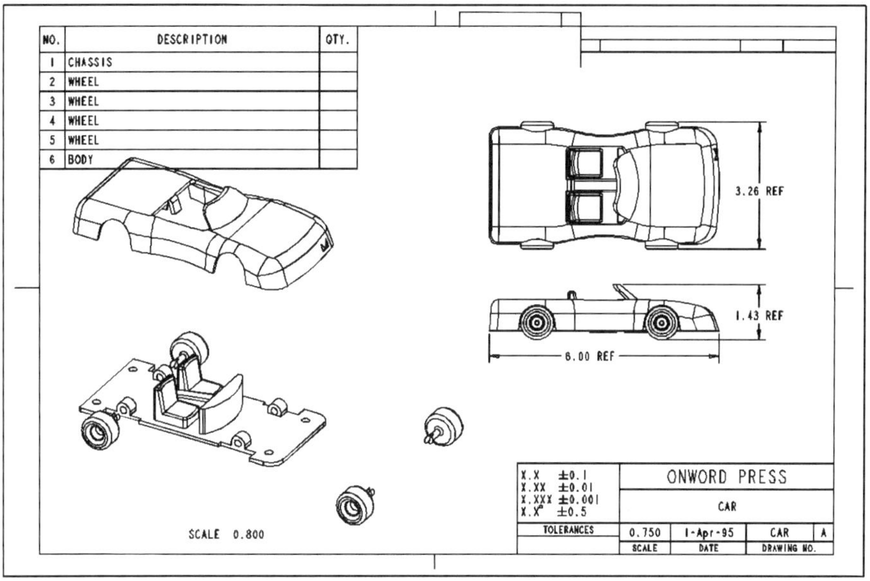

Figure 12-16

The BOM table shown in the assembly drawing with duplicate entries.

Modifying the Repeat Region's Attributes

To change the number of times the same part will display in the BOM table, change the attributes of the BOM table in the drawing format. Quit the drawing window and retrieve the *bassy* format. Change the table's attributes for this format by selecting **Table ➥ Repeat Region ➥ Attributes** from the TABLE menu. Then pick within the repeat region of the BOM table. Next, pick **No Duplicates | Flat | No Cbl Info | Done/Return**.

Retrieve the car assembly drawing. Use the same procedure as the last time you retrieved the car assembly drawing and replaced the existing format. When prompted whether to remove any tables from the drawing, tell Pro/ENGINEER to remove them. Now the BOM table looks correct—even the quantities are listed.

> **NOTE:** *Because you changed only the attributes in the format file, all drawings that retrieve this format in the future will display the new attribute settings. If you had changed only the attributes in the drawing, the format file would have remained unchanged.*

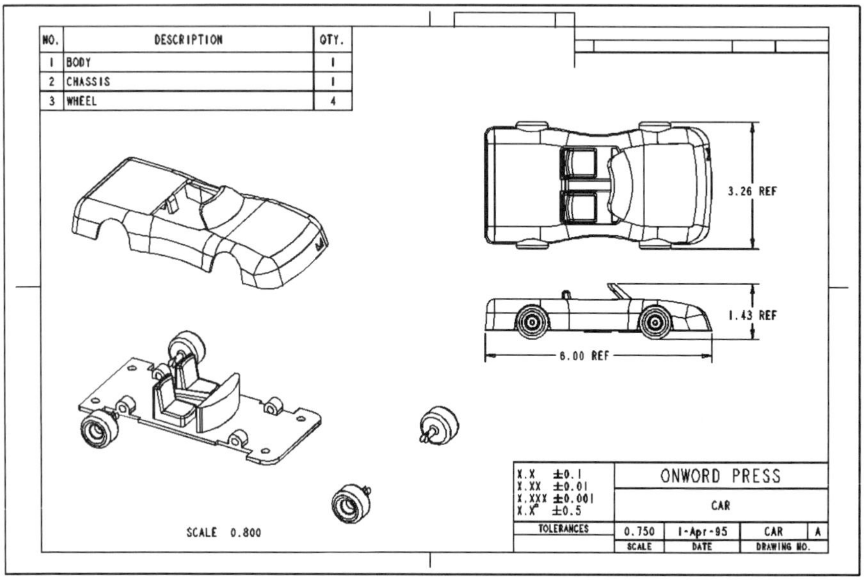

Figure 12-17
The BOM table shown in the assembly drawing.

Displaying Balloons for the Items in the BOM

Now that the BOM table appears correctly on the drawing, you need to place a balloon and leader for each of the components listed in the BOM table. The balloon should contain the number that corresponds to the index number listed for the component. From the DRAWING menu, pick the following commands:

Detail ➡ Show ➡ Balloon ➡ BOM ➡ Set Region ➡ Simple

Now pick within the repeat region of the BOM table and pick **Show** from the BOM BALLOONS menu. You need to move the displayed balloons or their attachment locations to new positions.

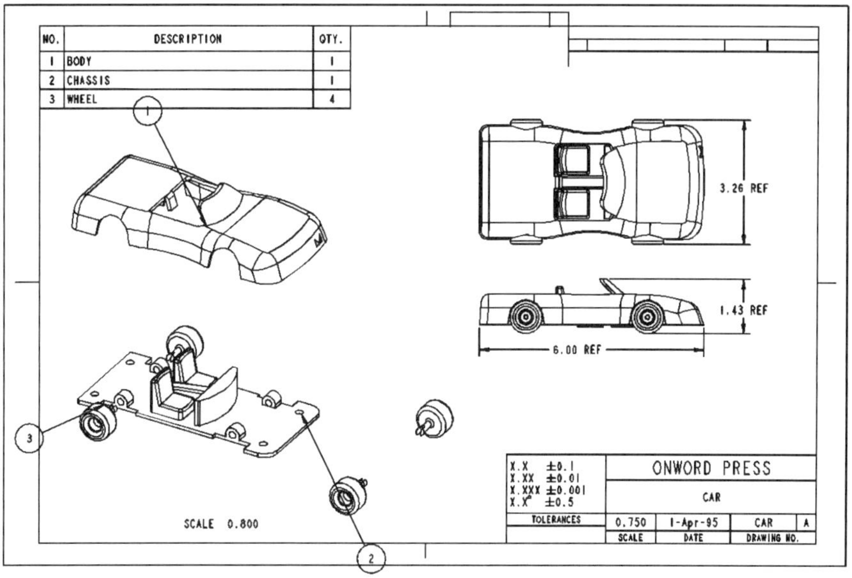

Figure 12-18
The car assembly with the BOM balloons in their initial position.

Relocating the BOM Balloons

From the DETAIL menu, pick **Move** and then the balloon you want to move. Pick the location to move the balloon to. To move the leader attachment location, pick **Move Attach**. Next, pick the balloon that needs the leader attachment moved. Next, pick **Change ref ➡ On Entity | Arrow Head** from the MOD OPTIONS menu. When you pick on an entity of the component, the leader attachment moves to that location. Move all of the balloons until the drawing looks like the following figure.

NOTE: *Note that Pro/ENGINEER will not let you attach a balloon to an invalid part. You cannot attach the balloon for a wheel to the chassis or body. But, you can attach it to another one of the wheels.*

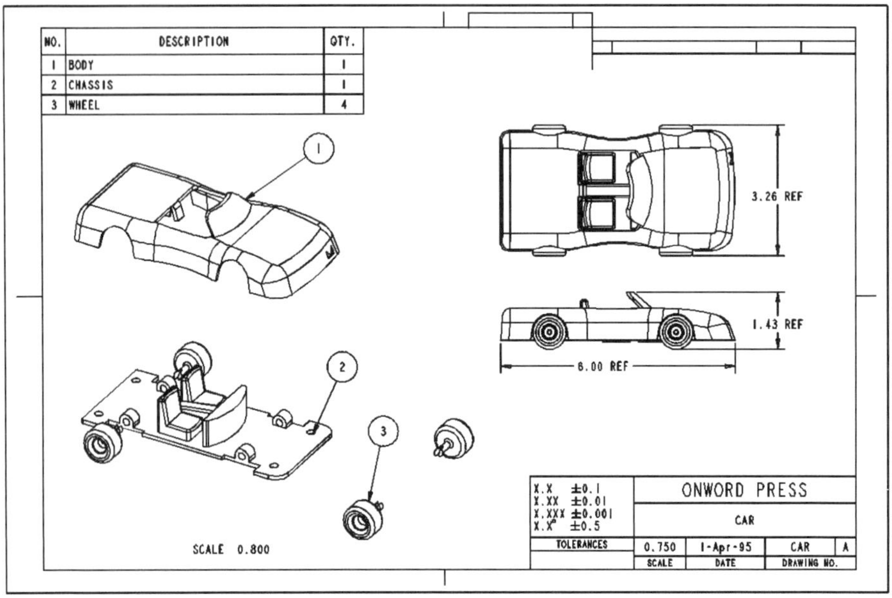

Figure 12-19
The car assembly with the BOM balloons in the final position.

SAVE

Conclusion

Drawing formats can be used to automate filling in the text of the title block. In addition, tables for BOM information can be used in assembly drawings to automatically reflect information about the components in the assembly file. In the next chapter, you will see how the information in the table of a tabulated drawing can be automatically provided.

Part 4
Managing Your Design Environment

Chapter 13

Managing Parts and Their Features

Relations, Family Tables, User-defined Features, and Programs

Introduction

This chapter introduces several methods for managing parts and their features. Relations are discussed in detail. You will constrain a part in an assembly so that it changes when another is changed. A family table will be created to show how similar parts can be tabulated into a single part file. You will learn how to create a *user-defined* feature that can be placed on any part. Finally, a parametric program will be created that will build a part based on user input. Concepts covered in this chapter include the following:

- Relations in Part and Assembly modes

- Relation comments
- Relation operators
- Basic family tables
- Feature suppression by family tables
- Creating formats for tabulated drawings
- Creating user-defined features
- Placing user-defined features
- Creating a program of a part

Understanding Relations

Relations are equations. By setting the value of a dimension with a relation, you can cause it to change whenever a variable in the equation changes. The variables used in a relation are the symbolic names for dimensions or parameter names. Each dimension is given a unique symbolic name such as `d57`. Even reference dimensions and tolerances have symbolic names that can be used in a relation. From now on, these dimension symbolic names will be referred to as dimension variables.

> **NOTE:** *The default dimension variables can be changed. Instead of using* `d18` *in your equations, you can change the dimension variable to* `width`. *Pick* ***Modify*** ➥ ***Dim Cosmetics*** ➥ ***Symbol****, pick the dimension text, and enter the new name.*

You can also create your own variables, called parameters, within Pro/ENGINEER. In the example of the mug from Chapter 1, the variables `OZ`, `VOL`, and `R` are parameters.

```
OZ=20
VOL=OZ*1.8047
R=DIA/2-THICK
HEIGHT=VOL/(PI*R^2)+THICK+.5
```

The variable `HEIGHT` is the dimension variable assigned to the cylinder height dimension of the mug. `DIA` is the dimension variable for the mug cylinder diameter, whereas `THICK` is the dimension variable for the mug wall thickness.

> **NOTE:** *The value solved for in a relation is the variable on the left of the equal (=) sign. Only a dimension variable or a parameter name can appear on the left of the equal (=) sign unless you are solving simultaneous equations. Simultaneous equations will be described later in this chapter.*

> ✓ **TIP:** *A parameter can be created by selecting* ***Add Param*** *from the RELATIONS menu. Unfortunately, a parameter entered this way can only be modified in the model file by deleting the parameter and adding it again, or by adding a relation that sets the parameter name equal to the new value. You can modify a parameter by placing it in a parametric note on a drawing if it was created by selecting* ***Add Param*** *and no relation file was made. However, this method requires you to always retrieve the drawing to change the parameter. A better way of creating parameters is to add a relation that sets the parameter equal to the desired value. Pro/ENGINEER realizes that it is simply a parameter, and it is easily modified by editing the relation. For example, add the relation* `COST=1`*: a parameter name of* `COST` *will be created. To change the parameter value, edit the relation. To delete the parameter, first delete the relation, and then use* ***Del Param*** *from the RELATIONS menu to remove the parameter name.*

Relations have many uses. Three examples follow.

- **Locating the Center of a Feature.** You have already seen how a relation can be used to center a feature in a part. This usage was shown by the relation in the chassis that set the offset dimension of DTM4 to be equal to half the part width.

- **Creating Equal Features.** A relation can also be used to align features without actually creating a parent/child relationship between them. The relationship between the location of the console and the dashboard in the chassis was an example of this concept. The dashboard feature can be deleted without affecting the console feature, yet the two features are always attached via the relation. Later in this chapter, features in two different parts will be aligned using a relation instead of creating a parent/child relationship between the features.

- **Engineering Calculations.** Relations can also be used to solve equations (such as the equations used to determine the mug height in Chapter 1).

Methods of Creating a Relation

Relations can be created in the following three ways:

- From the RELATION menu, select **Add**. You can enter several relations one after another. Just press ↵ when you are finished entering relations.

- From the RELATION menu, select **Edit Rel** to begin an editing session. You can then place your relations from within the editor.

- When modifying a dimension, a relation can be entered. Just answer `Y` to the prompt asking if you want to add the relation.

The first and second methods are used to add relations when all necessary features have already been created. The third method is handy for making quick relations on the fly when you know the variable names you want to use.

Relation Operators and Syntax

Pro/ENGINEER recognizes the use of standard arithmetic operators. The operators follow.

Arithmetic Operators

+ Plus

– Minus

/ Divide by

* Multiply by

^ Exponent

( ) Parentheses for grouping

= Equal to

Conditional statements such as IF, ELSE, and ENDIF are allowed in relations. In order to create these conditional statements, you can use the following comparison operators.

Comparison Operators

== Equal to

!=, <>, or ~= Not equal to

> Greater than

>= Greater than or equal to

< Less than

<= Less than or equal to

& And

| Or

~ or ! Not

> **NOTE:** *Do not forget that when you want to compare values to see if they are equal, use == instead of =. When you use a single equal sign, you are assigning a value.*

Many mathematical functions are also recognized by Pro/ENGINEER.

Mathematical Functions

`sqrt()` square root

`sin()` sine

`cos()` cosine

`tan()` tangent

`asin()` arc sine

`acos()` arc cosine

`atan()` arc tangent

`sinh()` hyperbolic sine

`cosh()` hyperbolic cosine

`tanh()` hyperbolic tangent

`log()` base 10 logarithm

`in()` natural logarithm

`exp()` e to an exponential degree

`abs()` absolute value

`ceil()` the smallest integer that is larger than the specified value

`floor()` the largest integer that is smaller than the specified value

> **NOTE:** *All trigonometric functions use degrees.*

Text or string variables can also be set in relations. Operators and functions available for these string variables follow.

= used to set a variable equal to a string

== used to compare strings as equal

!=, <>, or ~= used to compare strings as unequal

\+ used to concatenate strings

`itos(int)` used to convert an integer to a string

`search(string,substring)` used to search a string for a substring and return the location that the substring begins

`extract(string,position,length)` used to extract a portion of a string

Managing Relations

After entering relations, you can select **Show Rel** from the RELATIONS menu to display a window containing all relations and parameters. When you regenerate the model, each of the relations will be executed in the order that they appear in this window.

> **NOTE:** *When you create relations, the values are not included in the model until the model is regenerated.*

> ✗ **WARNING:** *Be careful with the order of relations.*

Do not use a dimension in a relation if the same dimension will be modified by a later relation. In that Pro/ENGINEER evaluates the relations in the order they are displayed, the dimension will be used in the relation with its old value before the new value is assigned. If you select **Sort Rel** from the RELATIONS menu, Pro/ENGINEER will automatically rearrange your relations to prevent this problem.

> **NOTE:** *If a dimension is assigned a value by more than one relation, the value of the dimension is determined by the last of those relations.*

Simultaneous Equations in Relations

Relations can also be used to solve simultaneous equations. The mug height equations could have been stated in the form of a simultaneous equation. For that case, the relations would have appeared as follows:

```
OZ=20
VOL=OZ*1.8047
SOLVE
  VOL=((DIA-2*THICK)^2*PI)/4*(HEIGHT-THICK-.5)
FOR HEIGHT
```

Variables equal to a constant value that are used in the simultaneous equations must be defined prior to the SOLVE statement. The VOL variable is an example. In this case there is only one unknown variable; thus, only one equation is required. When there are more variables and equations, then all of the variables to be solved for are listed on the line beginning with FOR.

> **NOTE:** *As you see in the example,* `PI` *is a parameter already defined by Pro/ENGINEER.*

In extreme cases, a series of relations can be circular in nature. For instance, consider

```
A=f(B)
C=f(A)
```

and

```
B=f(C).
```

If the value of B changes as a result of the third relation, the first relation will not be reevaluated to show the proper value. This scenario requires an iterative solution. Pro/ENGINEER will not iterate relations during the regeneration process, and these equations will not converge to a common solution. Try to avoid these situations when you write relations.

Creating a Relation in Assembly Mode

The obvious design goal for the car parts is for the body to attach to the chassis. To accomplish this, the bosses in the body must be aligned to the holes in the chassis. You should not create the chassis hole features in Assembly mode and make them children of the boss features on the body. It is usually best to create the individual part features in Part mode and constrain them with relations in Assembly mode, thereby avoiding the parent/child relationship.

Create a relation that equates the spacing of the bosses in the body to the spacing of the holes in the chassis. Retrieve the car assembly and pick **View ➥ Cosmetic ➥ Explode** to explode the assembly.

> ✓ **TIP:** *Seeing the features of each component is easier if you explode the assembly. In this way you can pick the feature that will display the desired dimension.*

Pick **Relations ➥ Assem Rel** so that you can create a relation with variables from different components. To see the variables that control the hole spacing, pick the surface that forms the wall of the hole shown in the following figure.

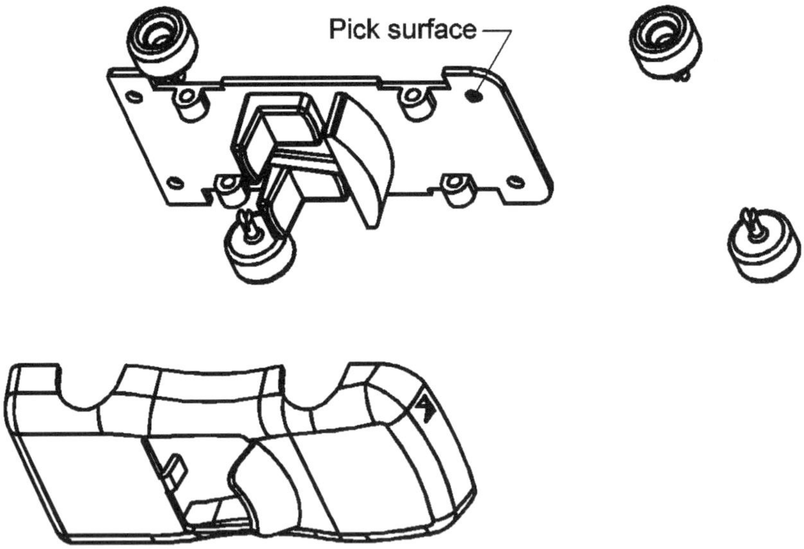

Figure 13-1
The surface to pick on the chassis for displaying the hole spacing dimensions.

After the dimensions that constrain the hole feature are displayed, make a note of the dimension variable used for the spacing from the back holes to the front holes. In this example, the dimension variable is `d47:0`.

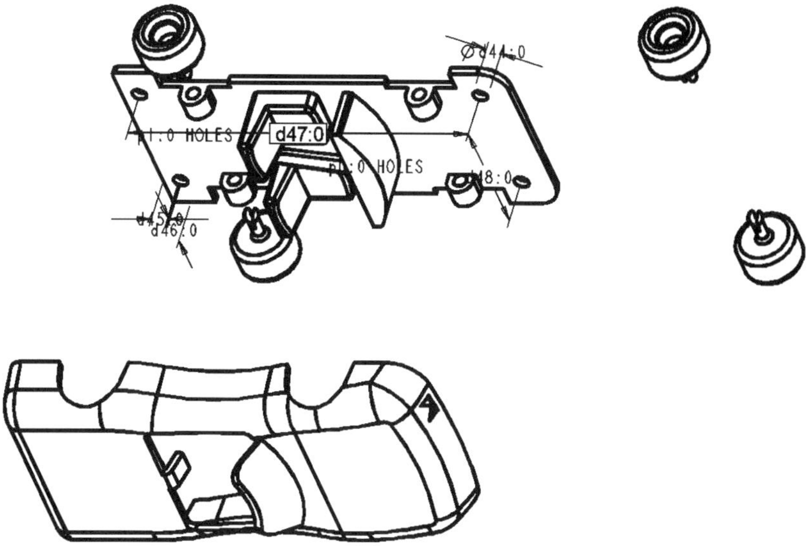

Figure 13-2
The dimension vaariables for the hole spacing in the chassis.

> **NOTE:** *The dimension variables may be different for your chassis because the variable numbers are assigned in the order you create your dimensions in Part mode.*

Notice how the variable is a bit different from those that were shown when you created a relation in Part mode. As you add a component to the assembly, the component is assigned a number. The number is then appended to each part's dimension variable. This makes each dimension variable unique for all parts in the assembly.

Now orient the assembly so that you can see the bosses on the bottom of the body. When you reorient the assembly, the displayed dimensions will disappear. Pick on the flat surface on the end of the front boss shown in the following figure.

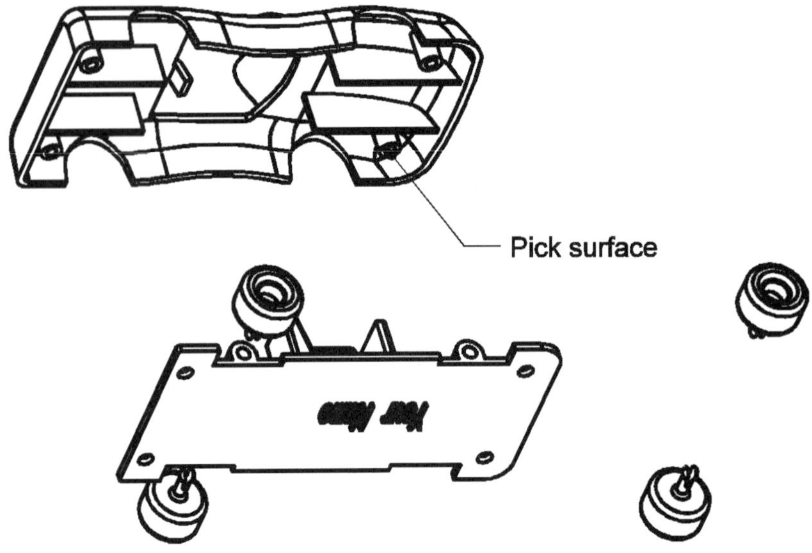

Figure 13-3
The surface to pick on the body for displaying the boss spacing dimension.

> ✓ **TIP:** *When you have added a draft feature to a surface, the draft feature's dimension will be displayed when you pick the surface to display dimensions. To display the dimensions for the feature that was created before the draft, use* **Query Sel.** *As you step through the features using* **Next,** *the type of feature and its feature number will display in the Message window. Pick* **Accept** *when the appropriate feature is displayed in the Message window.*

The dimension variable `d124:4` is the desired variable in this example.

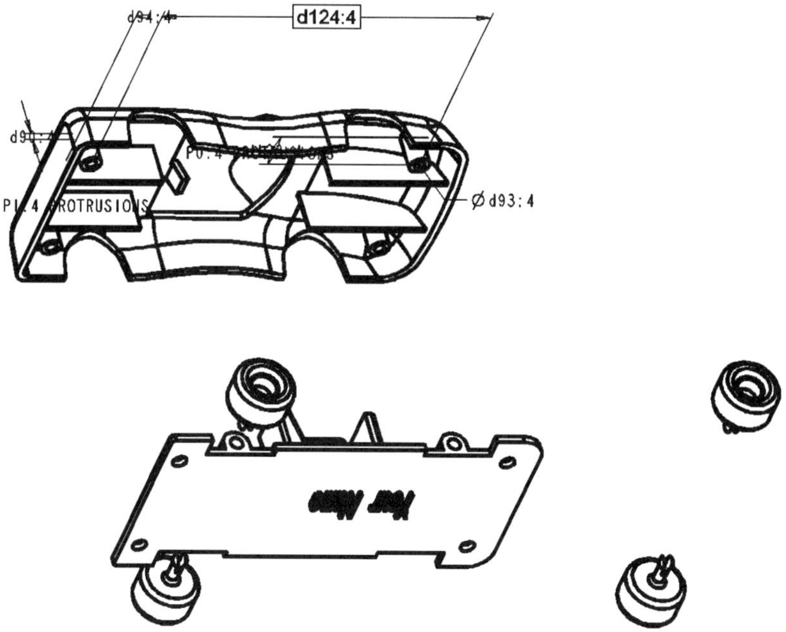

Figure 13-4
The dimensions for the spacing on the body.

Because all dimension variables are now known, the relation can be created. Pick **Add** and enter the relation that will make the spacing of the bosses on the body match the spacing of the holes in the chassis.

> ➾ ***NOTE:*** *You should try to place comments with each relation so that you will not forget the relation's purpose. In Pro/ENGINEER, you place comments by preceding the line of text with the characters / * . Comment lines should be placed before corresponding equations. The comments will then move with the equations they precede.*

Enter RELATION [QUIT]: **/*Equates chassis hole spacing to body boss spacing** ↵

Enter RELATION [QUIT]: **d124:4=d47:0** ↵

Enter RELATION [QUIT]: ↵

Pick **Switch Dim** until the dimensions are displayed as their numeric values again, and then pick **Done**.

Assembly Relation Behavior

The relation now makes the spacing of the bosses on the body change whenever the spacing of the holes in the chassis changes. If you try to modify the spacing of the bosses on the body from within the car assembly or with the car assembly *in session,* you will not be allowed to do so because that dimension is driven by a relation. If you modify the body boss spacing without the car assembly *in session,* the body bosses will move because Pro/ENGINEER does not know that this dimension is controlled by a relation. When you retrieve the assembly, the body boss spacing will change back to the value determined by the relation.

To view the effect of this relation, change the spacing for the holes in the chassis. From the ASSEMBLY menu, pick **Modify ➥ Mod Dim**. Pick the same hole in the chassis you did to display the dimension for the relation. Pick the `4.9` dimension value and change it to `4.5`. You are now ready to regenerate the appropriate parts. Pick **Regenerate ➥ Automatic** from the ASSEMBLY menu, and the modified components will begin regenerating.

> **NOTE:** *You do not have to select every component that has been modified when you regenerate an assembly. By picking* **Automatic**, *every component that has been modified, but not regenerated, will regenerate.*

Notice how the bosses in the body are automatically repositioned to maintain the same boss spacing in the body as the hole spacing in the chassis.

> **NOTE:** *Observe how this relation would not have been possible if not for the basic techniques you developed earlier on for part construction. If these holes and bosses had different controlling parametric dimensions, this relation would have been impossible to create.*

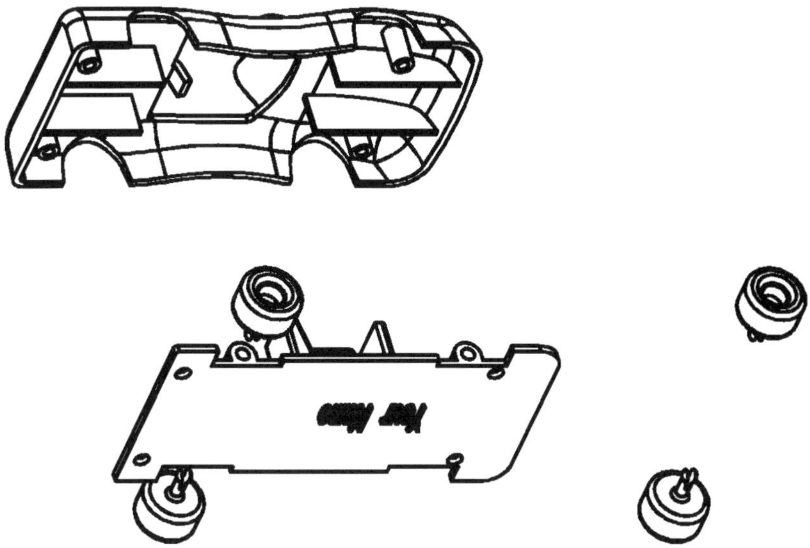

Figure 13-5
The bosses and holes are aligned because of the assembly relation.

Relations and Design Intent

Relations are used for situations where one part or feature *drives* another part or feature. In most mechanical designs, you will be able to decide which part or feature should be the driver of the relation. If your design uses a form of a standard component such as a bolt or an electrical connector, you should make the standard component the driver of the relation because its shape governs your design.

For situations where both parts involved in the relation are flexible, choose one of the parts to be the driver. You can always change your mind later. For example, imagine that a dimension in part A drives a corresponding dimension in part B. Your relation would be something like `d1 (in part B) = d3 (in part A)`. If a design change occurs that forces you to change part B, just flip the relation around so that part B becomes the driving component: `d3 (in part A) = d1 (in part B)`.

Relations are easy to use and maintain. Just make sure to keep them sorted and use comments so that six months later you will know what the relation controls.

Family Tables

As you have seen, modifying the dimensions of a Pro/ENGINEER part and changing its size are easy. You could make three different sizes of wheels by copying the existing wheel part to different file names. Modify each of the files as needed to have three separate part files, with each file containing a different wheel size. This functionality can be combined into a single part file.

What Is a Family Table?

If you have ever seen a tabulated drawing for a family of parts, you have seen the equivalent of a Pro/ENGINEER family table. A generic part represents your basic design shape. Some dimensions common to all parts are shown in views of the drawing. Other dimensions vary for each member of the family and will appear in the table on the drawing. A family table provides the ability to save many variations of a part within a single part file. To make a family of parts, you simply specify which dimensions will vary for each member of the family after you have completed the basic (generic) part.

Adding a Dimension to the Family Table

Make a family table for the wheel. Retrieve the wheel in Part mode. For this example, you will vary the tire width while leaving all other dimensions the same for each wheel. Pick **Family Tab** ➥ **Add Item** ➥ **Dimension** from the PART menu. Next, pick the feature on the wheel that will display the tire width dimension.

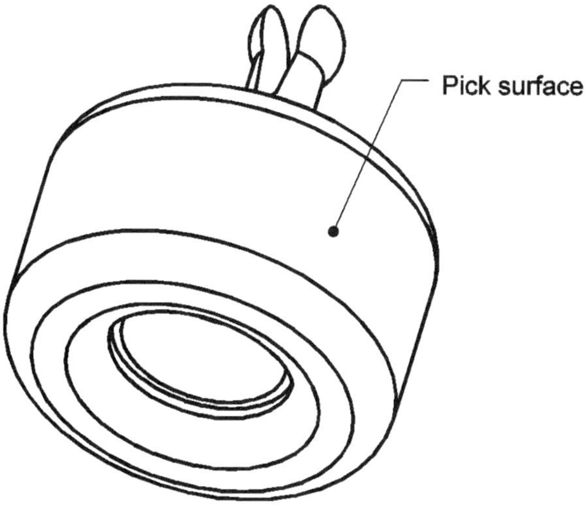

Figure 13-6
The surface to select for displaying the tire width dimension.

Now, pick on the `.50` dimension for the tire width. The dimension will highlight to indicate that it has been selected for the family table. A column for the tire width is now included in the family table.

> ➾ ***NOTE:*** *Valid dimensions for adding to the family table include part dimensions, pattern dimensions, geometric tolerances, and dimensional tolerances.*

Family Table Column Headings

To see the family table, pick **Show** from the FAMILY TABLE menu. A column has been added to the tabulated list for the selected dimension. The dimension's variable name is shown as a heading at the top of the column. As you can see, the default dimension variable names are not very distinctive. Change the name `d0` to `WIDTH`.

Exit from the family table viewer. Change the dimension's variable name by picking **Modify ➥ Dim Cosmetics ➥ Symbol** from the

PART menu. Then pick on the wheel width dimension text. At the prompt, enter the new name for the variable.

```
Enter symbol text: WIDTH ↵
```

Select **Relations** from the PART menu and you will see that `WIDTH` is the new dimension variable. Do not forget to pick **Switch Dim** before leaving the RELATIONS menu, so that the dimensions are displayed as their numeric values.

> ✓ **TIP:** *If you modify the dimension variable, you can change it from* `d23` *to another name such as* `A`*. When that happens, the family table column heading will become* `A`*. Next, instead of having a dimension number showing with the dimension leader lines when you create a tabulated drawing, you can have the dimension variable,* `A`*, displayed with the leader lines. This would identify the correct column,* `A`*, for finding that particular dimension.*

> **NOTE:** *When dimension variables are renamed, any relations using that dimension are automatically updated.*

Adding an Instance to the Family Table

Now edit the family table and add the correct values to the table for the wheels. Pick **Edit** from the FAMILY TABLE menu, and the editor will appear displaying the family table. Notice how the dimension variable name, `WIDTH`, is now shown.

```
!
!                    FAMILY TABLE EDITOR
!
!  1) Rows beginning with '@' will be saved as
!     comments.
!  2) Rows beginning with '!' and empty rows will be
!     '!' ignored.
!  3) Rows beginning with '$' contain locked instances.
!  4) The name of each part or assembly instance may
```

```
!     begin with a letter or a number and should be
!     unique within the entire family.
!  5) '*' can be used for the default value.
!  6) Values for the generic part cannot be changed.
!  7) Changes to instance values will, however, be
!     saved, if the instance is not locked.
!  8) Generic names of features if they appear are
!     enclosed in [].
!  9) You may add more entries to the bottom of the
!     table as needed.
! 10) Pro/TABLE formatting characters will also be
!     ignored.
! 11) Feature identifications are their internal ids.
!   Generic part name: WHEEL
!  Name                d0
!                   WIDTH
!
!  ========== ========== ==========
!  GENERIC          0.5000
```

Pro/ENGINEER refers to each part listed in a family table as an *instance*. The part you create a family table from is referred to as the *generic part*. When you first access the family table in the editor, there are no instances in the table. Scroll down to the bottom of the file in order to add the first instance. Leave the first column empty [below the exclamation (!) points]. In the second column, below the name `GENERIC`, enter the name for the first instance.

> ✓ **TIP:** *Many companies use the part number for the instance name. That name will be the one the instance is selected by when you want to retrieve it.*

Enter the name `narrow` for the first instance. Move to the next column below the heading `WIDTH` and enter `.4`, the tire width for the instance. Now, move down to the next line and do the same for an instance named `standard` with a width of `.5`. For the last line of the family table, add an instance named `wide` with a width value of `.6`. The family table now contains three instances.

```
!   Generic part name: WHEEL
!  Name                d0
!                    WIDTH
!
!  ========== ========== ==========
!  GENERIC          0.5000
   NARROW               .4
   STANDARD             .5
   WIDE                 .6
```

Save the family table and exit from the editor. See Appendix A for information on the use of the Pro/TABLE editor. After adding information to the family table, it is a good idea to pick **Verify** from the FAMILY TABLE menu. Pro/ENGINEER will then check each instance to verify that it can successfully regenerate. When you save the generic part, Pro/ENGINEER adds the instance names to an index file (*.idx* extension) so that they can be retrieved by their names like any part files.

Displaying a Family Table Instance

To see what one of the instances looks like, pick **Instance** from the FAMILY TABLE menu. Select **NARROW** from the INSTANCES menu. A new window shows the part named `narrow`. Erase the part named `narrow` from the session using the **DBMS ➡ Erase** command. Change back to the graphic window containing the generic part by picking **ChangeWindow**, and then picking in the window.

Adding a Parameter to an Instance

In addition to placing dimensions in the family table, you can place *parameters* in the family table. From the RELATIONS menu, you can define a variable for something such as the cost of a wheel. From the RELATIONS menu, pick **Add Param ➡ Number**.

```
Enter Parameter name: COST ↵
Enter value for COST [0.0]: 10 ↵
```

The parameter can then be added to the family table in the same way as the dimension was added. Pick **Add Item ➡ Parameter** from the FAMILY TABLE menu instead of **Add Item ➡ Dimension**, and then pick the parameter name, COST, from the PARAMETER menu. Pick **Done** to complete the selection.

> ✓ **TIP:** *If you are going to place a parameter in a family table, do not use a relation to create the parameter. The relation will govern the value of the parameter instead of the value entered in the family table.*

Edit the family table and enter the cost of each of the wheels in the column labeled `COST`. Use `10` for the narrow wheel, `12` for the standard wheel, and `15` for the wide wheel. You can associate attributes to instances in a family table in the same manner as to regular Pro/ENGINEER parts.

> ✓ **TIP:** *You can make a separate drawing for each instance instead of a tabulated drawing. In that the drawing format has the description determined by the value of the parameter* `descr`*, you can place a parameter by that name in the family table and have it relate to different model notes for each instance. You previously created a model note for the wheel, so a parameter named* `descr` *already exists in the part file. Just create a model note appropriate for each instance in the family table. Make sure you remember the ID number for each note you create so that you know which instance it applies to. Add the parameter,* `descr`*, to the family table. Then edit the column for that parameter and enter the appropriate ID number for the desired model note for each instance.*

Adding a Feature to an Instance

Families of parts will frequently use the same features. Sometimes one of the parts may have a feature that none of the others have. Instead of leaving it out of the family table, you can add the feature to the generic part and then add a *feature* column to the family table.

Put a protrusion of the letter `N` on the `narrow` instance of the wheel only. On the generic part, create the feature on the face of the hub so that it looks like the following figure.

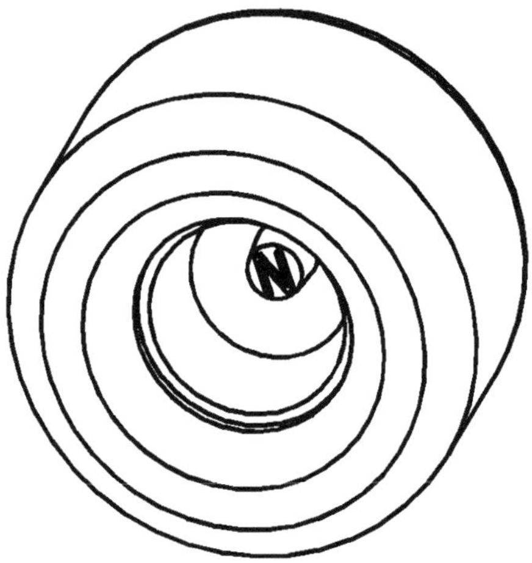

Figure 13-7
The generic part with the letter `N` added to the face of the hub.

After sketching the letter on the hub face, pick **Modify**, and pick the letter. Now select **Text Height** and change it to `.075`. Then change the text width factor to `.8`. Place a linear dimension of `.04` from the text to DTM2 and `.03` to DTM1. At the prompt for the protrusion depth, enter `.01`.

Now pick **Add Item ➡ Feature** from the FAMILY TABLE menu. After selecting the new feature, pick **Edit**, so that you can specify which instances should contain the new feature. On the row for the narrow instance, place Y in the last column. In the last column, place an N on all of the other rows that contain an instance. Notice how the column heading shows only the feature ID number. The heading can also be changed to something more descriptive. Save the family table file and exit from the editor. From the PART menu, pick **Set Up ➥ Name ➥ Feature**. Pick the last feature you created. At the prompt, rename it to LETTER_N. The next time you look at the family table, you will see that the column for the feature you just added is now labeled LETTER_N.

Another way to add the feature to only one instance is to retrieve the instance and add the feature directly to it instead of adding it to the generic part. This method will automatically add a column to the family table and place Y in the new column on the row for the instance the feature was created on. Place the letter W on the hub of the wide wheel. Retrieve the wide wheel instance by picking **Instance** from the FAMILY TABLE menu. Place the letter W on the wide wheel with a protrusion in the same way you placed the letter N on the generic wheel.

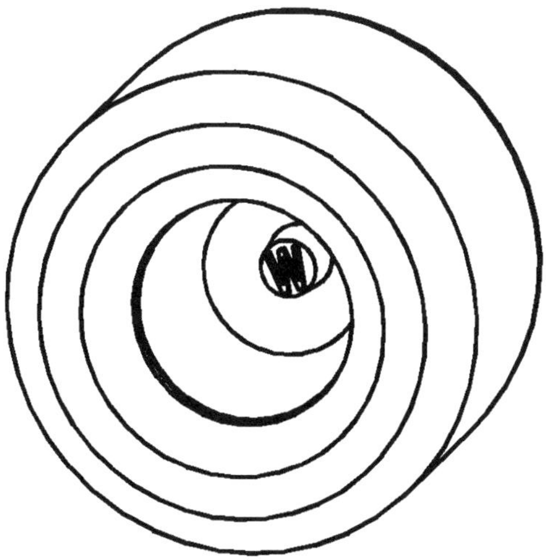

Figure 13-8
The wide instance with the letter W added to the face of the hub.

When you are finished, select **Quit Window**, and then change back to the generic graphic window by picking **ChangeWindow** followed by picking inside the generic graphic window. Notice that the letter `W` does not show as a feature on the generic. This is because it was created as a suppressed feature on the generic part. Show the family table and you will see that the row containing the `GENERIC` description has an `N`, to denote that the feature is suppressed, in the column for the new feature. The `wide` instance is the only one that has a `Y` for this last feature. That is exactly what you wanted. An asterisk (*) in a column means that it takes the value shown for the generic.

> ✓ ***TIP:*** *It is best to change the asterisk (*) to a value. If someone resumes the feature of the* `W`*, the asterisks (*) will cause each of those instances to contain the feature. Change them to* `N`*.*

```
!   Generic part name: WHEEL
!  Name               d0        COST        F362        F427
!                  WIDTH                LETTER_N    [PRTRSN]
!
!  ========= ========= ========= ========= =========
!  GENERIC      0.50000   10.0000           Y           N
   NARROW       0.40000   10.0000           Y           N
   STANDARD     0.50000   12.0000           N           N
   WIDE         0.60000   15.0000           N           Y
```

SAVE

> ✓ ***TIP:*** *Sometimes features that only pertain to one or two instances can conflict when you try to place them all in the generic part. This means you cannot have both features in the generic part at the same time unless at least one of them is suppressed. The easiest way to place both of these features into the family table (and into the generic part) is to create each feature on the instance where you want it associated.*

Modifying a Family Table Instance

The dimensions in the family table can be changed by editing the value in the family table file or by modifying the dimension in the part or drawing file like any other part dimension. However, you will be prompted to confirm that you want to modify a value in a family table. To edit the value in the family table, use the **Edit** command from the FAMILY TABLE menu.

> ➾ **NOTE:** *If you modify a generic dimension in an instance part, the generic part will automatically update to reflect the change. If you change to another existing graphics window containing a different instance part, you may have to regenerate the instance before the change is displayed.*

Retrieving a Family Table Instance into Assembly Mode

Now you know how to create a family table and retrieve an instance. Place an instance of the wheel on the car assembly. The wide wheel can be placed so that it replaces the two rear wheels. Retrieve the car assembly and pick **Component ➥ Replace**. Now pick one of the rear wheels. From the REPLACE WITH menu, pick **Family Table**. In this case, you could accept the default file name. Enter `?` instead, so that a menu of components is displayed.

```
Enter Name [WHEEL.PRT]: ? ↵
```

There are two ways to pick an instance to be retrieved into Assembly or Part mode. If you select **Verify** from the FAMILY TABLE menu after creating the family table and then saved the part file, each instance name will be shown on the menu that appears because they have been included in the directory index file. Each instance will be displayed as the instance name followed by the generic part name, `wide<wheel>.prt`. You can then select it by picking it directly off the menu. However, if you entered `wheel.prt` instead of `?`, or selected `wheel.prt` from the menu, you must answer the following prompts in order to place the wide wheel.

```
Do you want to assemble an instance of WHEEL? [Y]: ↵

Enter dim WIDTH value (<CR> for menu, > to skip, * to
name instance): * ↵
```

Pick **WIDE** from the INSTANCES menu.

Automatic Assembly Replacement of a Family Table Part

Pro/ENGINEER knows that the original wheel used in the assembly was the generic wheel. When a component of an assembly is a member of a family table (generic or instance), it can be replaced by any other member of the family table without again specifying the assembly constraints. Whenever Pro/ENGINEER identifies this situation, it displays the AUTO/MAN menu. When the AUTO/MAN menu is displayed, pick **Automatic**, so that the same assembly constraints are used for the wide wheel instance. The wide wheel is now placed in the assembly. Repeat the procedure for the other rear wheel.

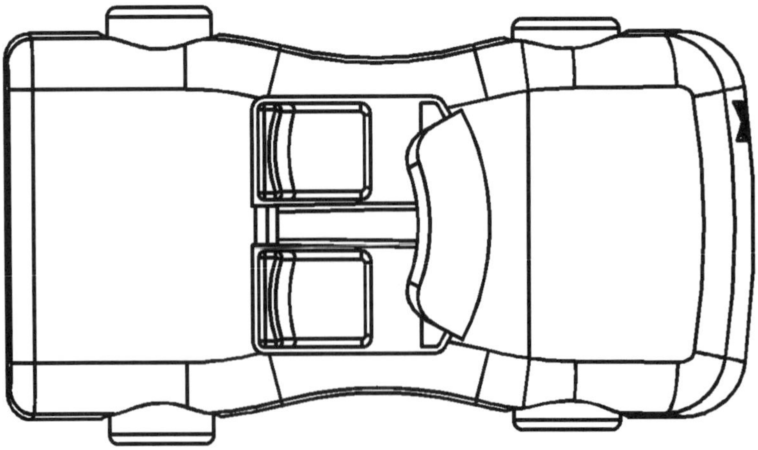

Figure 13-9
The car with the wide rear wheels.

➾ **NOTE:** *If geometry particular to one instance of a family table is used in an assembly constraint (you have features in one instance that locate it in the assembly, and those features are not included in another instance), you might not be able to automatically replace that instance in the family table with one of the others. In that case, the AUTO/MAN menu will not display.*

Making a Tabulated Part Drawing of a Family Table Part

You already know how to make a drawing of a part and how to make an assembly drawing that shows a BOM. A tabulated drawing is a combination of both types of drawings. You place the generic part on the drawing in the same way as when making a drawing of a part. You use a format in the drawing that contains a table in the same way as for an assembly drawing with a BOM. The table placed in the format is the legend for listing the tabulated information on the drawing.

Creating a Format Table for a Family Table Part Drawing

Create the format you will use for the tabulated drawing. Retrieve the "btitle" format and copy it to the new name "btab." Now retrieve the "btab" format. All title block information created in the "btitle" format is already included in this format. Only the tabulated dimensions table needs to be added. Create a table that contains two columns and two rows, as shown in the following figure. From the FORMAT menu, pick **Table** ➡ **Create** ➡ **Descending | Rightward | By Length** ➡ **Vertex**. Place the vertex at the top left corner of the border. Make the first column of cells 1.5" long and the second column 1" long. Make the first row of cells .5" high and the second row .375" high.

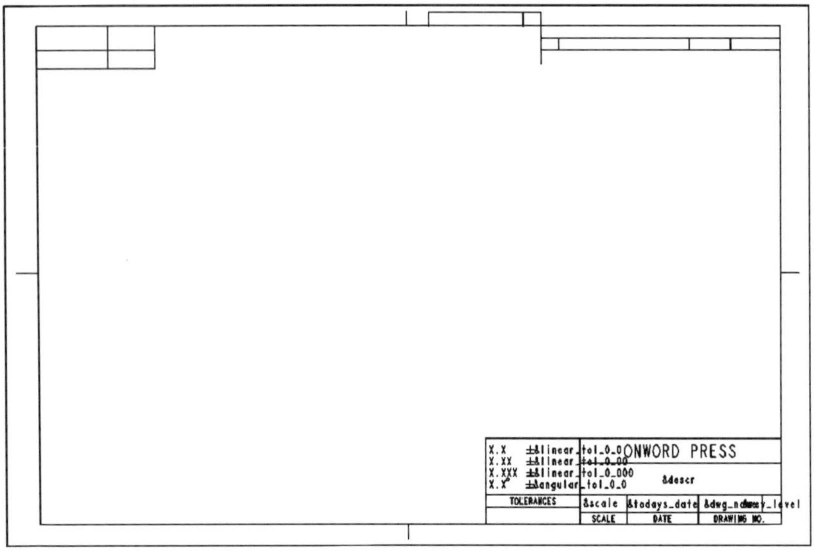

Figure 13-10
The table for the tabulated dimensions.

The information placed in the first column will be the part name for each instance. The next column will contain the wheel width dimensions from the family table `WIDTH` column. The title of the column will be the same as the title of the column for the family table. If there is more than one column of dimensions in the family table, the table will automatically add the appropriate number of columns so that all of the dimensions will display within the table. It will also add rows so that each instance of the family table will display in the table.

At this point, you should provide a title for the first column of the table. Enter the text `PART NAME` in the top left cell. The other cells will contain parametric text that varies depending on the information in the family table. The top row of each dimension column should contain the dimension variable name used in the family table. Enter the parametric text `&fam.inst. param.name` in the top right cell. The left cell of each row should contain the part name for each instance. Enter the parametric text `&fam.inst. name` in the bottom left cell. The last cell will contain the dimensions for each instance. Enter the parametric text `&fam.inst.param.value` in the bottom right cell. The table should look like the following figure.

> **NOTE:** *If you enter the text in the table prior to denoting the repeat region, you must enter the text for the report symbols. You do not have the option of selecting them from the REPORT SYM menu, as occurred when you created the format for the assembly drawing.*

PART NAME	&fam.inst.param.name
&fam.inst.name	&fam.inst.param.value

Figure 13-11
The table for the tabulated dimensions with the text placed.

Now we need to specify how the rows and columns of the table should be repeated so that the table will contain all instances and items from the family table. The row containing the part name for each instance should be repeated for each instance in the family table. Likewise, the column containing the dimension variable name should be repeated for each column of dimensions in the family table. This is done by specifying a repeat region. The repeat region for this table is a bit different from the one used for the assembly format's BOM table. The assembly BOM repeated a row only as another component was added to the assembly.

The table will repeat rows and columns. To create this type of repeat region, pick **Repeat Region ➥ Add ➥ Two-D**. Next, pick the cell in the highest row that will repeat (the top, right cell) and then the leftmost cell that will repeat (bottom, left cell). There will be the cells containing the names for each column and row. Finally, pick the cell common to both the row and column repeat regions. In this case, select the cell containing the parametric text `&fam.inst.param.value`. The tabulated dimension table is now complete.

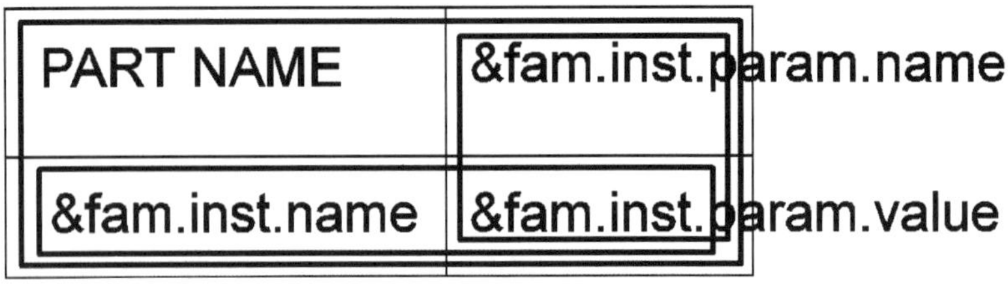

Figure 13-12
The completed tabulated dimension table in the format.

SAVE

Adjusting the Table Appearance of Your Drawing

To make the tabulated drawing, retrieve the wheel drawing. Replace the old drawing format with the new one. Pick **Sheets** ➥ **Format** ➥ **Add/Replace** ➥ **Unblank** and enter `btab` for the new format name. Remove all of the old format tables when you are prompted. The tabulated dimension table will automatically expand to display all family table part names and column information.

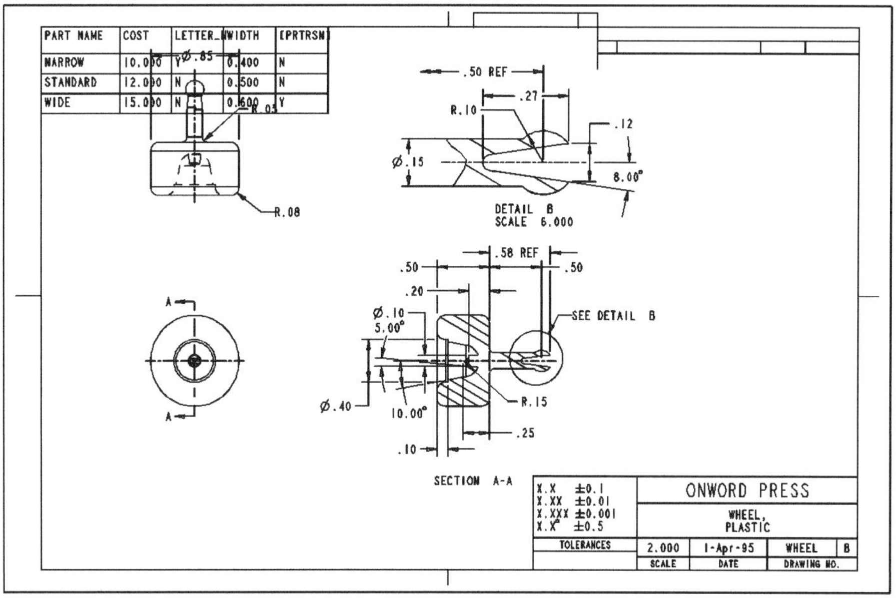

Figure 13-13
The tabulated drawing of the wheel.

Move the views so that they no longer overlap the table. The table contains all family table information. For the example, you want the column for the wheel width, so pick **Table ➥ Repeat Region ➥ Filters** from the DRAWING menu. Pick in any cell except the one containing the text `PART NAME`. Next, select **Add** from the FILTER REG menu. Now you are ready to enter the filter that will exclude from the table all family table columns except the one corresponding to the wheel width dimension.

```
Enter next filter: &fam.inst.param.name == WIDTH ↵
Enter next filter: ↵
```

After picking **Done**, the table shows only the wheel width column. The dimension on the drawing for the wheel width needs to be changed to display its variable name. Pick **Modify ➥ Text ➥ Text Line** from the DRAWING menu. Next, pick the `.50` dimension for the wheel width. Edit the text from `@D` to `@S`. Now the dimension on the drawing will always be displayed as the variable name.

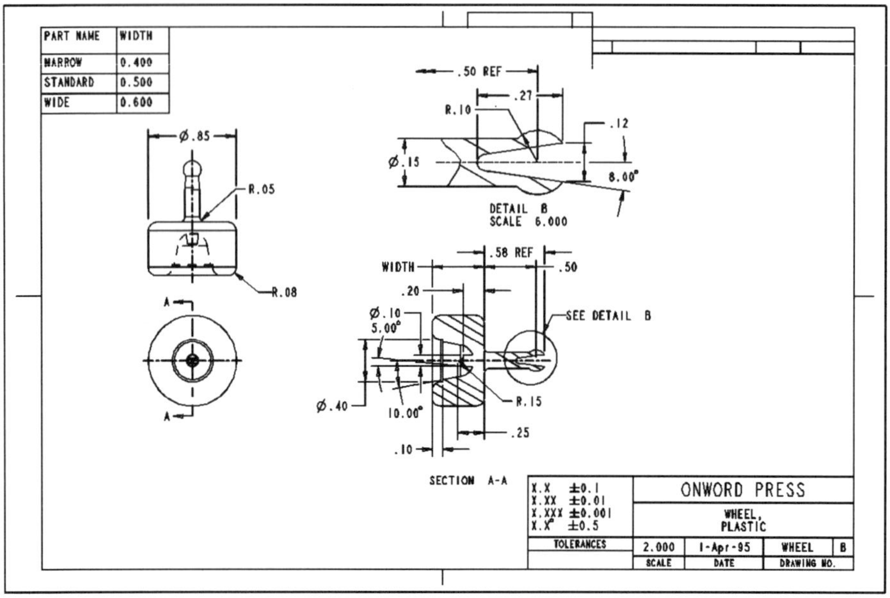

Figure 13-14

The tabulated drawing of the wheel with the variable name displayed for the wheel width.

➻ ***NOTE:*** *If the dimension variable name is not changed in Part mode, it can be changed in Drawing mode. Pick* **Modify ➥ Dim Params ➥ Dim Symbol** *from the DRAWING menu. Pick on the dimension text to change and enter the new variable name at the prompt. If the name is not* `WIDTH`*, the repeat region filter will not show the dimensions in the table.*

SAVE

Family Tables in Your Designs

Many companies design parts or products that can use Pro/ENGINEER's family table capabilities. Simple parts such as screws and bolts are good examples. More complex designs such as valves or electrical connectors can also take advantage of family tables. If you use family tables, try to take advantage of all capabilities. Use parameters for items such as cost, part weight, part description, and vendor name. Adding these items will help you create a complete 3D database with many downstream capabilities.

User-defined Features

While overall similarity between two parts may not exist, elements of a design are sometimes repeated from part to part. Whether a single or multiple features, these smaller elements can be captured for future use. Such elements can be placed on the same part again or any other part that requires the same element. Pro/ENGINEER provides a method for saving a group of features known as *user-defined feature*. User-defined features are a great way of standardizing certain elements of your designs. The size of the user-defined feature can be determined by any combination of the following methods:

- ❒ New features will have dimensions identical to the saved features' dimensions.
- ❒ Some dimensions of the new features will be prompted for during placement.
- ❒ Some dimensions will be determined by selecting an instance from a family table.

A reference part can also be saved to aid in the user-defined feature placement process. When Pro/ENGINEER prompts you to pick geometry, equivalent geometry on the reference part will be highlighted. After placing the user-defined feature, the dimensions can be modified like any other feature's dimensions, unless you refuse this option when you are retrieving the user-defined feature.

Selecting the Features to Include in a User-defined Feature

Create a user-defined feature of the logo on the front of the car body. A logo is a good example of a common feature or a common group of features that can be placed on several different parts. Retrieve the body part. From the PART menu, pick **Feature ➥ Group ➥ Define.** Enter a name for this feature.

```
Enter group name [QUIT]: logo ↵
```

Now you need to select the features to be included in the user-defined feature. In this case, there is only one feature that makes the logo. Pick the logo feature.

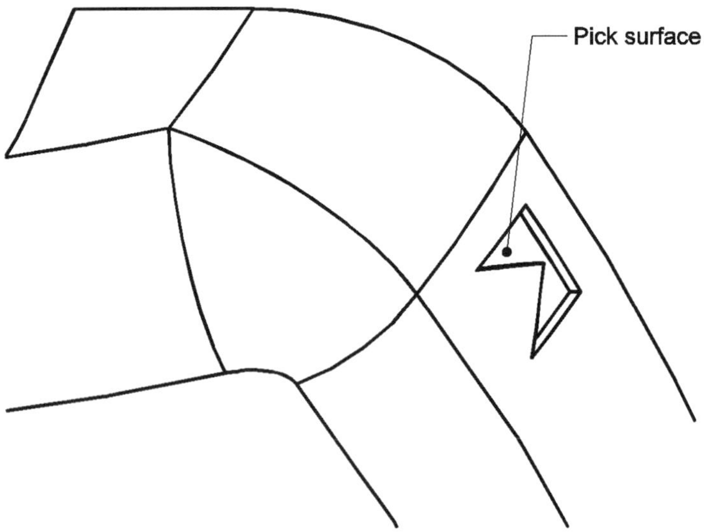

Figure 13-15
A surface to select to pick the logo feature.

After picking the logo feature, pick **Done** from the SELECT FEAT menu.

Prompts During User-defined Feature Placement

In this simple user-defined feature, you do not need to save instances of different logos. Thus, a family table is not required. Pick **Stand Alone | Done** from the DEFINE GROUP menu. In this example, the user-defined feature should be created so that the size of the logo will not vary, but the location can. Next, you need to pick the dimensions that can change when the feature is placed. Orient your view of the car body so that you are viewing the front. Pick **Redraw Dims**, so that the dimensions display again. Pick the dimensions A and B shown in the following figure.

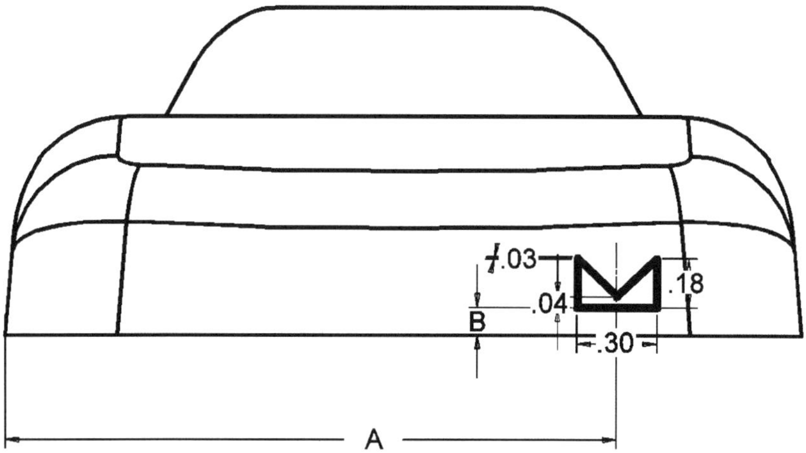

Figure 13-16
The dimensions of the logo feature that can change when the user-defined feature is placed.

After picking dimension A, you must specify the wording of the prompt that Pro/ENGINEER provides when it requests the value for this dimension. Pro/ENGINEER will begin the prompt with the word `Enter` and then append the text you provide.

`Enter prompt for the dimension value:` **`the distance from the side reference plane`** ↵

After picking the dimension `B`, enter the following:

`Enter prompt for the dimension value:` **`the distance from the bottom reference plane`** ↵

> **NOTE:** *Before selecting* ***Done****, you should orient the view to the default view because you will not be able to change the view during the subsequent prompts.*

Pick **Done** from the EXIT DIM PROMPT menu to signify that you do not want to be prompted for additional dimension values when you are placing the user-defined feature. The surfaces and edges referenced by the logo feature at the time of creation will now be highlighted. You need to provide the wording of the prompt to be used by Pro/ENGINEER when it is requesting that you pick a similar entity during placement. Pro/ENGINEER will begin the prompt with the word `Select`.

When DTM1 is highlighted, enter the following:

`Enter prompt for the surface in reference color:` **`the plane normal to the offset direction`** ↵

When DTM3 is highlighted, enter the following:

`Enter prompt for the surface in reference color:` **`the bottom reference plane`** ↵

Enter the following text when the surface shown in the next figure is highlighted:

`Enter prompt for the surface in reference color:` **`the surface to place the logo on`** ↵

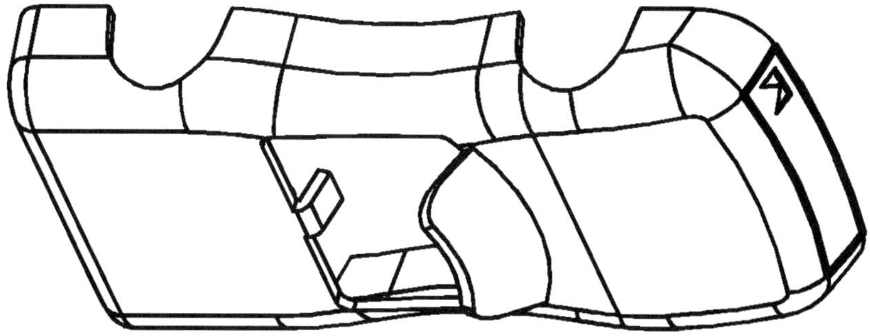

Figure 13-17
The surface the logo was placed on.

When DTM2 is highlighted, enter the following:

```
Enter prompt for the surface in reference color: the
side reference plane ↵
```

Now Pro/ENGINEER wants to know if it should save the part as the reference part. Typically, you will want the reference part to be simple. For this user-defined feature, you could simply use a box with the logo placed on one side as a reference part. Do not save the car body as the reference part.

```
Do you want to save a reference part? (Y/N): n ↵
```

The user-defined feature is automatically saved as *logo.gph* at this point.

Creating the Reference Part

Associate a simpler reference part with the logo user-defined feature. Create a rectangular, dimensioned boxlike part named *box* that contains default datum planes and a coordinate system. The part should resemble the following figure. Use the coordinate system as your guide to position the box in positive coordinate space: .5 in the positive X direction, 2 in the positive Y direction, and 1 in the positive Z direction.

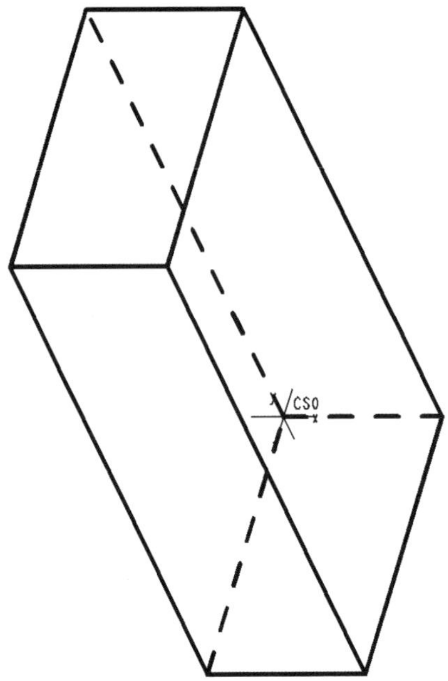

Figure 13-18
The box part.

Next, pick **Feature** ➡ **Group** ➡ **Modify** ➡ **Retrieve**. Select the logo user-defined feature.

```
Enter group name? [LOGO]: ↵
```

After picking **Ref Part** from the GRP MODIFY menu, select the part to be used as the reference part (the box you just created).

```
Enter reference part [-QUIT-]: box ↵
```

Now you will provide the information necessary to place the logo user-defined feature on the box. Pick **Same Dims** | **Done** from the SCALE menu.

```
Enter the distance from the side reference plane
[2.30]: .75 ↵
```

```
Enter the distance from the bottom reference plane
```

```
[0.10]: ↵
```

Pick **Normal** from the DISP OPTION menu. When you are prompted to select the plane normal to the offset direction, pick DTM1. Select DTM3 when you are prompted for the bottom reference plane. When prompted to select the side reference plane, select DTM2. Pick on the surface shown in the following figure when prompted to select the surface on which to place the logo.

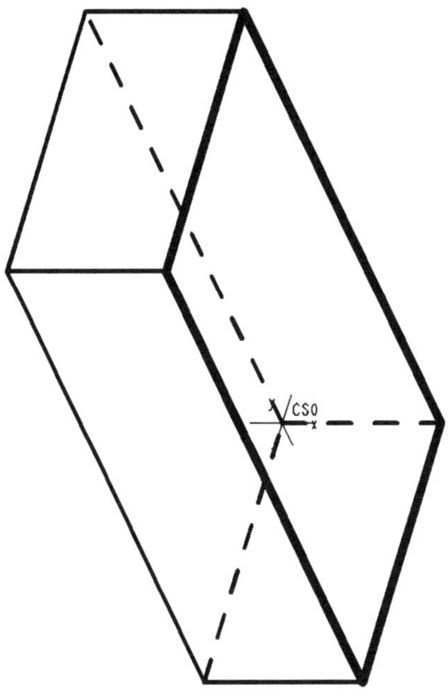

FIgure 13-19
The surface of the box on which to place the logo.

Pick **Done** from the MODIFY PREV menu. Pick **Okay** from the DIRECTION menu when the arrow points in the positive Z direction. Select **Done** from the GRP MODIFY menu. The reference part is then saved as *logo_gp.prt* and is now associated with the logo user-defined feature. Now you have a simple reference part for the logo user-defined feature that is not cluttered with unnecessary features.

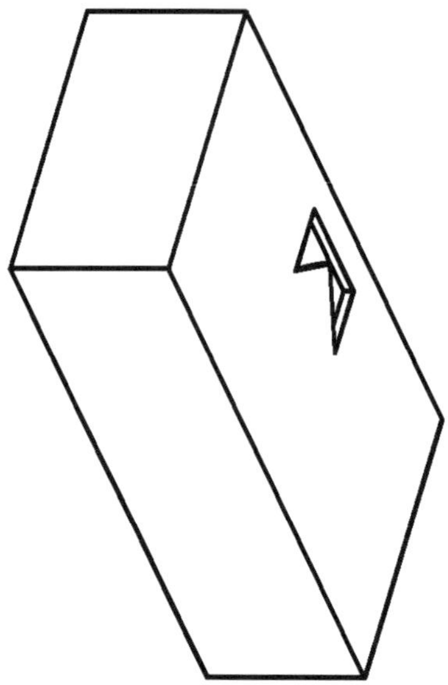

Figure 13-20
Reference part for the user-defined feature.

You can change user-defined feature prompts by selecting **Modify** from the GROUP menu and retrieving the desired group (user-defined feature) file. Pick **Prompt** from the GRP MODIFY menu and select whether you want to modify a prompt for a reference. If you want to modify an existing prompt for a dimension, select **Dim Type**, and then pick the dimension two times. After selecting the dimension the second time, you will be asked to enter a prompt for the dimension. If you no longer wish to be prompted for a dimension value or you want to be prompted for a dimension you currently are *not* prompted for, select **Dim Type** and pick the dimension once. If you wish to change the value for a dimension you are *not* being prompted for, select **Value** and then pick the dimension. To change the dimension units for the user-defined feature, select **Unit** and then pick the new units from the LENGTH menu.

> ✓ ***TIP:*** *You should place all user-defined feature files in a subdirectory of your standards directory (*/usr/standards*). Place the following option in your* config.pro *file to specify the location of*

the user-defined feature files: `pro_group_dir /usr/standards/udf`.

Placing a User-defined Feature

Now place the logo user-defined feature on the rear of the body part. To begin, pick the following commands:

Feature ➡ Create ➡ User-defined ➡ Search/Retr ➡ logo
Independent ➡ Same Dims | Done

Picking **Independent** for the dimensions causes them to be placed as if you were creating each feature individually. If you had chosen **UDF Driven**, the feature dimensions in a user-defined feature's family table would have been linked to the values in the family table. This means that you cannot modify the dimensions except by editing the family table of the user-defined feature. Editing the family table would cause all parts that are UDF driven by the family table to be changed as well. However, in this case, there was no family table in the user-defined feature. Therefore, **UDF Driven** would not be a valid selection.

Provide the offset values for the logo. Change the distance from the side reference plane to .5 and leave the distance from the bottom reference plane at .1.

```
Enter the distance from the side reference plane
[2.30]: .5 ↵
```

```
Enter the distance from the bottom reference plane
[0.10]: ↵
```

Next, you need to specify how to display the other dimensions used in the logo. You have three choices:

- ❒ **Normal** — Display and allow modification of the dimensions in the same way as any other Pro/ENGINEER dimension.

- **Read Only** — Display each dimension, but do not allow modification.
- **Blank** — Do not display the dimensions.

> ➾ **NOTE:** *If you select* **Read Only** *or* **Blank**, *you cannot modify the dimensions in the future. In addition, if you select* **Blank**, *you can never display the dimensions unless you delete the feature and place it again using another display option. If you previously selected* **UDF Driven**, **Read Only** *would be a valid selection, in that changing the user-defined feature family table is the method for modifying these dimensions.*

For this example, pick **Normal**. Now you need to specify references in the body that correspond to the surfaces referenced by the user-defined feature. Do not forget that the text of the prompt and the view of the reference part provide you with information to make the appropriate selection. When prompted to select the plane normal to the offset direction, pick DTM1. To select the bottom reference plane, pick DTM3. Use the following figure to make the correct surface selection when prompted to select the surface on which to place the logo.

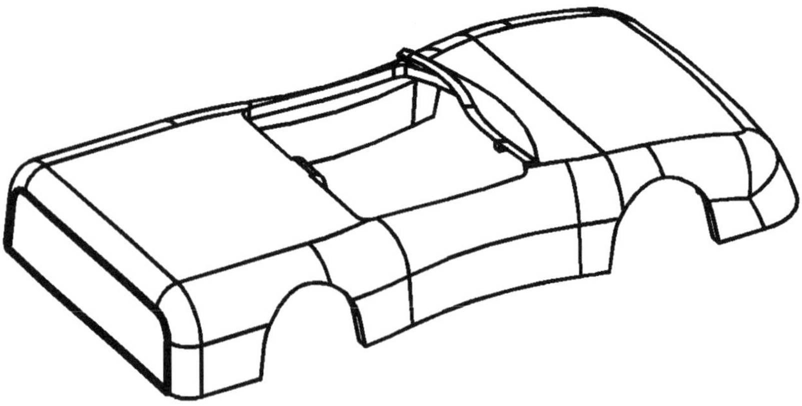

Figure 13-21
The surface on which to place the logo.

Finally, you need to select DTM2 when prompted to select the side reference plane. Pick **Done** from the MODIFY PREV menu when all references have been specified. From the DIRECTION menu, pick **Okay** when the arrow points in the positive Z direction. The user-defined feature should now be displayed on the rear of the body.

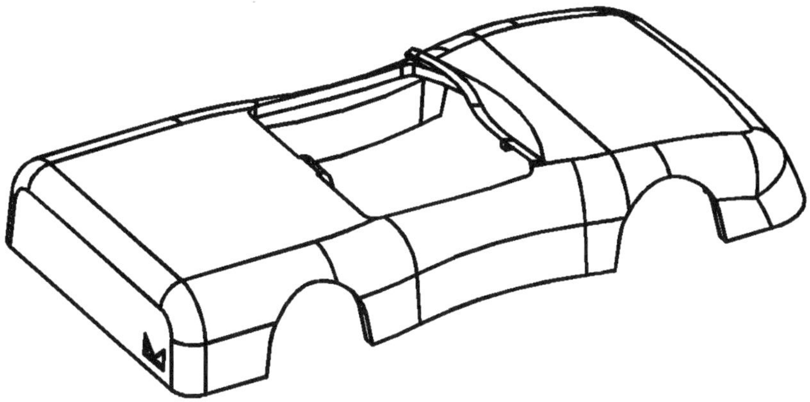

Figure 13-22
The logo on the rear of the car body.

Using User-defined Features in Your Designs

When you wish to standardize certain elements of your designs or create custom features, user-defined features are a solution. For molded parts, you can create a user-defined feature of a boss with a hole, gussets, and draft. Such features are an alternative to manually creating three or four features. User-defined features take a little time to set up, but they can save you a tremendous amount of time in the long run.

Creating Your Own Program for Making Parts and Assemblies

Pro/ENGINEER provides a utility called Pro/PROGRAM for creating a program that will make a parametric part or assembly. This utility can prompt you for part information at the time of execution. The completed program is saved within the model file. The program is not complete until you have successfully incorporated it into the model (successfully executed the program) with no errors. Whenever you want to execute the program, you simply retrieve the model into Pro/ENGINEER and then execute the program. Creating a custom part program in Pro/ENGINEER using Pro/PROGRAM is very easy.

A program is useful when you must make a custom part with varying size or optional features that are determined at the time of creation. When you receive the order for the part, execute the program and specify the information required to customize the part. The model will then regenerate according to your specifications.

Creating the Program to Duplicate an Existing Part

Make a program for the chassis part. In this program, you will specify whether or not to include the text protrusion feature on the bottom of the chassis. If you choose to include the feature, you will be prompted to specify the text protrusion depth. Retrieve the chassis in Part mode. Before having Pro/ENGINEER create the program, determine the feature number of the text protrusion on the bottom of the chassis. Next, you identify the dimension variable name for the protrusion depth of the text. Pick **Modify ➥ Query Sel** and select the feature.

> ✓ **TIP:** *When you use **Query Sel** to select a feature, its feature number is displayed in the message window. By default, if you use **Pick** to select the feature, nothing about the feature is*

displayed in the message window. However, if you add the `provide_pick_message_always yes` *option to the* config.pro *file, feature information is displayed in the message window regardless of how you select the feature.*

Make note of the feature number displayed in the Message window. In this example, the feature number is 32. Accept feature 32 so that its dimensions are displayed. Now, pick **Relations** from the PART menu so that the dimension variable names are displayed. Make note of the name for the text protrusion thickness. In this example, the name is `d106`.

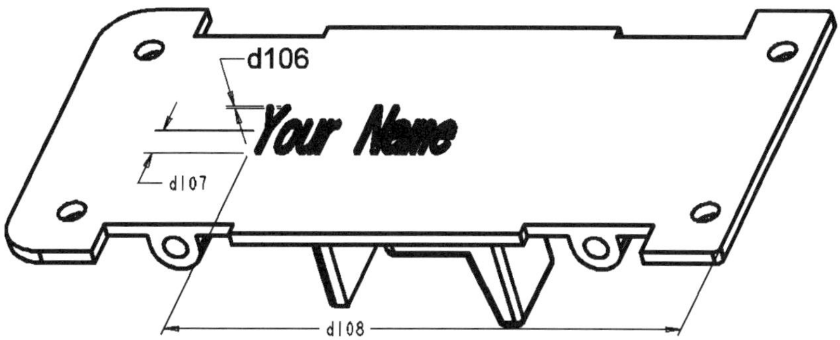

Figure 13-23
The dimensions for the text protrusion feature.

You are now ready to create the chassis program. Pick **Program ➥ Edit Design** from the PART menu. The program is displayed in an editor window ready for you to add your input information.

```
VERSION 17.0
REVNUM 2582
LISTING FOR PART CHASSIS

INPUT
END INPUT

RELATIONS
D5=D4/2
D87=D85-D88
END RELATIONS
```

```
ADD FEATURE (initial number 1)
INTERNAL FEATURE ID  33.
```

At some point during the editing session, you may want to quit before you are completely finished editing the program. In that case, Pro/ENGINEER will save your editing session in a file. In this example, the file would be named *chassis.pls*. For assemblies, the file extension would be `.als`. The file will be created whenever you exit from the program editor with errors in your program and choose to abort further editing of the program. The file will also be created if you choose not to incorporate your program into the model after you exit from the program editing session. The next time you edit your program, you will have to specify whether you want to edit this file or start over and use the program saved in the model (the last program to successfully execute).

> ✘ **WARNING:** *If you choose to edit the program from the model, your file from the previous incomplete editing session is overwritten with the last program that successfully executed.*

Providing the Prompts for the Program

The prompts and variables for the program are inserted between the line containing `INPUT` and the line containing `END INPUT`. Insert the following text between those two lines:

```
SHOW_NAME YES_NO
"DO YOU WANT YOUR NAME DISPLAYED ON THE CHASSIS"
IF SHOW_NAME == YES
    NAME_HEIGHT NUMBER
    "ENTER THE PROTRUSION DEPTH FOR THE TEXT"
ENDIF
```

These lines specify the variable names, variable types, and the prompts for the variables. The first line specifies that the variable `SHOW_NAME` will be a type that can only equal `YES` or `NO`. The allowable types for variables are `NUMBER`, `YES_NO`, or `STRING`.

The second line of text is the prompt that will be displayed for the variable. Conditional statements can also be used in the input section of the program. In this example, you will be prompted to provide a value for the text protrusion depth only if you specify that the part should contain the text protrusion.

Creating Relations in the Program

If you want a value from the input section to change a dimension value, you should place an equation in the relations section of the program. Add the following program lines after the line containing `RELATIONS:`

```
IF SHOW_NAME == YES
  IF NAME_HEIGHT > 0
    D106=NAME_HEIGHT
  ELSE
    SHOW_NAME = NO
  ENDIF
ENDIF
```

For this example, the dimension for the text protrusion depth, `d106`, will equal the `NAME_HEIGHT` variable if the user specifies that the text protrusion should be included in the part. However, a negative protrusion depth will not be allowed. If you provide a negative protrusion depth, the text protrusion feature will not be included in the part. As you can see, conditional statements are also allowed in the relations section of the program. The program lines at the beginning of the program should now look like the segment below.

```
VERSION 17.0
REVNUM 2582
LISTING FOR PART CHASSIS

INPUT
SHOW_NAME YES_NO
"DO YOU WANT TEXT DISPLAYED ON THE CHASSIS"
IF SHOW_NAME == YES
  NAME_HEIGHT NUMBER
```

```
  "ENTER THE PROTRUSION DEPTH FOR THE TEXT"
ENDIF
END INPUT

RELATIONS
IF SHOW_NAME == YES
  IF NAME_HEIGHT > 0
    D106=NAME_HEIGHT
  ELSE
    SHOW_NAME = NO
  ENDIF
ENDIF
D5=D4/2
D87=D85-D88
END RELATIONS
```

Including a Feature Based on an Input Response

To omit a feature from the part, place a conditional statement in the program that will skip the feature. Scroll to the end of the program where the text protrusion feature (feature 32) is listed. Place the conditional statement `IF SHOW_NAME == YES` before the line containing `ADD FEATURE (initial number 32).` After the line containing `END ADD` for feature 32, add `ENDIF`. Whenever the variable `SHOW_NAME` is set to `NO`, the text protrusion feature will be skipped. The program lines for feature 32 would then read as follows:

```
IF SHOW_NAME == YES
 ADD FEATURE (initial number 32)
 INTERNAL FEATURE ID  977
 PARENTS = 33(#1) 62(#6) 35(#2) 41(#5)
 TYPE = PROTRUSION
 FORM = EXTRUDED
 SECTION NAME = S2D0001
 DEPTH = FROM SKETCH TO BLIND
 FEATURE'S DIMENSIONS:
 d106 = .05
 d107 = .25
```

```
 d108 = 4.00
 END ADD
ENDIF
```

You have now completed program editing. Save the file and exit from the editor. Specify that the changes will be incorporated in the model when prompted as follows:

```
Do you want to incorporate your changes into the mod-
el? [Y]: ↵
```

> **NOTE:** *If you made a syntax error when editing the program, Pro/ENGINEER will notify you in the message window. Select* Edit *from the PROG ERROR menu to enter the editor again to fix the error. An error message will appear after the incorrect program line.*

Executing the Program

To execute the program and answer the prompts, pick **Enter** from the GET INPUT menu. Select each input variable you want to be prompted for by selecting it from the INPUT SEL menu by name. Once selected, a check mark will appear next to the variable name. Choose **Select All | Done Sel** if you want to be prompted for all variables. In this example, pick **Select All | Done**.

Answer the prompts so that you create the text protrusion feature with a protrusion depth of .02. Execute the program again. This time do not include the text protrusion feature. In order to execute the program again, regenerate the model. Because the model contains a program, a GET INPUT menu is displayed. Select **Enter** from the menu. Proceed as before, except specify that you do not want your name displayed. The chassis will now be created without the text protrusion feature. Notice that the text protrusion feature is not created or suppressed. In fact, it does not exist in the part.

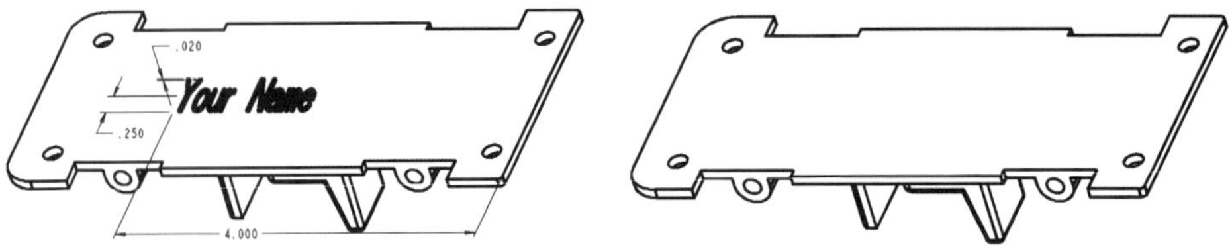

Figure 13-24
The chassis with the text feature and without the text feature.

Custom-sized parts that conform to a standard shape are easily created with Pro/PROGRAM. Custom assemblies that place different parts in the assembly according to user input can also be created. Programs are best used for automating regular changes to finished designs.

Conclusion

The following four design management topics were covered in this chapter:

- Relations
- Family tables
- User-defined features
- Pro/PROGRAM

When used properly, relations and user-defined features both will save you time and let you put more intelligence into your design. Both are excellent methods for managing features. Most companies can make use of family tables in one way or another. Even if your company does not have part families, you probably use standard hardware in your designs that falls into this category. If you need to create individual parts with user-customizable size, or individual assemblies with user-customizable components, you may wish to investigate Pro/PROGRAM.

Chapter 14
Plotting and File Translations

Importing and Exporting Designs

Introduction

This chapter introduces the basic method for plotting parts, assemblies, or drawings. Plotter configuration files are explained, as well as how to assign different pens in the pen table file for plotting visible lines, hidden lines, and centerlines. Next, the method for transferring files to and from both IGES and DXF file formats is shown. The method and settings for creating stereolithography (.STL) files is demonstrated. Commands and concepts covered in this chapter follow:

- ❒ How to create plotter configuration files

- ❒ How to make wireframe plots
- ❒ How to make shaded plots
- ❒ How to translate files to and from IGES or DXF
- ❒ How to export stereolithography files (.STL)

Plotting Configuration

Plots can be created in several formats. Vector plots can be created in HPGL or PostScript format without the use of special modules. Raster plots of shaded images can also be created in PostScript format without special modules.

A few options pertain to all plots you create. It is usually a good idea to set the following options in your *config.pro* file:

```
plot_file_dir   /usr/plot files
```

The configuration option above places all of your plots in a directory you specify. This practice will keep the plot files from cluttering up your working directory. In addition, when you decide to delete all of your plot files, you can easily remove all files in the plot file directory easily.

```
plot_names   yes
```

This option will place unique file name extensions on all of the plot files. For an HPGL plot, the extension is *.hp*, and for a PostScript plot, *.ps*.

```
pro_plot_config_dir   /usr/standards/plotter
```

Use the above configuration option to specify the directory where your plotter configuration files (*.pcf*) reside. Pro/ENGINEER will find your *.pcf* files if you place them in the specified directory.

✓ **TIP:** *The first two options can be placed inside the* .pcf *files. In that the shaded PostScript procedure does not use the* .pcf *files, those options would not be used with shaded output. To make these options apply to shaded output as well, place them only in your* config.pro *file.*

HPGL Plotter Configuration File

A plotter configuration file is an ASCII file containing configuration options for a specific printer or plotter. By using these configuration files, you can minimize the effort required to create and send a plot. Creating multiple plotter configuration files that help you plot to several different plotters is easy.

When you make HPGL plotter configuration files, new plotter choices will appear on the PLOTTERS menu. An option within the *.pcf* file specifies the menu name. However, the file name must be lower case, and the file extension must be *.pcf*.

✓ **TIP:** *Place all of your company custom files in subdirectories under a single main custom directory (*/usr/standards*). All users can then access them, and it is easier for a system administrator to update the files later. These examples use the* /usr/standards/plotter *directory as the directory for plotter-related files.*

This example plotter configuration file is for an HP7585B plotter located in the engineering department. The plotter has been given the name Hpplotter by the system administrator. Use the file name *eng_hpgl.pcf* for this plotter configuration file.

➾ **NOTE:** *Plotter configuration file names must be lower case. Pro/ENGINEER will display the names of all* .pcf *files on the menu, but only the* .pcf *files with lower-case letters are valid selections.*

Use your favorite text editor to create the *eng_hpgl.pcf* file. Because of the previous *config.pro* option, `pro_plot_config_dir /usr/standards/plotter`, this file should reside in the */usr/standards/plotter* directory. The first line should contain the name to be displayed on the PLOTTERS menu, from which you will select this configuration file. Add the following line to *eng_hpgl.pcf*:

```
button_name   ENGR HP
```

The second line should contain the text that will display in the message window when the cursor crosses over the menu name on the PLOTTERS menu. Enter the following:

```
button_help   Engineering Dept. HP plotter in room 101
```

The next line should contain the command Pro/ENGINEER issues to send the plot to the correct plotter. Request this command from your system administrator. If you use the `lp` print spooler, this line of the plotter configuration file would look like the following:

```
plotter_command   lp -dHPplotter
```

If you use the `lpr` print spooler, the line would be as follows:

```
plotter_command   lpr -PHPplotter
```

The next line in the file should provide the type of plotter. The list of supported plotter types is provided in the PTC Supported Hardware list.

> ✓ **TIP:** *If your plotter is not listed, consult your plotter manual to identify one of the listed plotters that can be emulated by your plotter.*

In that the example plotter configuration file is being created for an HP7585B plotter, the next line follows:

```
plotter   HP7585B
```

The `plotter` option is required in all plotter configuration files.

Pro/ENGINEER allows several plots to be appended to one plot file. When you send the file to the plotter, all plots are plotted. However, your usual configuration should place only one plot in each plot file. Add the following line so that only one plot is placed in each plot file:

```
plot_access   create
```

Pen mapping tables are a useful tool for controlling the line styles and line weights used in your plots. For Pro/ENGINEER to recognize a pen mapping table, the table must be specified in the plotter configuration file. Assuming a file containing the pen mapping table is named *pentable.pnt*, and is placed in your plotter *standards* directory, add the following line to specify the path and file name for the pen mapping table:

```
pen_table_file   /usr/standards/plotter/pentable.pnt
```

> ✓ **TIP:** *If you use plotters with other software that depends on the pens to be set in a specific manner, you can use a pen mapping table to reconfigure Pro/ENGINEER's default pen settings.*

You have covered the most common plotter configuration file entries used for an HPGL plotter. You have created a file called *eng_hpgl.pcf* containing the following text:

```
button_name   ENGR HP
button_help   Engineering Dept. HP plotter in room 101
plotter_command   lp -dHPplotter
plotter   HP7585B
plot_access   create
pen_table_file   /usr/standards/plotter/pentable.pnt
```

PostScript Plotter Configuration File

PostScript plotter configuration files are very similar to their HPGL plotter configuration counterparts. Because of variations between different PostScript printers, you need to have control over the paper orientation when you make a PostScript plot. Create two plotter configuration files for a color PostScript printer in the engineering department. This printer is capable of creating both A- and B-size drawings. The A-size plots are printed in landscape orientation, and the B-size plots in portrait orientation.

In that the two formats require different configuration options, you need to create two different *.pcf* files. Use the file names *eng_a_cps.pcf* and *eng_b_cps.pcf*. Start your text editor and create each of these files. The following lines should be the first two lines of *eng_a_cps.pcf*:

```
button_name    ENGR PS A
button_help    Engineering Dept. A size printer in
room 311
```

Use the following two lines in *eng_b_cps.pcf.:*

```
button_name    ENGR PS B
button_help    Engineering Dept. B-size printer in
room 311
```

The system administrator has set up the color printer on your workstation with the name *CPSprinter*. The following plotter command line will reference the printer name for both configuration files.

```
plotter_command   lp -dCPSprinter
```

The plotter type for both configuration files should be set so that Pro/ENGINEER knows that the plotter is a color PostScript printer. The appropriate line follows:

```
plotter   colorpostsc
```

Add the following line to both configuration files to make new plot files each time you plot:

```
plot_access   create
```

Specify the pen mapping table as you did for the HPGL plotter configuration file. In this example, the same pen mapping table can be used for both the HPGL and the PostScript plotter configuration files.

```
pen_table_file   /usr/standards/plotter/pentable.pnt
```

Pro/ENGINEER's default PostScript printing mode is landscape. The B-size plots use an option that will make Pro/ENGINEER print in portrait mode. Place the following line in the B-size plotter configuration file, *eng_b_cps.pcf*:

```
landscape_postscript_print   no
```

> ➻ **NOTE:** *In release 14, the syntax for the B-size plot option differs from the previous command line. If you are using release 14, insert the following line in the* eng_b_cps.pcf *file:*
>
> ```
> landscape_plotting_print   20no.
> ```

In that you already know the drawing size each file will be used for, there is no need to select it at the time you create the plot file. You can add one of the following lines to the plotter configuration file to set the paper size:

```
paper_size   a
```

or

```
paper_size   b
```

At this point, you need to select an appropriate name for the menu and the message window help line. You should have two color PostScript configuration files. The file for the A-size color PostScript output (*eng_a_cps.pcf*) will look like the following:

```
button_name    ENGR PS A
button_help    Eng Dept. A size printer in room 311
plotter_command    lp -dCPSprinter
plotter    colorpostsc
plot_access    create
pen_table_file    /usr/standards/plotter/pentable.pnt
paper_size    a
```

For the B-size color PostScript output, the (*eng_b_cps.pcf*) file follows:

```
button_name    ENGR PS B
button_help    Eng Dept. B size printer in room 311
plotter_command    lp -dCPSprinter
plotter    colorpostsc
plot_access    create
pen_table_file    /usr/standards/plotter/pentable.pnt
landscape_postscript_print    no
paper_size    b
```

Pen Mapping Table

The HPGL plotter configuration file and both PostScript plotter configuration files refers to a pen mapping table file. The HPGL plotter configuration file refers to the table file so that you can change the default pens for the plot file. The PostScript plotter configuration files refer to the table file so that different line thicknesses can be specified for each pen number.

> **NOTE:** *The HPGL plotter configuration file ignores the line thickness value in the pen mapping table. The line thickness is determined by the pen number at the plotter.*

Pro/ENGINEER entities are assigned to one of four pens. However, if you have a pen plotter that supports eight pens, insert *use_8_plotter_pens yes* in your *config.pro* file so that certain entities can be plotted to the additional four pens. This option cannot be placed in a plotter configuration file. Pro/ENGINEER uses generic width values for each of these pen settings. Line width values range between 1 and 16, with 16 being the thickest. If the plotter listed in your configuration file does not support eight pens, this option is ignored. For an eight-pen plotter, the pens and corresponding entities follow.

Pen 1: Width value = 4 — Visible geometry and drawing sheet borders

Pen 2: Width value = 1 — Dimension lines, text, yellow sides of datum planes, and axes

Pen 3: Width value = 2 — Hidden lines

Pen 4: Width value = 3 — Red sides of datum planes and surface mesh lines

Pen 5: Width value = 2 — Sheetmetal entities

Pen 6: Width value = 3 — Sketcher section entities

Pen 7: Width value = 1 — Toggled sections

Pen 8: Width value = 4 — Spline surface grid

If you do not like the previous settings, they can be reconfigured by using pen mapping tables. Changes from the default pen settings may be required if the plotted line thickness for visible geometry is too narrow when compared with the hidden lines, dimension lines, or text. Line thickness values in a pen mapping table are entered in inches or centimeters instead of the generic width values. The syntax for an entry in a pen mapping table follows:

```
pen # pattern pvalues: thickness tvalue: color cval-
ues entity_type
```

The five values you specify follow:

— The pen number.

pvalues — These values control the line style by specifying the line length and gap length. The values are followed by units (in.or cm).

tvalue — The thickness value followed by units (in.or m).

cvalues — The RGB values ranging from 0 to 1.

entity_type — The type of entity that will be plotted by his pen.

For this example, create a pen mapping table file that will use Pen to draw visible geometry and drawing sheet borders. You should change pen 1 to create hidden lines. For the PostScript files, you ill specify a line thickness of .02" for visible geometry and drawing sheet borders. The line thickness for hidden lines will be 01".

As specified in the plotter configuration files, this file will be located in the */usr/standards/plotter* directory and the file name will be *pent-able.pnt*. Using your text editor, create the file and add the following lines:

```
pen 1 thickness 0.01 in: half_tone_color

pen 3 thickness 0.02 in: drawing_color
```

As you can see, the text used to specify the entity type is not very intuitive. Pro/ENGINEER uses the following nomenclature when specifying entity types:

- `drawing_color` — Visible geometry, drawing borders, and other entities that normally display in white
- `half_tone` — Hidden lines
- `letter_color` — Text and other entities that normally display in yellow
- `section_color` — Sketched entities while in a Sketcher window

Now all of the plotter configuration files have been created, as well as the pen mapping table file. You are ready to start Pro/ENGINEER and create a plot.

Wireframe Plot Creation

Wireframe (vector) plots can be created from a sketched section, part, assembly, or drawing. The method of creating the vector plot is the same for each.

Plotting from Part Mode

For this example, you will create a plot of the wheel you created in Chapter 6. Enter Part mode and retrieve the wheel. Using the ENVIRONMENT menu, turn off the datum planes, points, point tags, axes, coordinate systems, and hidden lines.

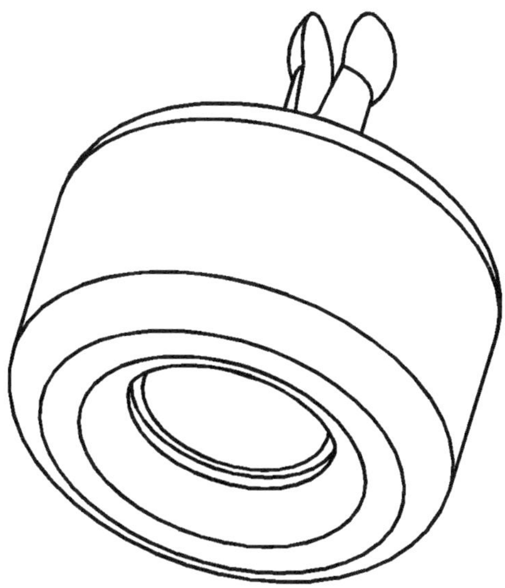

Figure 14-1
The wheel in the orientation to plot.

Pick **Interface ➥ Export ➥ Plotter** from the PART menu. Select **ENGR HP** from the PLOTTER options in the dialog window (this is the menu name specified for the plotter configuration file you created for the HPGL plotter). Now pick **Plot Setup** so that you can see the available options when creating a plot file. Use the selections at the top of this dialog window and the previous (PLOT) dialog window to gain access to the following options:

Xon/Xoff — To initiate software handshaking

Hardware — To initiate hardware handshaking for plotting to a local device

Based on Zoom — To plot an image based on the current zoom factor

Full Plot — To plot all of the entities

Clipped and **Plot Area** — To plot an area of the graphics window

Scale Model — Scales the plot by a factor relative to the model scale

Scale Plot — Allows you to input a final plot scaling factor

Quality 0 — Plots with no line check

Quality 1 — Plots with no overlap check

Quality 2 — Plots with simple overlap checking

Quality 3 — Plots with complex overlap checking

With Label — Deactivated, it does not add a label

With Label — Adds a label to the left side that contains your name, the object name, and the date

Set Slewing — Allows you to input pen speed parameters

Select the paper size from the list of available sheet sizes displayed on the SIZE window. Select **A**. Select **OK** from the dialog menu to accept the default options.

```
Enter Plot File name [WHEEL]: ↵
```

The plot file will now be created with the file name of *wheel.hp* and be placed in the */usr/plot files* directory. If you want to send the plot file to the specified plotter, pick **Yes** from the PLOT COMMAND window. You can select **Yes** again if you want to make more than one plot. Pick **Quit** to indicate that you are finished plotting.

Plotting from Drawing Mode

Enter Drawing mode and retrieve the drawing of the wheel. Pick **Export ➥ Plotter ➥ ENGR HP**. When you select **Options** from the dialog window, you will notice that a few more commands from the OPTIONS menu are available.

- ❒ **With Format** — Plots the drawing border
- ❒ **With Format** — Deactivated, does not plot the drawing border
- ❒ **Curr Layers** — Plots everything currently displayed
- ❒ **By Layer Name** — Plots specific layers only
- ❒ **Segmented** — Plots a large drawing on multiple sheets of a smaller size paper

Select the following commands from the **Plot Setup** windows:

With Format | Full Plot | Current Layers | Quality 3 | No Label | No Slewing | OK

> ✓ ***TIP:*** *When you create a plot of a drawing containing a drawing format, and the size of the format drawing sheet equals the selected*

plot size, Pro/ENGINEER will warn you because part of the border may not plot on the paper. If part of your drawing border does not plot, you have two choices. First, you can use **Scale** *and enter a value less that 1.0 at the prompt. This reduces the size of the entire drawing so that the entire border will plot. Second, if you really want the drawing to plot at a scale of 1:1, you need to edit the drawing format file and move your drawing border lines farther from the edges of the paper. You will still receive the warning about the format equaling the size of the selected paper, but the drawing border should not be clipped.*

Provide a value for the plot scale.

```
Enter a factor between 0.1 and 10.0 for plotting
[1.0]: ↵
```

Shaded Plot Creation

Shaded plots can also be created from your part or assembly. However, the plotter configuration files have no effect on these plot files. Next, everything displayed in the graphics window is plotted. Retrieve the assembly of the model car, and shade it by picking **View ➥ Cosmetic ➥ Shade ➥ Display** from the MAIN menu.

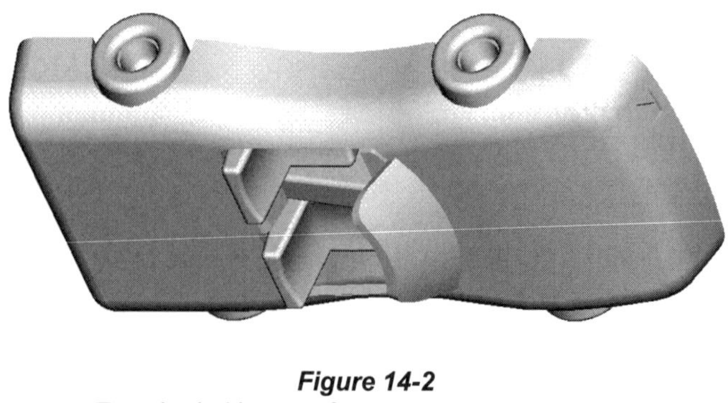

Figure 14-2
The shaded image of the car that will be plotted.

Immediately after shading the car, pick **PostScript** from the SHADE menu. A list of supported PostScript printers will be displayed on the PS PRINTERS menu. The supported printers are Phaser II PX, Phaser III PX, Seiko, and CC 6613 PS. Select the appropriate printer from this list.

> ✓ **TIP:** *If your PostScript printer is not on the list, pick one of the listed printers anyway. The PostScript output will probably work with your PostScript printer. The only differences between some of the listed printers are available output sizes.*

Pick **Resolution** from the POSTSCRIPT menu and set the resolution to one of those listed on the RESOLUTION menu. The allowable resolutions range from 100 dpi to 400 dpi.

> ✓ **TIP:** *The higher the resolution, the larger the plot file, and the greater the plotting time.*

Select **Image Size** from the POSTSCRIPT menu and pick an A-size sheet from the list. The list of sizes depends on the PostScript printer you previously selected. Now select **Output to PS** from the POSTSCRIPT menu and answer the following prompts.

```
Maximum size of plot file will be 978k. Proceed? [Y]: ↵
Enter Plot File name [CAR]: ↵
```

A PostScript file with the file extension of *.ps* will be placed in the */usr/plot files* directory. From a UNIX window, send this plot file to your PostScript printer by entering `lp -dCPSprinter /usr/plot files/car.ps` or `lpr -PCPSprinter /usr/plot files/car.ps`, depending on the print spooler your system uses. You should now have a shaded plot. You can now exit the UNIX window and return to Pro/ENGINEER.

IGES File Transfers

IGES (Initial Graphics Exchange Specification) is commonly used as a vehicle for transferring 3D geometry between applications. Pro/ENGINEER supports IGES import and export of 2D and 3D geometry.

Exporting 3D Wireframe

Pro/ENGINEER can export a wireframe model of your solid parts through IGES. This is sometimes necessary if the application that will read this IGES file only supports wireframe data. For this example, begin a part mode session and retrieve the chassis part. Pick the following commands from the PART menu:

Interface ➥ Export ➥ IGES

Pro/ENGINEER will prompt you for the file name.

`Enter IGES file name [CHASSIS]:` **`chas_wf`** ↵

To tell Pro/ENGINEER that you want to output only wireframe data, check **Wframe Only** from the IGES OPTS menu and pick **Done**. In Pro/ENGINEER, you are not required to have a coordinate system. The IGES file needs a 3D reference point for its data. If you do not have a datum coordinate system, you have the opportunity to make one. The chassis part should already have a datum coordinate system. Pick this coordinate system (CS0) to use as the reference for the IGES file. After you select the coordinate system, the IGES file *chas_wf.igs* will be created.

> **NOTE:** *If the target CAD system supports entity type 126 (B-SPLINE), you may need to use the following configuration option in your* config.pro *file:* `iges_out_spl_crvs_as_126 yes`. *Set this option to* `no` *if you need to output entity type 112 (SPLINE).*

Exporting 3D Surfaces

IGES does not support parametric solid models. The best transfer of data possible is a complete, 3D surface model of your part. To output a surface model to IGES, pick the same commands from the PART menu.

Interface ➥ Export ➥ IGES

Pro/ENGINEER will prompt you for the file name.

`Enter IGES file name [CHASSIS]:` **`chas_srf`** ↵

Now pick **Trimmed Srf** from the IGES OPTS menu and pick **Done**. Select the coordinate system, and the IGES file *chas_srf.igs* will be created.

> ➾ **NOTE:** *You should probably set the configuration option* `iges_out_spl_srfs_as_128 yes`. *Most CAD systems that can read IGES surface models accept entity type 128. In some cases, you may need to use the option* `iges_out_all_srfs_as 128`.

Importing 3D Geometry

You can import 3D IGES files into Part mode. To import the IGES file, pick the following commands from the PART menu:

Interface ➥ Import ➥ IGES

Pro/ENGINEER will prompt you for a file name and a coordinate system. The IGES file will be imported as a single feature called IGES PART. This feature will contain all curves and surfaces within the IGES file.

✓ **TIP:** *You can import an IGES file when you first create a part at the ENTERPART menu. The IGES file will be imported relative to the default coordinate system.*

If you import 3D wireframe geometry, you can construct surfaces through the wireframe and join them together to make a solid. You can also use the wireframe geometry in a section and create solid features. If you import a 3D surface model, you can join these surfaces together with the **Redefine ➡ Attributes ➡ Join Srfs** command to create a merged surface. After creating the merged surface, you can use **Protrusion ➡ Use Srfs** to convert it into a solid. In both cases (wireframe or surfaces), your resultant part will be nonparametric. The only way to make a parametric model with IGES is by importing 2D geometry into the sketcher. This would be very time-consuming for a large part.

NOTE: *A nonparametric solid behaves as a single feature. Other parametric features can be added to it, and those features can then be modified like any other parametric feature.*

DXF File Transfers

DXF (Autodesk's Data eXchange Format) files can be used to exchange 2D geometry with Pro/ENGINEER. The most common uses of DXF files are exporting or importing a drawing.

Exporting a Drawing

Exporting a drawing to DXF format is very simple. Begin a Drawing mode session and retrieve the wheel drawing. Pick the following commands to export the drawing to DXF format:

Interface ➡ Export ➡ DXF

Pro/ENGINEER will prompt you for the file name. You may enter a new file name at this point or let the system default.

```
Enter export file name [WHEEL]:
```

If the default name is used it will automatically create a DXF file using the name of the drawing. If you have elements on layers that need to maintain their layer status in the DXF file, you must use the **Layer ➥ Setup Layer ➥ Specify Id** command to assign an interface layer ID number to those particular layers that need to be output. Items residing on multiple layers in Pro/ENGINEER will be output on the interface layer ID that has the highest number.

Importing a Drawing

Pro/ENGINEER can also import DXF files. When DXF drawings are imported into Drawing mode, they are translated into draft entities. These drawing files can be modified using the standard drafting commands. To import a DXF file, select **Interface ➥ Import ➥ DXF** when you are in Drawing mode. Pro/ENGINEER will prompt you for the file name.

SLA File Creation

Pro/ENGINEER can create an SLA (StereoLithography Apparatus) file from a part or assembly. This SLA file can be used to create a prototype part via stereolithography or other rapid prototyping technique.

Creating an SLA File of a Component

Now you should create an SLA file of the wheel. Retrieve the wheel into Part mode, and then pick **Interface ➥ Export ➥ SLA**. The SLA file will contain triangles to approximate the surfaces of the part. The menu choices of **Chord Height** and **AngleControl** add control over how closely the part surfaces are approximated by the triangles.

Chord Height Value

The chord height value establishes a global control over the size of the triangles. The smaller the chord height value, the more accurate the approximation. However, the number of triangles required will increase. As the number of triangles increases, so does the size of the SLA file. The following figure shows how the size of the chord height tolerance relates the SLA surface to the part surface.

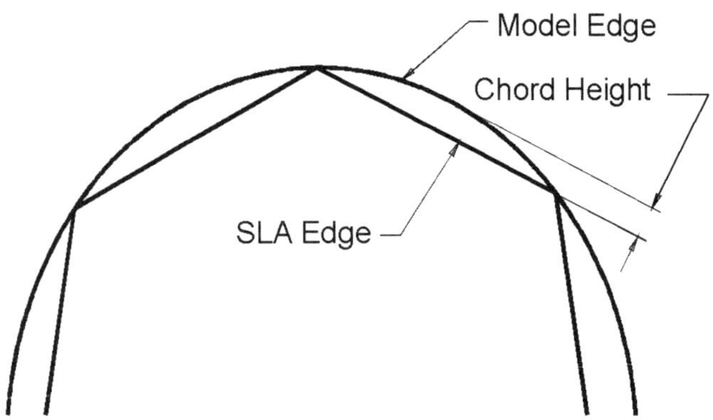

Figure 14-3
The relationship of the chord height to the part surface and SLA surface.

Angle Control Value

The angle control value is used to specify a local chord height value for small radii. A radius that is very small when compared with the overall size of the part may not be accurately represented in the SLA file. A smaller chord height can be specified for this case by specifying an angle control value greater than 0. As you increase the angle control value, the SLA approximation of the small radii will become more accurate. Use a value of 1 to achieve maximum accuracy on small radii.

Outputting the SLA File

The size of a binary SLA file is smaller than an ASCII SLA file. Most rapid prototyping systems will accept binary SLA files. Choose **SLA Binary** and pick **Output**. When the GET COORD S menu appears, pick **Select** and then pick the datum coordinate system in the wheel.

```
Enter triangulation file name [WHEEL]: ↵
```

> ➾ ***NOTE:*** *When an SLA file will not create, you may need to add a coordinate system to the part, instead of letting the default coordinate system take over. To do this, pick* ***CREATE*** *from the GET COORD menu. The OPTIONS menu will appear. Pick the type of coordinate system best suited for the part. Once the parameters have been defined, the coordinate system will then be displayed. Another cause may be related to a specific feature within a part. Pro/ENGINEER will highlight these features and abort the SLA file creation. You may even need to suppress these features for the SLA file to run. Remember, however, that this is a last resort and should only be used if the file will not run, by adjusting* ***Chord Height*** *or* ***Angle Control****.*

In the past, stereolithography applications required the datum coordinate system to be located so that no SLA entities would have a nonpositive coordinate value. This is no longer a requirement for most applications. If the application that will use your SLA file still has this requirement, you should create a new datum coordinate system that places your model in positive coordinate space. For the datum coordinate system in the wheel, you will be warned that the SLA file will contain nonpositive entities.

```
Some points have non-positive coordinates. Continue?
[N]: y ↵
```

The SLA file will now be created in the current directory, with the file name *wheel.stl*.

Creating an SLA File of an Assembly

An SLA file can be created for an assembly in the same manner as for a part. The only difference is that before picking **Output**, you must select the parts of the assembly to include in the file. Pick **Include** from the SLA menu and pick each component you want to output. Pick **Output** from the SLA menu to continue with the process, as described earlier.

Conclusion

You should try to use plotter configuration files for all of your plotting needs. These will save you time by automating many of the menu choices you would normally have to make. They also save you time by automatically sending the plot to the plotter. Importing and exporting geometry is very straightforward. The procedures shown here are typical of what a regular user performs. Pro/ENGINEER has one of the most reliable and robust IGES translators when compared with other 3D CAD products. If you have problems with IGES file transfers, it is usually not a result of Pro/ENGINEER's translation.

Chapter 15

Customizing Your Environment

Mapkeys, Trail Files, Additional Menu Entries, and Configuration Options

Introduction

Several methods can be used to customize the Pro/ENGINEER work session to your preferences. These methods include creating mapkeys (hot keys), using trail files, adding menu entries, and placing configuration file options within the *config.pro* file.

In order to make configuration changes, you should use the Pro/TABLE editor. In reality, any text editor can be used, but if you use Pro/TABLE you have access to on-line assistance for configuration commands and respective options. The Pro/TABLE is a spreadsheet-style editor that is a part of Pro/ENGINEER. Appendix A covers the use of Pro/TABLE. Editing trail files and menu files should be performed with a basic ASCII text editor. In that there are many different editors, commands for starting, editing, saving, and quitting your text editor are not provided. Ask your system administrator if you are uncertain about the text editor available for your use.

Mapkeys

Mapkeys provide a method to create a *macro,* or a means to execute one or several commands in sequence automatically. The macro is assigned to keys of your choice on the keyboard. To create a mapkey, make note of the wording of the commands as you select them from the menus. By adding these menu picks to a `mapkey` entry in your *config.pro* file , you can automate many common functions in Pro/ENGINEER.

Mapkey Macro Creation

If you want to shade your part on the screen, select the following commands:

View ➡ Cosmetic ➡ Shade ➡ Display

If you also want the view menus to disappear after shading the view, select **Done-Return** from the MAIN VIEW menu. For this mapkey example, you will create a mapkey macro that duplicates these menu picks when you type the characters `sv`.

The easiest way to edit your *config.pro* file is to pick **Misc ➡ Edit Config** from the MAIN menu. Pro/ENGINEER will prompt you for the file name.

```
Enter Configuration file name [config.pro]: ↵
```

Pro/ENGINEER will retrieve the *config.pro* file from the current directory. If the file does not exist in your current directory, Pro/ENGINEER will create the file.

> ➾ **NOTE:** *If you are not in the directory where the* config.pro *exists, you can provide a path, along with the file name at the prompt (i.e.,* ../config.pro*).*

Enter the following line in your *config.pro* file to perform the shading function.

```
mapkey sv #view;#cosmetic;#shade;#display;#done-re-
turn;
```

> ➾ **NOTE:** *The last semicolon (:) in the mapkey definition is optional.*

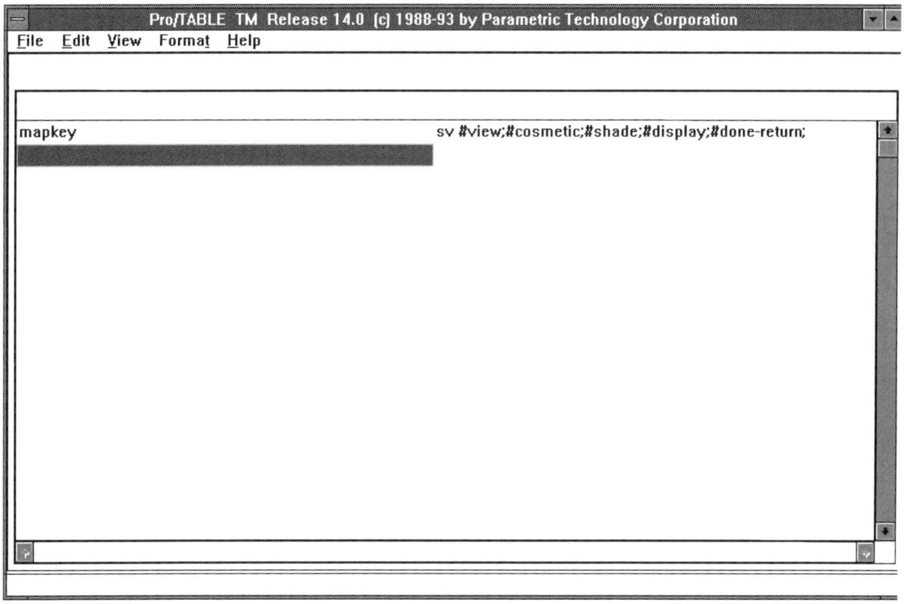

Figure 15-1
A pro/TABLE session with the mapkey entered on the first line.

> ✓ **TIP:** *If the text for the configuration seems too long, or is cut off in the value column, the column width is too narrow for the characters. The top line of the window shows the real value of the cell. Either ignore the dropped characters or reconfigure the column width.*

Exit the editing session.

> ➾ **NOTE:** *Pro/ENGINEER will save your file automatically if you select* ***Exit*** *from within the Pro/TABLE editor.*

If you have a part on the screen now, go ahead and type "sv" to see what happens.

Are you still waiting? Well, you could wait a long time. Changes to the *config.pro* file will take effect only if you start Pro/ENGINEER over again or pick **Misc ➥ Load Config** from the MAIN menu.

> ➾ **NOTE:** *Mapkeys will always reinitialize when you use* ***Load Config****. Some configuration options, such as hardware graphics settings, will take effect only upon start-up of Pro/ENGINEER.*

Now try typing "sv." Your part on the screen should be shaded.

Mapkey with Response to a Prompt

The pound sign character (#) precedes each Pro/ENGINEER menu command. A semicolon separates commands. If a user response to a keyboard prompt is normally required after a Pro/ENGINEER command, and you wish to provide that input in your mapkey, you can create a mapkey like the following one, which assigns the material properties as saved in the *steel.mat* material property file when you enter `st`.

```
mapkey st #set up;#material;#assign;#from
file;steel; #done;
```

Before you use the mapkey for the first time, select **Set Up ➡ Material ➡ Define**. Then you will create a material named `steel`.

```
Enter material name [ ]: steel ↵
```

An editor session will start and you should place some values in the file and then exit the editor. Pick **Write ➡ From Part ➡ STEEL ➡ Accept**. Now provide the file name to save the material properties to.

```
Enter material name [STEEL]: ↵
```

As you see in this mapkey macro, `steel` is the entry you would key in if choosing the sequence of commands from the menus. Any mapkey entry between semicolons or after the last semicolon, and which does not begin with a pound sign (#), is used as an entry to a Pro/ENGINEER prompt. Before using this mapkey, change menus until the PART menu is visible. Now try the mapkey and you will see that the information stored in the mapkey file is now loaded into the current part file.

Mapkey Which Takes the Default in Response to a Prompt

If the prompt has a default choice, the default choice can be selected in the mapkey by repeating the semicolon. An example is the following mapkey, assigned to the characters `ls`, which is used to list the contents of the current working directory.

```
mapkey ls #misc;#list dir;;
```

This mapkey executes the **Misc ➡ List** command and then provides a <Return> character to answer the last prompt.

Mapkey User Key-in as Response to a Prompt

If you desire to respond to keyboard prompts at the time of executing the mapkey, do not place a response to the prompt in the mapkey. The following mapkey illustrates this when you save an object by entering the mapkey `sn` and then providing an in-session object name.

```
mapkey sn #dbms;#save;#done-return;
```

In that no entry is provided in the mapkey macro in response to the Pro/ENGINEER prompt for the object name to save, you will be prompted to enter it when you execute the mapkey. After entering the name, the DBMS menu is removed because of the `#done-return:` entry.

Mapkey Assigned to a Function Key

Mapkeys can also be mapped to function keys. A function key must be preceded by a dollar sign character ($) when defining the mapkey. The following mapkey macro saves the default object name when you press the function key `F2`.

```
mapkey $F2 #dbms;#save;;#done-return;
```

Mapkey Length Reduction

Each mapkey entry in the *config.pro* file should be no more than 80 characters in length. If you require more characters, have the mapkey macro call other mapkeys so that the length of the mapkey macro is shortened. The following mapkeys are used to fit the object to the screen and illustrate how to shorten the macro by ten characters through nesting the mapkeys.

```
mapkey dr #done-return;
```

```
mapkey fv #view;#pan/zoom;#reset;%dr
```

> **NOTE 1:** *As you may have noticed, the example mapkeys are composed of two characters. When executing a mapkey, key in the characters assigned to the mapkey macro. As soon as you enter the last mapkey character, the macro is executed. No <Return> is required. If a mapkey is composed of only one character, that character cannot be the first character of any other mapkey. The one-character mapkey would always execute before you could enter a second character. However, one-character mapkeys are desirable because they are easy to enter. But you must ensure that no other mapkey starts with that character.*

> **NOTE 2:** *The example mapkeys, except for the one assigned to the characters "st," include all of the menu selections starting from the MAIN menu. This is done to ensure that the submenus containing each of the macro's commands are displayed when that macro command is executed. Select* ***Feature*** *from the PART menu and then enter "st." At this point, you change back to the PART menu because the only command found on the FEAT menu was* ***Done****. Mapkeys skip any command beginning with a pound sign (#) if the command is not contained on a currently displayed menu. They then try to execute the next command beginning with a pound sign.*

Mapkey Restrictions

Entries in a mapkey are not case sensitive, and spaces are ignored unless the entry is for keyboard input of text. Mapkey execution will stop and no more commands will execute whenever a prompt requires a mouse pick in a graphics window. Mapkeys are limited to 80 characters per line.

Sample Mapkeys

Mapkeys you can try for yourself follow.

```
mapkey dr #done-return;
mapkey dv #view;#default;%dr
mapkey fv #view;#pan/zoom;#reset;%dr
mapkey pv #view;#pan/zoom;#pan;
mapkey rv #view;#repaint;%dr
mapkey sv #view;#cosmetic;#shade;#display;%dr
mapkey sp #view;#orientation;#spin;
mapkey vn #view;#names;
mapkey zi #view;#pan/zoom;
mapkey zo #view;#pan/zoom;#zoom out;%dr
mapkey fr #view;#names;#Front;%dr
mapkey bk #view;#names;#Back;%dr
mapkey rt #view;;#names;#Right;%dr
mapkey lt #view;#names;#Left;%dr
mapkey tp #view;#names;#Top;%dr
mapkey bt #view;#names;#Bottom;%dr
```

Trail Files

Pro/ENGINEER keeps track of everything you select and enter during each session and saves that information in a file called *trail.txt*. These trail files have unique version numbers, similar to part files, when saved. Thus, the first trail file created in a directory will have a file name of *trail.txt.1*; the second one will be *trail.txt.2*. If you exit Pro/ENGINEER without saving your file, you can replay the newest trail file to recreate everything you did during that session of Pro/ENGINEER. Trail files can also be used for future playback of many commands at one time. To create a trail file for this use, start up a new Pro/ENGINEER session.

> ✓ **TIP:** *You can specify that all trail files will be placed in one directory no matter which directory you start a Pro/ENGINEER session from. This will enable you to remove all trail files easily when you need more disk space. Place the following option in your* config.pro *file and create the specified directory with write privileges for everyone:* `trail_dir   /usr/trailfiles`.

Part Start-up Trail File Creation

Select Part mode and create a part with the file name *master*. Create the default datum planes and a datum coordinate system. By selecting the appropriate datum planes, orient these planes and create view names for the front, back, left, right, top, and bottom views. After orienting and naming each of these views, save the part and exit Pro/ENGINEER. Now edit the *trail.txt* file with the highest version number. Remove the line containing #DBMS and all subsequent lines. The following listing of the end of the trail file shows the lines to delete from the file.

```
#NAMES
#SAVE
!Enter NAME:
bottom
!View is saved.
#DBMS
#SAVE
!Enter object to save [MASTER.PRT]:   |
!MASTER has been saved                |  Remove
#EXIT                                 |  these lines.
!Do you really want to exit? [N]:     |
YES                                   |
```

Save this trail file as *view.txt*. Now start Pro/ENGINEER and replay the trail file. Pick **Misc ➥ Trail** from the MAIN menu. Then select the trail file to replay.

```
Enter TRAIL FILE name [QUIT]: view ↵
```

A Better Part Start-up File

As you may have noticed when you replayed the trail file, *view.txt*, every menu and command selected when the trail file was created is displayed and executed again. Instead of replaying the *view.txt* trail file each time you want to start a new part, it is faster to retrieve the part file named *master*. Before starting Pro/ENGINEER, you should write-protect the *master.prt* file. Check your operating system guides or ask your system administrator if you do not know how to write-protect a file.

Move the file to your standards directory. Now start Pro/ENGINEER again. Pick **Mode ➥ Part ➥ Retrieve** and enter the part file name.

```
Enter Part name [QUIT]: /usr/standards/master ↵
```

Rename this part file. Pick **Dbms ➥ Rename**.

```
Enter Object To Save [MASTER.PRT]: ↵
Enter TO: new ↵
```

Now select **Backup** from the DBMS menu.

```
Which directory ? [/usr/people/pro/]: ↵
Enter object to save [NEW.PRT]: ↵
```

Select **Erase ➥ Confirm** from the DBMS menu. Then select **Part ➥ Retrieve** from the MODE menu.

```
Enter Part name [-Quit-]: new ↵
```

Finally, select **Dbms ➥Rename**.

```
Enter Object To Save [NEW.PRT]: ↵
Enter TO: new1 ↵
```

Now exit from Pro/ENGINEER and remove the part file *new1.prt.1* from your directory. Edit the trail file from this session. Remove the lines following the last prompt for the new name for the file. Rename the trail file to *master.txt*. The resulting trail file should contain the following lines:

```
!trail file version No. 744
Pro/ENGINEER TM Release 14.0 (c) 1988-93 by Paramet-
ric Technology Corporation  All Rights Reserved.
!Select a menu item.
#MODE
#PART
#RETRIEVE
!Enter Part name  [-Quit-]:
/usr/standards/master
!9-Mar-95 10:35:41  Start /usr/standards/mas-
ter.prt.1
!3-Mar-95 10:35:44  End   /usr/standards/master.prt.1
#DBMS
#RENAME
!Enter Object To Save [MASTER.PRT]:

!Enter TO:
new
@ ok
!MASTER has been renamed to NEW.
#BACKUP
!Which directory ? [/usr/people/pro/]:

!Enter object to save [NEW.PRT]:

!NEW has been saved
#ERASE
!Confirm request to clear NEW.PRT from workstation
memory
#CONFIRM
!NEW has been cleared
#PART
#RETRIEVE
!Enter Part name [-Quit-]:
new
!9-Mar-95 10:36:21  Start /usr/people/pro/NEW.PRT.1
!3-Mar-95 10:36:24  End   /usr/people/pro/NEW.PRT.1
```

```
#DBMS<R>#RENAME
!Enter Object To Save [NEW.PRT]:
```

> **NOTE:** *A blank line must be saved at the end of the trail file so that Pro/ENGINEER will know that the default part file name is accepted.*

Now whenever you are ready to start a new part file, replay the trail file *master.txt* first. Then, in response to the last prompt from replaying the trail file, key in the file name for the part you are about to create. You will already have your default datum planes and a datum coordinate system. View names will already be defined for front, back, left, right, top, and bottom views.

Menu Additions

Items can be added to an existing menu. To add an item, create a *menu_def.pro* file in your home directory. Each line in this file contains information to add a command to a menu. As an example, you can place a command to replay the *master.txt* trail file on the MAIN menu. Enter the following text on the first line of the file:

```
@setbutton MAIN Part#Start-up "#misc;#trail:master"\
"Creates a new part with default datum planes and
view names"
```

The entry is divided into five parts, with a blank space entered between each part.

- `@setbutton` denotes that the following line will define a new item on a menu.
- The second entry is the name of the menu the new command will be added to. You will be adding a command to the MAIN menu in the example. If the menu name had been more than one word, a pound sign (#) would have been used in place of the space between the words of the menu name.

- The third entry is the new command name that will be added to the menu. Your new command name will be Part Start-up. If the command name is to be more than one word, substitute pound signs (#) for spaces between words.

- The fourth entry is the action to occur after you select the command from the menu. It should be placed within quotes (""), and its syntax is the same as the macro entry in a mapkey.

  NOTE: *Place a backslash (\) after the action text string to tell Pro/ENGINEER that the rest of the menu information will follow on the next line.*

- The last text string is used to provide a one-line help message that will appear in the Message window whenever the new command is highlighted. If no help text string is provided for the new menu command, the help message displayed when the menu item is highlighted will be "User-defined action."

Now when you start Pro/ENGINEER, the command **Part Start-up** will appear at the bottom of the MAIN menu. You will be able to pick that command instead of playing back the *master.txt* trail file when you want to start a new part file.

Configuration File Options for Your *Config.pro* File

As mentioned previously, mapkey information is placed in the *config.pro* file. Information that can be set from the ENVIRONMENT submenu can also be set in the *config.pro* file. Many other custom settings can be placed in the *config.pro* file so that they are set for every Pro/ENGINEER session. Some suggested entries for the *config.pro* file are grouped according to use and provided in this section.

Graphic Display Customization

The following options are useful for customizing the graphic display. The default value for the option is shown first.

❒ `graphics gl | x_windows | starbase | pex | xgl | gpx | win32_gdi`

Sets the graphics environment. If your hardware supports enhanced 3D capabilities (dynamic spinning of shaded images), select the graphics option appropriate for your workstation. Refer to the Pro/ENGINEER Installation Guide to determine the correct setting for your hardware.

❒ `save_triangles_flag no | yes`

Sets whether the shaded surfaces are saved in the software so that you can spin a shaded image.

> ✓ ***TIP:*** *Set this option to* `yes` *only if you want to spin shaded images and your hardware does not support enhanced 3D capabilities. In other words, if you try to spin a shaded model and it changes back to wireframe, you can set* `save_triangles_flag` *to* `yes` *and the model will stay shaded while spinning. The model will slowly spin with this option set.*

Menu and Window Display Customization

The following options are useful for customizing the display of the menus.

❒ `header_menu no | yes`

Sets the location for the MAIN menu. The MAIN menu can be placed horizontally above the graphics window. This will provide more room down the side of the graphics window for menu windows to appear. When the MAIN menu is not placed horizontally above the graphics window, some menu windows will overlap and hide some of the menu commands.

> ✓ **TIP:** *At first glance, the option* `header_menu   yes` *seems desirable. However, when you make a selection from the MAIN menu, another menu window appears on top of the graphics window. When that window is dismissed, the graphics window repaints. For instance, if you shade the graphics window, the graphics window will shade a second time as the menu disappears.*

❐ `header_menu_alignment   left | right`

Sets the order of the commands on the MAIN menu if the menu is placed horizontally above the graphics window. You are specifying the end of the horizontal menu that will start with the **Mode** command.

❐ `menu_horizontal_hint   left | right`

Sets the side of the MAIN menu to place its submenus if the MAIN menu is not placed horizontally above the graphics window. If the submenus are placed on the left side of the MAIN menu, they will overlap the graphics window. When that happens, the graphics window display will be slowed because it will be repainted again when the menu window is removed.

> ✓ **TIP 1:** *If you set* `menu_horizontal_hint   right`*, the menus will not appear over the graphics window. Graphics speed is most efficient with this setting. However, there is a drawback. Unless you provide enough room on the right side of the MAIN menu for the other menus to appear, they will still display on the left of the MAIN menu. The only way to provide this room is to make the graphics window smaller by changing the windows_scale option to less than 1.*

> ✓ **TIP 2:** *If you do not want to make your graphics window smaller, it is better to place the MAIN menu horizontally on top of the graphics window. With the MAIN menu to the side of the graphics window, all other menus are shifted to a lower position. This causes some menus to overlap others at the bottom of the screen because there is not enough room. You will probably notice this problem only if you are using a small monitor and your graphics resolution is 1024 by 768.*

❐ `windows_scale value`

Sets the size of the windows with respect to the default size. The allowable values are .5 to 1.

> ✓ **TIP:** *In order to have adequate room to place the submenus of the MAIN menu on the right side, the windows scale usually needs to be reduced. Select the largest value that still allows the submenus to be placed to the right of the MAIN menu. This configuration setting should not be used if you place the MAIN menu horizontally above the graphics window because all you accomplish is making your graphics window smaller.*

❐ `thermo_position_hint window_overlap| no_window_overlap`

Sets the location for the thermometer-type scale, which is displayed when you are spinning an object. Overlapping the graphics window with the thermometer scale has the same drawback as overlapping the graphics window with a Menu window. The graphics display takes longer because the window is repainted an additional time when the thermometer-type scale is removed. In addition, you cannot see part of the graphics window if you do not move it.

❐ `forced_cascading_level value`

Selects the minimum number of menu windows to place down the side of the graphics window before automatically starting to overlap them. A negative number specifies the number of menus to be cascaded when there is inadequate room to display all menus. The negative number specifies how many menus before and including the menu that first runs out of space will be cascaded. Pro/ENGINEER sometimes starts to overlap menu windows even when there is sufficient room to display them. Using `forced_cascading_level -1` will instruct Pro/ENGINEER to begin cascading menus, beginning with the first one that has insufficient room to be displayed below the other menus.

❐ `visible_message_lines   value`

Sets the number of visible lines in the message window. The allowable range is 1 to 5.

❐ `iconify_entire_pro   yes | no`

Specifies whether the entire Pro/ENGINEER session will be iconified when you select to iconify a graphics window. If the `no` option is specified, only the graphics window be iconified.

Color Customization

The colors displayed in Pro/ENGINEER can be customized. To make the new colors your default, put the red-green-blue (rgb) values for the system color name in your *config.pro* file. For example, to set the window background color, define the rgb values for the system color name as follows:

```
system_background_color   value value value
```

The first value specifies the amount of red; the second, green; the third, blue. The numbers 0 to 100 will give you ratios of each color.

> ✓ **TIP:** *Adjusting* `system_background_color` *to black allows entities to be more easily seen than with the default blue background. Add* `system_ background_color 0 0 0` *to* config.pro *to make this color change your default. Do not make the letter color and the background color the same, because your menu items will not be visible.*

To experiment with the range of colors (and their value numbers), use the following commands, starting at the MAIN menu: **Misc** ➥ **SystemColors** ➥ **Redefine** ➥ **Background**.

A small SYSTEM COLOR window will appear. Using the arrow to pick on each number below the color will give you various shades along the color spectrum. To select a color you have mixed, put the cursor on the color bar above the word *Red*. To return to the default blue background color, and all other default color selections, select **Default** in the SYSTEMCOLORS menu.

Object Display Customization

The options that follow are useful for customizing the display of objects.

❐ `display wireframe | hiddenvis | hiddeninvis | shade`

Sets how hidden lines are displayed.

> ➾ **NOTE:** *Wireframe display allows the fastest display speed. Other settings allow the display to be more easily interpreted.*

❐ `orientation trimetric | isometric`

Establishes the orientation of the default view.

Dimension Display Customization

The following options are useful for customizing dimensioning.

❐ `default_dec_places    value`

Sets the default decimal places used for part dimensions that are not created in the sketcher.

❐ `sketcher_dec_places   value`

Sets the number of displayed decimal places for all dimensions created in the sketcher.

> ✓ ***TIP:*** *Set the value to the number of decimal places you most often show on your drawings. You can then manually modify exceptions to the default.*

❐ `parenthesize_ref_dim   no | yes`

Sets the reference dimension display. The option `parenthesize_ref_dim   yes` displays the reference dimension-with parentheses instead of following the dimension with the text "REF."

❐ `create_fraction_dim   no | yes`

Sets the method to display dimensions as either decimal or fractional.

❐ `dim_fraction_denominator   value`

Sets the largest allowed denominator to be used with a dimension displayed as a fraction. For instance, use `64` if you do not want fractions smaller than 1/64.

❐ `linear_tol   value value`

Sets linear tolerance as a function of the number of decimals in the part linear dimension. The first value specifies the number of decimal places in the linear dimension. The second value specifies the tolerance for the dimension.

❒ `angular_tol   value value`

Sets angular tolerance as a function of the number of decimals in the part angular dimension. The first value specifies the number of decimal places in the angular dimension. The second value specifies the tolerance for that dimension.

❒ `tol_display   no | yes`

Sets whether tolerances will be displayed with the part dimension.

> **NOTE:** *This configuration setting does not affect tolerance display in Drawing mode. In order to display tolerances in drawings, tolerances need to be turned on in the drawing standards information for the drawing file.*

❒ `tol_mode   limits | nominal | plusminus | plusminussym`

Sets how tolerances will appear when displayed.

> ✓ **TIP:** *Set the tolerance to the style used most often in your drawings. You can manually modify the exceptions to the default.*

Object File Storage and Retrieval Customization

The following options are useful for customizing how object files are saved and retrieved.

❒ `store_display   no | yes`

Saves the screen display with the object file. The model is saved with current orientation information so that hidden lines do not have to be calculated when the model is retrieved.

> ➾ **NOTE:** *Using* `store_display    yes` *allows the file to be retrieved faster. However, a larger file is created.*

❒ `compress_output_files    no | yes`

Specifies whether to compress the object file when saving it to disk.

> ➾ **NOTE:** *Compressing when saving a file requires less hard disk space. However, saving and retrieving the file are more time-consuming.*

❒ `save_object_in_current    no | yes`

Saves a file in the current directory if you do not have write permission to the directory the file was retrieved from.

❒ `search_path    text`

Lists directories in the order they will be searched to retrieve an entered file name.

> ➾ **NOTE:** *Enter several paths on a single line or use this setting several times in your* config.pro *file if you want more than one directory searched.*

> ✓ **TIP 1:** *If you have directories in which you place files after they have been approved, you can specify that these directories be searched. Files will then be retrieved from those directories as if they resided in your current working directory.*

> ✓ **TIP 2:** *The fastest file retrieval is achieved by using the following configuration setting:*

```
store_display   yes
compress_output_files   no
```

✓ **TIP 3:** *To reduce the amount of hard disk space used when saving a file, use the following configuration settings:*

```
store_display   no
compress_output_files   yes
```

Miscellaneous Configuration Settings

The following entries can be placed in the *config.pro* file to make your design session more productive:

- `bell   yes | no`

Sets whether the bell should ring each time Pro/ENGINEER prompts you for information.

- `pro_unit_length   unit_inch | unit_foot | unit_mm | unit_cm | unit_m`

Sets the default units.

- `pro_unit_mass   unit_pound | unit_ounce | unit_ton | unit_gram | unit_kilogram | unit_tonne`

Sets the unit of mass for mass property calculations.

- `provide_pick_message_always   no | yes`

Provides information about the feature selected when using the **Pick** option.

> **NOTE:** *Pro/ENGINEER provides information in the Message window about the selected feature when you are using* ***Query Sel****, but not when using* ***Pick****. The*

`provide_pick_message_always` `yes` *line will provide feature information when using* ***Pick*** *to ensure that you selected the correct feature.*

`use_dimensioned_edges   yes | no`

/hen set to `no`, the previous line restricts Pro/ENGINEER from sing an edge that has been dimensioned to, as if it had been selected an alignment reference. If you want to align to that entity, you ust specifically select it as an alignment entity.

`flip_arrow_scale   value`

ts the scale at which the flip arrow appears after some sketcher ssions. This is used if the displayed arrow is too small.

`repeat_datum_create   no | yes`

llows datum features to continue being placed without reselecting e datum feature after every placement.

`drawing_file_editor   editor | protab`

akes Pro/ENGINEER use its editor or the system editor when iting the drawing standards information in a drawing file.

`relation_file_editor   editor | protab`

akes Pro/ENGINEER use its editor or the system editor when iting relations.

`pro_editor_command   editor`

ovides the command for Pro/ENGINEER to use when you specify system editor in `drawing_file_editor` or `lation_file_editor`. The default is to use the `vi` editor.

> ✓ **TIP:** *For beginning UNIX users, selecting a system editor command other than the default is highly recommended. For SGI workstations, use* `jot -f`*. If you do not have a system editor other than* `vi`*, change the* `drawing_file_editor` *and* `relation_file_editor` *options to* `protab`.

Drawing Mode Configuration Settings

The following entries can be placed in the *config.pro* file to affect the function of Drawing mode:

❒ `auto_regen_view   yes | no`

Determines whether view regeneration should occur automatically or only when you select it.

> ➾ **NOTE:** *Pro/ENGINEER will regenerate a drawing view as soon as a change has been made. If view regeneration is time-consuming and you want to make several changes, you should consider specifying that you will control when regeneration occurs. The following line is used for this purpose:* `auto_regen_view   no`.

❒ `drawing_setup_file   text`

Specifies where to find the file that specifies drawing configuration information such as arrow size and text size.

❒ `highlight_new_dims   no | yes`

Specifies whether new dimensions displayed in Drawing mode will display in the highlight color (red) until they are moved or the screen is repainted.

> ✓ **TIP:** *Setting* `highlight_new_dims` *to* `yes` *will make it easier to identify newly displayed dimensions when you show dimensions.*

The sample *config.pro* file below contains suggested configuration options.

```
GRAPHICS DISPLAY
raphics   gl
MENU & WINDOW DISPLAY
enu_horizontal_hint   right
indows_scale   .98
hermo_position_hint   no_window_overlap
orced_cascading_level   -1
isible_message_lines   5
conify_entire_pro   no
COLOR CUSTOMIZATION
ystem_background_color   0 0 0
OBJECT DISPLAY
isplay   hiddenvis
DIMENSION DISPLAY
efault_dec_places   3
ketcher_dec_places   3
arenthesize_ref_dim   yes
im_fraction_denominator   64
near_tol   1 0.1
near_tol   2 0.03
near_tol   3 0.005
gular_tol   1 0.5
l_display   yes
l_mode   nominal
BJECT FILE STORING AND RETRIEVING
ore_display   yes
arch_path   /usr/prt:/usr/asm
ISCELLANEOUS SETTINGS
ll   no
 o_unit_length   unit_inch
 o_unit_mass   unit_pound
 ovide_pick_message_always   yes
e_dimensioned_edges   no
peat_datum_create   yes
awing_file_editor   protab
lation_file_editor   protab
RAWING MODE SETTINGS
to_regen_views   no
awing_setup_file   /usr/standards/custom.dtl
ghlight_new_dims   yes
LOTTER SETTINGS
ot_names   yes
```

```
!DEFAULT DIRECTORIES
pro_format_dir   /usr/standards/formats
pro_plot_config_dir   /usr/standards/plotter
pro_group_dir   /usr/standards/udf
trail_dir   /usr/trailfiles
plot_file_dir   /usr/plotfiles
!MAPKEY MACROS
mapkey dr #done-return:
mapkey dv #view:#default:%dr
mapkey fv #view:#pan/zoom:#reset:%dr
mapkey pv #view:#pan/zoom:#pan:
mapkey rv #view:#repaint:%dr
mapkey sv #view:#cosmetic:#shade:#display:%dr
mapkey sp #view:#orientation:#spin:
mapkey vn #view:#names:
mapkey zi #view:#pan/zoom:
mapkey zo #view:#pan/zoom:#zoom out:%dr
mapkey fr #view:#names:#Front:%dr
mapkey bk #view:#names:#Back:%dr
mapkey rt #view:#names:#Right:%dr
mapkey lt #view:#names:#Left:%dr
mapkey tp #view:#names:#Top:%dr
mapkey bt #view:#names:#Bottom:%dr
```

Conclusion

If you are comfortable with attempting custom configurations, you have achieved skills beyond those of a beginning Pro/ENGINEER user. Continued practice and refinement of the skills you now possess will prepare you for upcoming advances in your field.

Appendix A
Pro/TABLE Commands

Commonly Used Commands for the Pro/TABLE Editor

The Pro/TABLE editor is used for editing relations, family tables, and configuration files. The editor resembles a spreadsheet more than an average text editor. You see columns and rows, and you manipulate data in individual cells. The Pro/TABLE editor looks like the following illustration.

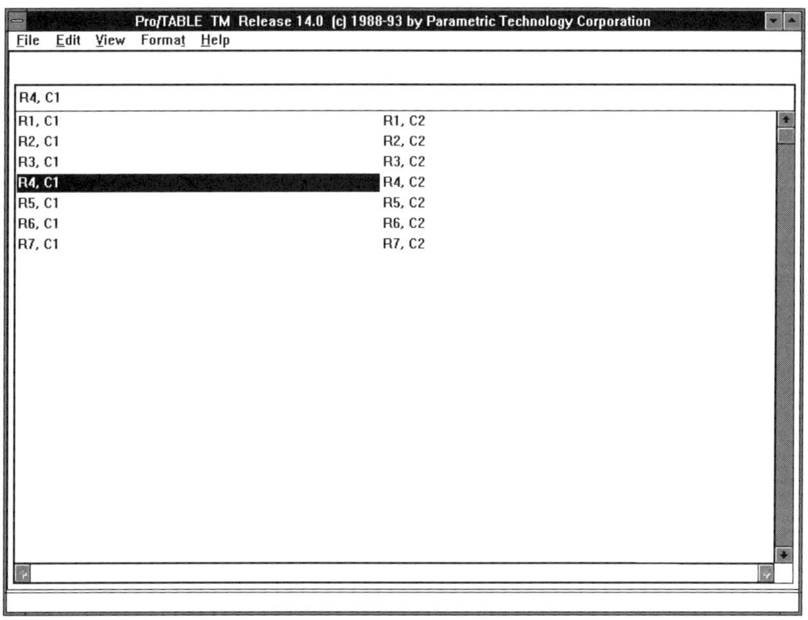

To move the cursor around in the window, use the arrow keys or the mouse. When you want to edit the data in a cell, highlight the cell and start to type. Your typing will appear in the line just below the pull-down menus. Press <Return> to place the new data in the cell.

The Pull-down Menus

Pro/TABLE commands are made available via pull-down menus. Point to the desired heading with the cursor and press the left mouse button to activate the menu. Pull-down menu headings follow.

File — Commands for saving files and quitting the editor

Edit — Commands for manipulating cells, rows, and columns

View — Commands for cursor positioning

Format — Commands for setting column width

Help — Commands for getting help with Pro/TABLE

The File Menu

New — Starts editing a new file

Open — Opens an existing file for editing

Save — Saves the current file

Save As — Saves the current file with a different name

Exit — Saves the current file and quits the Pro/TABLE editor

Quit — Quits the Pro/TABLE editor without saving the file

The Edit Menu

Undo — Undoes the last editing sequence

Cut — Erases the data in the highlighted cells and saves the contents to a buffer

Copy — Copies the data in the highlighted cells to the buffer

Paste — Places the contents of the buffer at the cursor location

Erase — Erases the data in the highlighted cells

Sort — Alphabetically sorts the highlighted cell data

Delete — Deletes the highlighted rows or columns

Insert — Inserts rows or columns

Choose Keywords — Provides keywords and options when editing configuration files

The View Menu

Home — Returns the cursor to the row 1, column 1 position

Goto — Places the cursor at the specified row and column

The Format Menu

Global Width — Controls the default column width setting

Column Width — Allows you to change the width of the highlighted columns

Reset Width — Resets the highlighted columns' width to the default setting

The Help Menu

Contents — Starts Pro/USERGUIDE and displays Pro/TABLE help

Search for Help on — Search through Pro/USERGUIDE by keyword

How To Use Help — Help for Pro/USERGUIDE

About Pro/TABLE — Displays the version number and license agreement

Pro/TABLE can be executed from the command prompt by typing `protab` ↵. This will allow you to edit configuration files or family tables from outside Pro/ENGINEER.

> ➾ **NOTE:** *The* ***Choose Keywords*** *command may not be available if you start Pro/TABLE from the command prompt.*

Appendix B
Product Descriptions

Pro/ENGINEER's Add-on Modules

Most users do not require all Pro/ENGINEER software. Consequently, the software is packaged in modules so that you buy only the modules you require. Many modules have prerequisites, which are noted in the following descriptions. Some Pro/ENGINEER modules can be defined as "floating," meaning that the module is available for use on several different workstations, as long as you do not exceed the maximum number of licenses. Modules that do not require Pro/ENGINEER function as stand-alone licenses. Floating modules can be attached to stand-alone licenses as appropriate during a design session. The following product descriptions also indicate which modules can float.

Pro/ENGINEER

Pro/ENGINEER is the base module, which provides capabilities for creation of parts, assemblies, and engineering drawings. The base module supports IGES, HPGL, and PostScript output. It provides the ability to dynamically spin, rotate, zoom, and pan both wireframe and shaded models. Full markup capability in drawing mode is also provided to allow redlining.

Pro/ASSEMBLY

❒ Required modules: Pro/ENGINEER

- ❐ Floating supported: No

Pro/ASSEMBLY is a module that provides the capability to manage an assembly of parts. Parts can be created in assembly mode. Features can be created on a part that reference other parts. Assembly features can be created that affect several parts.

Pro/CABLING

- ❐ Required modules: Pro/ENGINEER
- ❐ Floating supported: Yes

Pro/CABLING provides the capability for placement and modification of cables, wires, and bundles in 3D assemblies. Cable and bundle routes can be denoted by picking surfaces, holes, edges, points, and connectors. As the assembly changes, these parametric cable and bundle routes are automatically updated. Connector parameters such as pin configuration and internal jumpering are supported. BOM information is provided, as is cable and bundle straight lengths. Bundle splitting and joining are supported, and 3D harness drawings from routed assemblies can be generated. Signal assignment to pins and conductors is supported.

Pro/CASTING

- ❐ Required modules: Pro/ENGINEER
- ❐ Floating supported: Yes

Pro/CASTING provides a method of generating die cavities and creating casting geometry by directly referencing the model of the final cast model. Compensation for shrinkage of the cast part is provided. Evaluation of draft requirements can automatically be determined. Sand cores can be created and complex surfaces used for the parting surfaces. Die components can be automatically created after defining the parting surfaces. Interference checks can be conducted during the simulated die opening process. Part thickness can be analyzed to determine casting suitability.

Pro/CAT

- Required modules: Pro/ENGINEER or Pro/DETAIL
- Floating supported: Yes

The Pro/CAT module provides the capability to exchange geometry between Pro/ENGINEER and CATIA, a CAD/CAM tool from Dassault Systems. Solid models from Pro/ENGINEER can be exported to CATIA. CATIA surface models can be imported into Pro/ENGINEER. Parametric features can then be added to these surfaces as if the surfaces had been created from within Pro/ENGINEER. The imported surfaces can be converted to a solid, and parametric solid features can be added to the surfaces.

Pro/CDT

- Required modules: Pro/ENGINEER
- Floating supported: Yes

The Pro/CDT module provides the capability to transfer PROFESSIONAL CADAM 2D drawing files into Pro/ENGINEER. These files can be modified from within Pro/DETAIL.

Pro/COMPOSITE

- Required modules: Pro/ENGINEER, Pro/SURFACE
- Floating supported: No

The Pro/COMPOSITE module provides the capability to design, manufacture, and document ply lay-up composite panels. Design engineers are allowed to reference nominal panel surfaces and add beads, flanges, and additional mold-line geometry. Ply and core features that simulate the lay-up process are used to create a 3D solid model. Trim and stretch allowances can be added to the referenced mold-line geometry. Flat patterns of individual plies can be generated, as well as dimensioned part drawings and sequence drawings.

Pro/DESIGN

- Required modules: Pro/ENGINEER
- Floating supported: No

Pro/DESIGN provides a method for top-down assembly design. Global and local dimensions, relations, and datums can be placed in a layout. Instead of modifying dimensions or relations in each part of an assembly, only the layout, which contains the dimension and relation information for all parts, needs to be modified. 2D sketches and engineering notes that document the design can be included in the layout. Automatic assembly of components can be achieved by relating the appropriate part axes, datums, and planar surfaces to axes and datums within the layout. The capability to assemble parts in absolute 3D space coordinates is provided.

Pro/DETAIL

- Required modules: None
- Floating supported: Yes

Pro/DETAIL provides the tools for creating fully associative production drawings from Pro/ENGINEER models. Tools are provided to make drawings that conform to ANSI, ISO, DIN, and JIS standards. These drawings can include multiple sheets and views.

Text and 2D geometry specific to the drawing and not reflected in the model can be added. Drawings in IGES, DXF, and SET formats can be imported. Drawings can be exported by IGES, PostScript, and plotting.

Pro/DEVELOP

- Required modules: Pro/ENGINEER
- Floating supported: No

Pro/DEVELOP is a programming interface to Pro/ENGINEER that provides the means for software developers to create applications that directly integrate into the Pro/ENGINEER environment and obtain direct access to the Pro/ENGINEER database.

Pro/DIAGRAM

- Required modules: Pro/ENGINEER
- Floating supported: Yes

Pro/DIAGRAM provides the method for documenting schematic information in diagrams for applications such as electrical wiring, piping, HVAC, process flow, and project management. For wiring applications, fully detailed wiring and interconnect diagrams can be created. Information—including components, connectors, wires, and pins—can be graphically depicted. Output such as component lists, netlists, and wirelists can be created graphically or as text files. Wirelists can be used in conjunction with Pro/CABLING.

Pro/DIEFACE

- Required modules: Pro/ENGINEER, Pro/SURFACE
- Floating supported: No

Pro/DIEFACE provides functionality to define complex die surfaces used in the manufacture of auto body panels and other deep-drawn products.

Pro/ECAD

- Required modules: Pro/ENGINEER
- Floating supported: Yes

Pro/ECAD provides a method to parametrically place components on printed circuit boards (PCB). Components to place on the PCB can be from Pro/ENGINEER component libraries, or the component profiles can be transferred to Pro/ENGINEER and extruded into solid components. Data supported from the ECAD systems' standard neutral file include the following:

- PCB outline
- Board edge connectors
- Mounting and alignment hole position and size
- Cutout areas for large components or housings
- Keep-out areas where routing will not occur
- Height restriction areas

The PCB and components can be assembled, visualized, and analyzed like any other solid within Pro/ENGINEER. Currently supported ECAD systems include the following:

- CADENCE/VALID Allegro 5.1
- MENTOR GRAPHICS BoardStation 8.0 and Mechanical Interface
- RACAL REDAC Visula PCB Rev. 6.0
- DAZIX/INTERGRAPH CDX CAD (Cadnetix) Version F1.1A and above

- INTERGRAPH PCB Engineer

Pro/FEATURE

- Required modules: Pro/ENGINEER
- Floating supported: No

Pro/FEATURE provides an extended set of features to the basic Pro/ENGINEER. These features include the following:

- Shell
- Lip
- Complex domes
- 3D sweeps
- Thin features
- Complex blends
- Merge part
- Blend/sweep
- Multi-trajectory sweep
- Groove feature
- Offset area
- Parting lines
- Piping

Feature management tools included are the following:

- User-defined features
- Local groups
- Feature copying
- Feature mirroring

Pro/FEM-POST

- Required modules: Pro/ENGINEER, Pro/MESH
- Floating supported: No

Pro/FEM-POST integrates several FEA analysis packages into the Pro/ENGINEER environment. After creating loads and the mesh using the Pro/MESH module, only one menu pick initiates the solver without exiting Pro/ENGINEER. When the analysis is completed, tools are provided to graphically display the results. Solvers supported follow:

- NASTRAN
- ANSYS
- COSMOS/M
- PDA FEA

Pro/HARNESS-MFG

- Required modules: Pro/ENGINEER
- Floating supported: Yes

Pro/HARNESS-MFG provides a method for creating manufacturing data used in the production of electrical wire and cable harnesses. It is used in conjunction with Pro/DIAGRAM and Pro/CABLING. The harness can be used to create 3D semi-straightened or 2D fully flat harnesses. Any changes in shape or length of the 3D harness automatically update the flattened harness. The harness can be represented as stick-figure drawings or nail board drawings. Parts lists and from-to wire lists are provided. Connectors, tie wraps, and markers can be displayed. Nail board fixtures can be created and assembled.

Pro/INTERFACE

- Required modules: Pro/ENGINEER
- Floating supported: Yes

Pro/INTERFACE is a set of industry standard file translators for importing and exporting geometry with Pro/ENGINEER. Translators for importing files include the following:

- CATIA
- CATIA IIF
- CGM
- DXF
- IGES
- SET
- TIFF
- VDA

Translators for exporting files include the following:

- CGM
- COSMOS geometry
- DXF
- ECAD
- INVENTOR
- NEUTRAL
- PATRAN geometry
- RENDER
- SET
- SLA
- SUPERTAB geometry

- TIFF
- VDA

Pro/LANGUAGE

- Required modules: Pro/ENGINEER or Pro/DETAIL
- Floating supported: No

The Pro/LANGUAGE module translates Pro/ENGINEER menus and help files into French, German, and Japanese (Kanji). If the hardware platforms support Kanji characters and Kanji keyboards, notes in Kanji can be added to drawings. Various Pro/ENGINEER User Guide documents are also provided in German or Kanji.

Pro/LEGACY

- Required modules: Pro/ENGINEER, Pro/SURFACE (for surface legacy data)
- Floating supported: Yes

Pro/LEGACY provides a set of tools for manipulating 2D drawings and 3D data. These tools allow the creation, modification, and deletion of geometry. Pro/INTERFACE provides the formats for importing the legacy information into Pro/ENGINEER. Legacy 3D parts and assemblies can be used in Pro/ENGINEER assemblies in the same way as Pro/ENGINEER parts and assemblies. Manipulation of 3D surface data requires the Pro/SURFACE module.

Pro/LIBRARYACCESS

- Required modules: Pro/ENGINEER
- Floating supported: Yes

Pro/LIBRARYACCESS provides access to libraries of parametric parts, tools, and features. Families of user-defined parts, features, tools, connectors, and mold bases can be added to the standard libraries provided within the Pro/ENGINEER-supplied libraries. The following Pro/ENGINEER-supplied libraries are available:

- The BASIC Library provides common fasteners that correspond to ANSI, metric ANSI/ISO, USAS, and other standards. Standard features such as countersinks, counterbores, and corrugations are also provided in the form of user-defined features.
- The TOOLING Library provides standard tools and tool holders, as well as fixture components.
- The MOLD BASE Library provides parts and assemblies from DME and HASCO in standard and metric series.
- The CONNECTOR Library provides families of connectors that conform to MIL-C specifications.
- The ELECTRICAL SYMBOL Library provides electrical symbols that conform to ANSI/Canadian/IEEE standards. These symbols can be used with Pro/DETAIL, Pro/DESIGN, and Pro/DIAGRAM.
- The PIPING & HEATING SYMBOL Library provides piping, HVAC, and heating apparatus symbols that conform to ANSI/ASME standards. These symbols can be used with Pro/DETAIL, Pro/DESIGN, and Pro/DIAGRAM.
- The PIPE FITTING Library provides standard SAE and ANSI/ASME fittings such as hydraulic, automotive, refrigeration, and pipe fittings.

- The Human Factors Library provides models of both male and female bodies with 5th-, 50th-, and 95th-percentile body dimensions. Body dimensions can be modified, and predefined joint-angle settings are provided for positions such as standing, walking, and driving.

Pro/MANUFACTURING

- Required modules: Pro/ENGINEER

- Floating supported: Yes

Pro/MANUFACTURING provides a method to parametrically define the numerical control toolpaths required to manufacture a model. The toolpaths are parametrically linked with both the design model and the stock model. Toolpaths are generated for the following machining operations:

Hole Making

- Drill cycle
- Bore cycle
- Tapping cycle
- Deep drilling cycle with user-defined peck depth
- Centerdrill
- Spotface
- Reaming
- Hole pattern recognition

Milling

- Face milling
- 2.5-axis roughing
- Surface machine in 3, 4, and 5 axis
- Thread milling

- Identify and mill material remaining from previous operation

Turning

- Rough and finish turning
- Rough and finish facing
- Rough and boring
- Threading
- Grooving
- 4-axis lathes with simultaneous operations
- Mill / turn centers
- All hole making cycles

Wire Electric Discharge Machining

- 2- and 4-axis control
- WEDM corner conditions

Pro/MESH

- Required modules: Pro / ENGINEER
- Floating supported: Yes

Pro / MESH provides a method to create a finite element mesh of solid models and thin-walled solid models. Loads, boundary conditions, and the mesh are associatively defined to the design model so that they are modified as the model changes. Automatic mesh generation uses solid tetrahedral elements and triangular shell elements. The thickness of the shell element is determined from the model thickness. Mixed element meshing is supported for both parts and assemblies. Capabilities are provided for local mesh refinement. Shell elements can be generated at the mid-surface of the model or at the inner / outer surface. Mid-surface geometry can be output via

IGES 5.0. Output can be provided to the following analysis applications: ANSYS, COSMOS/M, NASTRAN, PATRAN, and SUPERTAB.

Pro/MOLDESIGN

- Required modules: Pro/ENGINEER
- Floating supported: Yes

Pro/MOLDESIGN provides the tools required to design mold components and mold base assemblies. The mold impression is automatically created and parametrically based on the design model. Compensation for shrinkage of the mold part is provided, as well as simulation of mold opening and molding ejection with interference checking between parts. Complex multiple plate and insert molds can easily be created. AC Technology's C-FLOW/EZ analysis software is included in Pro/MOLDESIGN to provide cavity fill simulation in addition to air trapping, flow front, ram speed, weld lines, and flow velocity. Information such as fill volume and impression surface area is available. User-defined mold features such as gates, runners, and sprues can be created. The MOLDBASE Library of standard mold base assemblies and components can be accessed with the use of Pro/LIBRARYACCESS.

Pro/NC-CHECK

- Required modules: Pro/MANUFACTURING or Pro/SHEETMETAL
- Floating supported: Yes

Pro/NC-CHECK provides a method to graphically simulate the removal of material due to milling and drilling operations. Various visualization tools are provided to better identify any interference problems.

Pro/PIPING

- ❒ Required modules: Pro/ENGINEER
- ❒ Floating supported: Yes

Pro/PIPING is used to parametrically route fitted pipe, bent tubing, or flexible hose. Components such as clamps and support brackets can be created by referencing the pipeline. Changes in the pipe diameter or minimum bend radius will also be reflected in these components.

Pro/PLOT

- ❒ Required modules: Pro/ENGINEER, Pro/DETAIL, or Pro/REVIEW
- ❒ Floating supported: No

Pro/PLOT provides the following plotting formats in addition to Pro/ENGINEER's standard HPGL and PostScript file output capabilities: Calcomp, Gerber, HPGL2, and Versatec.

Pro/REPORT

- ❒ Required modules: Pro/ENGINEER or Pro/DETAIL
- ❒ Floating supported: No

The Pro/REPORT module provides the ability to create associative customized reports. Customized bills of material (BOM) and family tables can be added to drawings so that they automatically update as information is added to the drawing. Tools are provided to filter and sort information. Duplicate entries can be toggled between show and no show. Balloons associated with an assembly BOM can be displayed.

Pro/REVIEW

- Required modules: none
- Floating supported: Yes

Pro/REVIEW is designed to be used without Pro/ENGINEER to view and inspect parts, assemblies, drawings, and manufacturing models. Viewing capabilities include spin, shade, explode, zoom, and change view. Plotting capabilities include HPGL and PostScript. The Pro/PLOT module can extend these plotting capabilities. Pro/REVIEW provides for a method to extract information such as measurements, mass properties, and BOM. Markup capabilities to sketch revisions and add comments to drawings and 3D models are provided.

Pro/SCAN-TOOLS

- Required modules: Pro/ENGINEER, Pro/SURFACE
- Floating supported: Yes

Pro/SCAN-TOOLS provide a method for smoothly fitting complex curves and surfaces through scanned data.

Pro/SHEETMETAL

- Required modules: Pro/ENGINEER
- Floating supported: Yes

Pro/SHEETMETAL provides automated feature-based design and manufacture of sheet metal parts. Features are included to address the bending and unbending of models through the use of bend table libraries that base the bend allowance upon the bend radius, angle, material type, and thickness. Features can be added with the model in the bent or unbent configuration. Form features can easily be saved as a user-defined feature. Manufacturing capabilities include the ability to generate toolpaths and tool index tables. Manufacturing operations such as punching, notching, nibbling, contour cutting, and metal forming are provided. Interference checking and gouge avoidance checking can be performed. Dynamic simulation of tools and paths can be visualized.

Pro/STEP-INTERFACE

- Required modules: Pro/ENGINEER
- Floating supported: Yes

Pro/STEP-INTERFACE enables the transfer of data to and from Pro/ENGINEER using ISO 10303 (the international standard known as STEP).

Pro/SURFACE

- Required modules: Pro/ENGINEER
- Floating supported: Yes

Pro/SURFACE allows the creation and editing of complex parametric surfaces. Surfaces can be joined to create surface patches. Surfaces can be trimmed at the intersection with other surfaces. Surfaces can be added to a wireframe model or a partially surfaced model and then stitched together to form a solid model. Surfaces can be used to perform cuts or protrusions to solid models.

Index

Symbols

A

B

C

D

E

F

G

H

I

J

K

L

M

N

O

Q

R

S

T

U

More OnWord Press Titles

Computing/Business

Lotus Notes for Web Workgroups
$34.95

Mapping with Microsoft Office
$29.95 Includes Disk

Geographic Information Systems (GIS)

GIS: A Visual Approach
$39.95

The GIS Book, 3E
$34.95

INSIDE MapInfo Professional
$49.95 Includes CD-ROM

Raster Imagery in Geographic Information Systems
$59.95

INSIDE ArcView
$39.95 Includes CD-ROM

ArcView Exercise Book
$49.95 Includes CD-ROM

ArcView/Avenue Developer's Guide
$49.95

ArcView/Avenue Programmer's Reference
$49.95

101 ArcView/Avenue Scripts: The Disk
Disk $101.00

ARC/INFO Quick Reference
$24.95

MicroStation

INSIDE MicroStation 95, 4E
$39.95 Includes Disk

MicroStation 95 Exercise Book
$39.95 Includes Disk
Optional Instructor's Guide $14.95

MicroStation 95 Quick Reference
$24.95

MicroStation 95 Productivity Book
$49.95

Adventures in MicroStation 3D
$49.95 Includes CD-ROM

MicroStation for AutoCAD Users, 2E
$34.95

MicroStation Exercise Book 5.X
$34.95 Includes Disk
Optional Instructor's Guide $14.95

MicroStation Reference Guide 5.X
$18.95

Build Cell for 5.X
Software $69.95

101 MDL Commands (5.X and 95)
Executable Disk $101.00
Source Disks (6) $259.95

Pro/ENGINEER and Pro/JR.

Automating Design in Pro/ENGINEER with Pro/PROGRAM
$59.95 Includes CD-ROM

Pro/ENGINEER Exercise Book, 2E
$39.95 Includes Disk

Pro/ENGINEER Quick Reference, 2E
$24.95

Thinking Pro/ENGINEER
$49.95

Pro/ENGINEER Tips and Techniques
$59.95

INSIDE Pro/JR.
$49.95

Softdesk

INSIDE Softdesk Architectural
$49.95 Includes Disk

Softdesk Architecture 1 Certified Courseware
$34.95 Includes CD-ROM

Softdesk Architecture 2 Certified Courseware
$34.95 Includes CD-ROM

INSIDE Softdesk Civil
$49.95 Includes Disk

Softdesk Civil 1 Certified Courseware
$34.95 Includes CD-ROM

Softdesk Civil 2 Certified Courseware
$34.95 Includes CD-ROM

Interleaf

INSIDE Interleaf (v. 6)
$49.95 Includes Disk

Interleaf Quick Reference (v. 6)
$24.95

Interleaf Exercise Book (v. 5)
$39.95 Includes Disk

Interleaf Tips and Tricks (v. 5)
$49.95 Includes Disk

Adventurer's Guide to Interleaf LISP
$49.95 Includes Disk

Other CAD

Manager's Guide to Computer-Aided Engineering
$49.95

Fallingwater in 3D Studio
$39.95 Includes Disk

Windows NT

Windows NT for the Technical Professional
$39.95

SunSoft Solaris

SunSoft Solaris 2. for Managers and Administrators*
$34.95

SunSoft Solaris 2. User's Guide*
$29.95 Includes Disk

SunSoft Solaris 2. Quick Reference*
$18.95

*Five Steps to SunSoft Solaris 2.**
$24.95 Includes Disk

SunSoft Solaris 2. for Windows Users*
$24.95

HP-UX

HP-UX User's Guide
$29.95

HP-UX Quick Reference
$18.95

Five Steps to HP-UX
$24.95 Includes Disk

OnWord Press Distribution

End Users/User Groups/Corporate Sales

OnWord Press books are available worldwide to end users, user groups, and corporate accounts from local booksellers or from Softstore/CADNEWS Bookstore: call 1-800-CADNEWS (1-800-223-6397) or 505-474-5120; fax 505-474-5020; write to SoftStore, Inc., 2530 Camino Entrada, Santa Fe, NM 87505-4835, USA or e-mail orders@hmp.com. SoftStore, Inc., is a High Mountain Press Company.

Wholesale, Including Overseas Distribution

High Mountain Press distributes OnWord Press books internationally. For terms call 1-800-4-ONWORD (1-800-466-9673) or 505-474-5130; fax to 505-474-5030; e-mail orders@hmp.com; or write to High Mountain Press, 2530 Camino Entrada, Santa Fe, NM 87505-4835, USA.

On the Internet: http://www.hmp.com

OnWord Press, 2530 Camino Entrada, Santa Fe, NM 87505-4835 USA